Rishabha Malviya, Sonali Sundram, Sathvik Belagodu Sridhar, Shivam Rajput (Eds.)

Artificial Intelligence in Polymer Science and Nanotechnology

Also of interest

Smart and Functional Coatings.
Nanotechnology in Coatings
Arno Schut, 2025
ISBN 978-3-11-132644-3, e-ISBN (PDF) 978-3-11-132649-8,
e-ISBN (EPUB) 978-3-11-132755-6

Physical Chemistry of Polymers.
A Conceptual Introduction
Sebastian Seiffert, 2nd Edition 2023
ISBN 978-3-11-071327-5, e-ISBN (PDF) 978-3-11-071326-8,
e-ISBN (EPUB) 978-3-11-071339-8

Polymer Matrix Composite Materials.
Structural and Functional Applications
Debdatta Ratna, Bikash Chandra Chakraborty, 2023
ISBN 978-3-11-078148-9, e-ISBN (PDF) 978-3-11-078157-1,
e-ISBN (EPUB) 978-3-11-078170-0

Smart and Functional Textiles
Bapan Adak, Samrat Mukhopadhyay (Eds.), 2023
ISBN 978-3-11-075972-3, e-ISBN (PDF) 978-3-11-075974-7,
e-ISBN (EPUB) 978-3-11-075993-8

Artificial Intelligence in Polymer Science and Nanotechnology

---

Autonomous Labs, Polymer Design, Materials Simulation

Edited by
Rishabha Malviya, Sonali Sundram, Sathvik Belagodu Sridhar
and Shivam Rajput

DE GRUYTER

Editors
Dr. Rishabha Malviya
Department of Pharmacy
Galgotias University
201310 Greater Noida
Uttar Pradesh
India
and
Galgotias Multi-Disciplinary Research &
Development Cell (G-MRDC)
Galgotias University
201310 Greater Noida
Uttar Pradesh
India
rishabhamalviya19@gmail.com

Prof. Sonali Sundram
Department of Pharmacy
School of Medical and Allied Sciences
Galgotias University
201310 Greater Noida
Uttar Pradesh
India
sonaliaim13@gmail.com

Prof. Dr. Sathvik Belagodu Sridhar
RAK Medical and Health Sciences University
RAK College of Pharmacy
11172 Ras Al Khaimah
United Arab Emirates
sathvik@rakmhsu.ac.ae

Shivam Rajput
Galgotias University
201310 Greater Noida
Uttar Pradesh
India
shivamrajput909@gmail.com

ISBN 978-3-11-168182-5
e-ISBN (PDF) 978-3-11-168187-0
e-ISBN (EPUB) 978-3-11-168238-9

Library of Congress Control Number: 2025942110

Bibliographic information published by the Deutsche Nationalbibliothek
The Deutsche Nationalbibliothek lists this publication in the Deutsche Nationalbibliografie; detailed bibliographic data are available on the Internet at http://dnb.dnb.de.

Cover image: wildpixel/iStock/Getty Images Plus
Typesetting: Integra Software Services Pvt. Ltd.
Printing and binding: CPI books GmbH, Leck

www.degruyterbrill.com
Questions about General Product Safety Regulation:
productsafety@degruyterbrill.com

Preface

In many scientific fields, including polymer science and nanotechnology, artificial intelligence (AI) is transforming research approaches. This interdisciplinary collaboration promises faster discovery, material design, a better knowledge of structure-property correlations, and greater customization and innovation potential. This book explores the role of AI in polymer design and characterization. The chapters discuss a wide range of subjects, including autonomous self-driving labs for material innovation and AI-enhanced polymeric drug delivery systems. To address difficult design problems, forecast polymer properties, enhance material performance, and uncover underlying relationships between structure and functioning, the book additionally discusses the use of machine learning in materials science.

The aims of the book are: to inspire future research at the intersection of data science, material informatics, nanotechnology, and polymer engineering; to introduce readers the basic and emerging applications of AI in polymer and nanomaterial research; to outline the technologies, methodologies, and frameworks reshaping material discovery; and to demonstrate the role of AI to develop autonomous, predictive, and scalable research models.

The book is aimed at a wide range of readers, including research scientists, graduate and postgraduate students, academic scholars, industry engineers, and technology developers, who are interested in the exciting convergence of AI and materials science. It provides useful perspectives and practical insights for academics and industrial experts.

https://doi.org/10.1515/9783111681870-202

Foreword

The potential of artificial intelligence (AI) and machine learning has triggered a revolutionary shift in the fields of materials science and polymer technology. AI has accelerated innovation by identifying hidden patterns, forecasting the characteristics of materials, creating molecular structures, and automating entire labs. This book is a significant scientific contribution in the field of polymer science. The book, edited by Dr. Rishabha Malviya and his team, provides an in-depth discussion of the role of AI technologies in nanoscience and polymer technology. The chapters in this book demonstrate a careful balance between theoretical frameworks, state-of-the-art research, and practical applications.

Since polymeric materials are essential to consumer technology, biomedical applications, environmental sustainability, and drug delivery, the book focuses on the applications and societal aspects of these developments. AI integration enables faster, smarter, safer, and more sustainable solutions that prioritize society and the environment.

The book invites researchers, engineers, academics, and students to adopt a new paradigm of data-driven, autonomous, and intelligent material discovery. It is not simply an academic book but a blueprint for the future. Through his efforts and leadership in bringing together an excellent group of contributors, Dr. Rishabha Malviya has compiled a book that will be an essential resource for present and future scholars working at the convergence of AI, polymer science, and nanotechnology.

Happy reading and best wishes,

Dr. Dhruv Galgotia
CEO, Galgotias University
Greater Noida, India

https://doi.org/10.1515/9783111681870-203

Contents

List of contributing authors

Wan Nurhidayah Wan Hanaffi
Laboratoire de Chimie Physique
Macromoléculaire
Université de Lorraine
CNRS, F-54000 Nancy
France
and
Department of Immunology
School of Medical Sciences
Universiti Sains Malaysia
Pulau Pinang, Malaysia

Khan Aejaz Ahmed
Faculty of Pharmacy
Integral University
Kursi Road
Lucknow 226026, Uttar Pradesh, India

Usama Ahmad
Faculty of Pharmacy
Integral University
Kursi Road
Lucknow 226026, Uttar Pradesh, India

Leandro Rodrigues Oviedo
Applied Nanomaterials Research Group (GPNAp)
Franciscan University (UFN)
Santa Maria, Rio Grande do Sul, Brazil

Maurício Dalla Costa Rodrigues da Silva
Federal University of Health
Sciences of Porto Alegre (UFCSPA)
Porto Alegre, Rio Grande do Sul, Brazil

Pâmela Cristine Ladwig Muraro Brasil
Northern Regional Technological Institute
Technological University of Uruguay (UTEC)
Rivera, Uruguay

Raisi Lenz Baldez
Northern Regional Technological Institute
Technological University of Uruguay (UTEC)
Rivera, Uruguay

Tiago Moreno Volkmer
Federal University of Pelotas (UFPEL)
Pelotas, Rio Grande do Sul, Brazil

William Leonardo da Silva
Applied Nanomaterials Research Group (GPNAp)
Franciscan University (UFN)
Santa Maria, Rio Grande do Sul, Brazil

Shikha Kumari
Department of Physics
School of Basic and Applied Sciences
IILM University
Greater Noida, Uttar Pradesh, India

Pratima Parashar Pandey
Department of Physics
School of Basic and Applied Sciences
IILM University
Greater Noida, Uttar Pradesh, India

Neha Minocha
School of Medical and Allied Sciences
K. R. Mangalam University
Gurgaon 122103, Haryana, India

Parijat Pandey
Department of Pharmaceutical Sciences
Gurugram University
Gurugram 122018, Haryana, India

Syed Salman Ali
Lloyd Institute of Management and Technology
Plot No. 11, Knowledge Park-II
Greater Noida 201306, Uttar Pradesh, India

Hrithik Dey
Lloyd Institute of Management and Technology
Plot No. 11, Knowledge Park-II
Greater Noida 201306, Uttar Pradesh, India

Atifa Khan
Lloyd Institute of Management and Technology
Plot No. 11, Knowledge Park-II
Greater Noida 201306, Uttar Pradesh, India

https://doi.org/10.1515/9783111681870-205

Bhoomika Chaudhary
Charak School of Pharmacy
Chaudhary Charan Singh University
Meerut, Uttar Pradesh, India

Yarun
Lloyd Institute of Management and Technology
Plot No. 11, Knowledge Park-II
Greater Noida 201306, Uttar Pradesh, India

Mohammad Asif
Faculty of Pharmacy
Lachoo Memorial College of Science and Technology
Sector A, Shastri Nagar
Jodhpur 342001, Rajasthan, India

Rahul Sharma
Department of Pharmacology
School of Medical and Allied Sciences
Galgotias University
Greater Noida, Uttar Pradesh, India

Sonali Sundram
Department of Pharmacology
School of Medical and Allied Sciences
Galgotias University
Greater Noida, Uttar Pradesh, India

Naga Rani Kagithala
Department of Pharmacology
School of Medical and Allied Sciences
Galgotias University
Greater Noida, Uttar Pradesh, India

Ruby Srivastava
Centre for Cellular and Molecular Biology – CSIR
Hyderabad, Telangana, India

Isha Kalbhor
Department of Pharmaceutical Sciences
School of Health Sciences and Technology
Dr. Vishwanath Karad MIT World Peace University
Pune 411038, Maharashtra, India

Shantanu Chakote
Department of Pharmaceutical Sciences
School of Health Sciences and Technology
Dr. Vishwanath Karad MIT World Peace University
Pune 411038, Maharashtra, India

Rushikesh Dabhade
Department of Pharmaceutical Sciences
School of Health Sciences and Technology
Dr. Vishwanath Karad MIT World Peace University
Pune 411038, Maharashtra, India

Pranay Sharma
Department of Pharmaceutical Sciences
School of Health Sciences and Technology
Dr. Vishwanath Karad MIT World Peace University
Pune 411038, Maharashtra, India

Sushant Gaikwad
Department of Pharmaceutical Sciences
School of Health Sciences and Technology
Dr. Vishwanath Karad MIT World Peace University
Pune 411038, Maharashtra, India

Raj Agarwal
Department of Pharmaceutical Sciences
School of Health Sciences and Technology
Dr. Vishwanath Karad MIT World Peace University
Pune 411038, Maharashtra, India

Satish Polshettiwar
Department of Pharmaceutical Sciences
School of Health Sciences and Technology
Dr. Vishwanath Karad MIT World Peace University
Pune 411038, Maharashtra, India

Kanaka Durga Devi Nelluri
KVSR Siddhartha College of Pharmaceutical Sciences
Vijayawada, Andhra Pradesh, India

S.K. Abdul Rahaman
School of Medical and Allied Sciences
Galgotias University
Greater Noida, Uttar Pradesh, India

Minakshi Kommoju
KVSR Siddhartha College of Pharmaceutical Sciences
Vijayawada, Andhra Pradesh, India

Bhavana Alaparthi
KVSR Siddhartha College of Pharmaceutical Sciences
Vijayawada, Andhra Pradesh, India

Jatinder Singh Dhaliwal
Indian Oil Corporation Ltd.
R&D Centre, Faridabad 121007
Haryana, India

Pravesh Kumar
Indian Oil Corporation Ltd.
R&D Centre, Faridabad 121007
Haryana, India

Rajesh M. Badhe
Indian Oil Corporation Ltd.
R&D Centre, Faridabad 121007
Haryana, India

Alex C. Pulikottil
Indian Oil Corporation Ltd.
R&D Centre, Faridabad 121007
Haryana, India

Alok Sharma
Indian Oil Corporation Ltd.
R&D Centre, Faridabad 121007
Haryana, India

Tamanna Rai
United College of Engineering and Research
Greater Noida, Uttar Pradesh, India
and
Department of Pharmacy
School of Medical and Allied Sciences
Galgotias University
Greater Noida, Uttar Pradesh, India

Rishabha Malviya
School of Medical and Allied Sciences Galgotias University
Greater Noida, Uttar Pradesh, India

Harshita Shakya
Department of Chemistry
Gurugram University
Mayfield Garden, Sector 51
Gurugram 122003, Haryana, India

Sona Gandhi
School of Engineering and Technology
Manav Rachna International Institute of Research and Studies
Faridabad 121004, Haryana, India

Fatima Haider
Saulat Institute of Pharmaceutical Sciences and Drug Research
Quaid-i-Azam University
Islamabad 45320, Pakistan

Uswa Mansoor
Saulat Institute of Pharmaceutical Sciences and Drug Research
Quaid-i-Azam University
Islamabad 45320, Pakistan

Maham Naeem
Saulat Institute of Pharmaceutical Sciences and Drug Research
Quaid-i-Azam University
Islamabad 45320, Pakistan

Gul Shahnaz
Saulat Institute of Pharmaceutical Sciences and Drug Research
Quaid-i-Azam University
Islamabad 45320, Pakistan

Komal Gupta
Department of Pharmacy
School of Medical and Allied Sciences
Galgotias University
Greater Noida, Uttar Pradesh, India
and
Galgotias College of Pharmacy
Greater Noida, Uttar Pradesh, India

Niranjan Kaushik
Department of Pharmacy
School of Medical and Allied Sciences
Galgotias University
Greater Noida, Uttar Pradesh, India

Amit Singh
School of Pharmacy
Monad University
Hapur, Uttar Pradesh, India

Vikram Sharma
Galgotias College of Pharmacy
Greater Noida, Uttar Pradesh, India

Tanya Biswa
Galgotias College of Pharmacy
Greater Noida, Uttar Pradesh, India
and
Delhi Institute of Pharmaceutical Sciences and Research
New Delhi, Delhi 110017, India

Wan Nurhidayah Wan Hanaffi, Khan Aejaz Ahmed, and
Usama Ahmad*

Chapter 1
Nanotechnology innovation: AI-enhanced polymer drug delivery systems

Abstract: The confluence of polymer nanotechnology with artificial intelligence (AI) is revolutionizing drug development and targeted drug delivery, ensuring unprecedented precision in therapeutics. By coupling data analytics with AI-driven polymeric nanocarrier systems, the technology surpasses the constraints of traditional drug delivery, including bioavailability and patient-specificity, by virtue of superior design, encapsulation efficiency, and controlled release mechanisms for the drug. Predictive modeling and machine learning algorithms aid AI in optimizing drug–polymer interaction and optimizing the physicochemical characteristics of polymer nanocarriers so that therapeutics are released accurately and controllably at target sites to reduce side effects. AI's capability to forecast the interaction of disease-specific, biocompatible, and responsive polymers for patient profiles advances the principle of precision medicine. Improved encapsulation efficiency and optimized kinetics by AI-driven processes maximize the direct delivery of therapeutics to the target site with optimal doses for enhanced bioavailability and therapeutic efficacy. The amalgamation of AI with polymeric nanotechnology thus enables drug discovery in the rapid prototype and optimization of nanocarrier systems to deliver precise, individualized drug delivery with high accuracy to diverse therapeutic domains. This chapter discusses the development potential of AI-fortified polymer-based drug delivery systems, their ability to maximize the efficacy and selectivity of treatment regimens. With AI-driven innovation, polymer nanotechnology has the ability to improve patient treatment through highly personalized and potent medical options, beginning a new era of tailored medicine and new paths of drug innovation.

Keywords: Artificial intelligence, targeted drug delivery, polymeric nanocarriers, nanotechnology, precision medicine, machine learning, personalized treatment

***Corresponding author: Dr. Usama Ahmad**, Associate Professor, Faculty of Pharmacy, Integral University, Kursi Road, Lucknow 226026, Uttar Pradesh, India, e-mail: usamaahmad.10@outlook.com
Wan Nurhidayah Wan Hanaffi, Laboratoire de Chimie Physique Macromoléculaire, Université de Lorraine, CNRS, F-54000 Nancy, France; Department of Immunology, School of Medical Sciences, Universiti Sains Malaysia
Khan Aejaz Ahmed, Faculty of Pharmacy, Integral University, Kursi Road, Lucknow 226026, India

https://doi.org/10.1515/9783111681870-001

1.1 Introduction to AI and polymeric nanotechnology in drug delivery

Artificial intelligence (AI) is transforming contemporary drug delivery systems by increasing formulation and optimization of therapeutic systems. AI finds the most appropriate drug delivery systems tailored to patients' specific requirements, improving treatment efficacy [1]. In pharmaceutical technology, AI-driven models predict drug–polymer interaction, improving stability, controlled release, and efficiency of the drug carrier [2]. Machine learning allows quick innovation, with the ability to design nanoparticles with desirable characteristics, optimizing targeted drug delivery with reduced development time and expense [3]. AI optimizes bioavailability by forecasting drug behavior in the biological system, resulting in improved absorption and therapeutic action [4]. AI utilizes patient-specific data to enable treatment individualization on the basis of genetic profiles to optimize clinical gain [5]. This technological advancement redrew pharmaceutical sciences, offering smart and efficient drug delivery strategies. Design and development of targeted drug delivery systems are presented in Figure 1.1, with AI integration.

The combination of AI and polymer nanotechnology has revolutionized materiomics, biomedicine, and drug delivery by improving the design, optimization, and functionalization of polymeric nanocarriers. Machine learning and deep learning in AI speed up the identification of polymer membranes for use in carbon capture and cancer therapy by predicting polymer composition, drug uptake, and controlled release, streamlining timelines and accuracy [6]. Polymeric nanosystems such as nanoparticles and vesicles have shown greater stability, tunability, and capacity to entrap more than one therapeutic agent, which is useful for addressing complicated diseases such as triple-negative breast cancer and resistant cancers [7–9]. AI has facilitated the development of polymeric systems that can be designed to emulate biological cells, allowing cellular behavior to be accessed with ease and drug delivery to be optimized [10]. Advanced models have maximized functionalization of nanostructured lipid carriers to enhance cellular recognition, which is crucial for targeting diseases such as chronic inflammation and chemotherapy-resistant cancers [11].

AI algorithms predict the self-organization and spatial arrangement of polymer-modified nanoparticles, which maximize the controlled release of drugs and cell targeting [12]. Emerging trends involve the combination of MXenes and polymers in nanocomposites for drug delivery and diagnostics, where AI maximizes design and therapeutic effectiveness [13]. AI-aided polymer nanoparticles have been synthesized via processes like double emulsion and pH-triggered synthesis, which enhance bioavailability and treat bacterial infections [14]. AI-aided polymeric systems are being studied in neuromorphic computing, where polymer devices exhibit synaptic behavior for brain-inspired computational modeling [15, 16]. These advancements show the way AI-polymer nanocarriers offer possibilities for precision medicine, targeted therapy, and other biomedical progress.

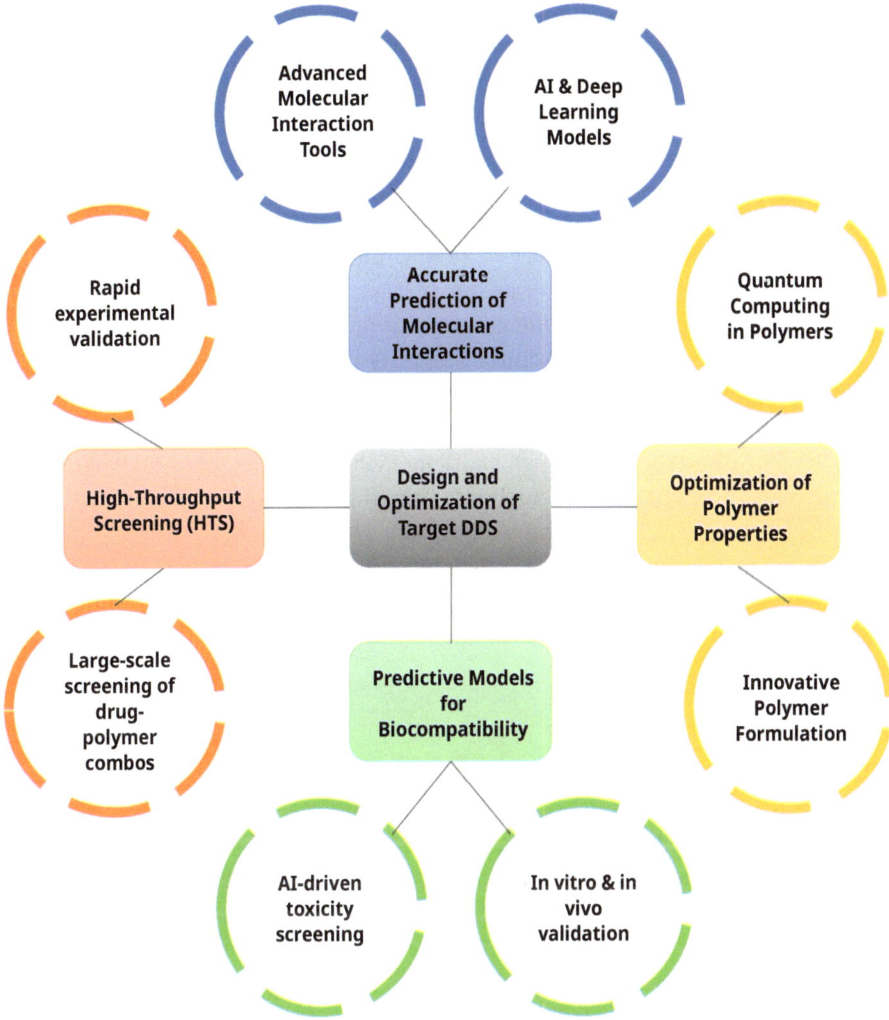

Figure 1.1: Design and optimization of targeted drug delivery system with integration of AI.

1.2 Design and optimization of targeted drug delivery systems

1.2.1 Accurate prediction of molecular interactions

Latest advances in computational algorithms enhanced predictive capability to engage in interactions with higher precision and effectiveness when higher accuracy and efficacy are needed. Some models, such as the SPVec-SGCN-CPI model, use a simplification of graph convolutional networks to predict compound–protein interactions using neighborhood information and topological features, outperforming other approaches in imbalanced data [17]. The SSGraphCPI model utilizes a three-channel deep learning model, integrating sequential and structural features of 1D SMILES strings and 2D molecular graphs, showing outstanding performance in datasets such as GPCR [18]. The LINE network representation approach utilizes a heterogeneous multi-molecular information platform to predict interactions with high precision [19]. Integration of molecular and interactome information through sophisticated deep learning architectures enhances the accuracy of prediction by integrating views beyond sequence homology and structural similarity [20]. Machine learning force fields (MLFFs) have been developed to simulate molecular dynamics with ab initio accuracy, solving issues such as nonlocal interatomic interactions [21]. Approaches to predicting binding free energies through end-state molecular dynamics simulations have shown strong correlation with actual data, which offers a lower-cost alternative to classical approaches [22]. Advances such as AlphaFold2 for protein–protein interaction prediction [23] and the CREST tool for exploring conformational spaces in receptor–ligand binding [24] enhance the accuracy of molecular interaction prediction and hence improve drug discovery and advance the knowledge of biological processes.

1.2.2 Optimization of polymer properties

Polymer property optimization involves an integrated method that includes advanced computational methods to maximize polymer performance and application. Multimodal deep-learning models are a forefront method that combines physical and chemical data to forecast properties of advanced polymer composites across extensive compositional ranges. The method facilitates handling of high-dimensional data in inverse material design with optimized properties [25]. Transfer learning of injection molding processes has been employed to optimize machine parameters based on limited data, demonstrating AI capability to enhance process efficiency and product quality in polymer processing [26]. Quantum computing facilitates optimization of polymer attributes by sampling dense polymer mixtures at equilibrium using quantum annealing technology, allowing polymers with designed structural properties [27]. AI application in designing multifunctional composites of MXene and graphene blends displays innovation in advanced multifunctional composites, indicating AI capability in designing materials with improved thermal, electrical, and mechanical properties for smart electronics [28]. In silico methodologies, such as molecular mechanics-driven multi-monomer concurrent docking, facilitate the optimization of molecularly imprinted polymers through the assessment of monomer combinations to enhance binding capacities, thereby improving the sensitivity and selectivity of artificial receptors [29]. Such methods greatly facilitate optimization of polymer properties, enabling advanced material design with improved performance and functionality.

1.2.3 Development of predictive models for biocompatibility

Synthesis of polymeric biocompatibility predictive models requires understanding the interactions between biological systems and polymeric substrates. Recent research has revealed key determinants of biocompatibility, which have assisted predictive modeling. Microporous scaffolds by stereolithography cellular compatibility is improved by postprocessing methods, including ultraviolet irradiation and Soxhlet extraction, eliminating residual toxic monomers [30]. Similarly, swelling behavior and crosslinking density in ionic liquid-containing hydrogels affect biocompatibility, with evidence from cell seeding experiments using L929 mouse fibroblasts [31]. Polysaccharides, naturally occurring and biodegradable, are of interest in drug delivery due to low immune reactivity and efficient release kinetics [32]. Copolymers containing thioester functionality, degrading amid physiological thiols, are of considerable significance in the realm of pharmaceutical delivery, owing to noncytotoxic breakdown products [33]. Electron beam-irradiated carboxymethyl chitosan hydrogels in wound healing display excellent biocompatibility in vitro and in vivo [34]. Addition of natural biopolymers, including starch and gelatin, in composite membranes improves compatibility with biological tissues, as evidenced by in vivo implantation studies [35]. Bioinspired polymeric materials that replicate cell membranes demonstrate enhanced biocompatibility in biomedical devices [36]. Hydrophilic polyphosphoesters show minimal interaction with blood components, demonstrating the contribution of polymeric chemistry to hemocompatibility [37]. Parameters such as polymeric architecture, crosslinking density, degradation behavior, and postprocessing techniques are indicated by these findings as determinants of significant influence on biocompatibility. These variables, when included with predictive models, improve the design of safer and more efficient biomedical materials while reducing extensive biological tests.

1.2.4 High-throughput screening

The technique of high-throughput screening (HTS) of polymers poses high importance, mostly in industries ranging from materials science and pharmaceuticals to environmental science, with emphasis on rapid identification and evaluation of polymer properties and potential applications. HTS has screened biopolymers such as chitosan and cationized cellulose nanofibrils to be screen printed on polypropylene substrates for facial masks, enhancing antiviral activity, while retaining viral filtration efficiency [38]. HTS accelerates automation of simulations for new material discovery, such as using the many-body perturbation theory to surpass density functional theory limitations for increased-level simulation convergence in materials science computations [39]. Screening through HTS for antimicrobial agents, accelerating the discovery of small molecules stimulating antimicrobial peptide synthesis, offers a platform for new therapies [40]. HTS screens single-particle acidity of the catalyst, giving insight into activity and compatibility with fluid catalytic cracking processes [41]. HTS is also leading in drug discovery, which is evident through the identification of small molecule inhibitors of proteins from pathogens like *P. falciparum*, which is critical to the development of antimalarial therapy [42]. Utilization of HTS in such applications speaks volumes about its usefulness and significance in speeding up compound discovery and optimization across a variety of scientific disciplines.

1.3 Enhancing encapsulation and controlled release

1.3.1 Enhancing encapsulation

The encapsulation within a suitable polymer involves diverse factors, which includes chemical structure of the polymer, molecular weight, and functional groups, influencing encapsulation efficiency and stability of the encapsulated compound. For instance, in encapsulation of cold-pressed hickory oil, β-cyclodextrin (β-CD) and porous starch (PS) possessed great encapsulation efficiencies of 80.40% and 75.52%, respectively, compared to maltodextrin and hydroxylpropyl-β-cyclodextrin, owing to conformational stability of the structure and heat stability, which are critical for oil oxidative stability maintenance [43]. The replacement of poly(ethylene glycol) (PEG) with linear polyglycerol (LPG), in the presence of Interferon-α2a, strongly regulates the pharmacokinetics and efficacy of bioconjugates, with PEGylated variants showing increased levels of exposure [44]. The physicochemical properties of poly-vinyl polymers, such as vinyl pyrrolidone-to-vinyl acetate ratio, regulate the efficiency of amorphous solid dispersions, modulating the dissolution and solid form stability [45]. The balance of cationic-hydrophobic functionality is important in the optimization of antibacterial efficacy and selectivity and reflects how the application of specific functional groups optimizes encapsulation efficiency through enhanced interaction with target molecules [46]. Entrapment of hydrophilic agents in extracellular vesicles demonstrates the significance of encapsulation procedures, where freeze-thaw and osmotic shock procedures are capable of achieving high drug encapsulation while preserving vesicle properties [47]. These examples demonstrate that the choice of a polymer is a complex process requiring careful consideration of the purpose of application, stability requirements, and interaction dynamics with the encapsulated entity to introduce increases in efficiency and performance.

Machine learning models and AI-based simulations maximize nanoformulation processes through optimization of comprehension and regulation of the properties of polymers. Machine learning models forecast the gas-separation behavior of polymers, illustrated by Barnett et al., who forecasted performance of over 11,000 homopolymers from a small dataset, developing membranes beyond the state-of-the-art performance for CO_2/CH_4 separation [48]. AI-based digital twins, as illustrated by Davidopoulou and Ouranidis, offer realistic in silico simulation of the manufacturing process, enabling optimization of nanosuspension formulations by forecasting and controlling the output from experimental data [49]. The combination of machine learning and infrared thermography facilitates automatic forecasting of the properties of polymer composites, required for the optimization of formulation [50]. Transfer learning, as illustrated by Tsubaki and Mizoguchi, facilitates forecasting of polymer properties by utilizing pre-trained models with small molecules, gaining insight into quantum-chemical properties of polymers [51]. Utilizing machine learning models, combined with Bayesian optimization, as illustrated by Lampe et al., facilitates rapid optimization of nanocrystal syntheses, providing potential to enhance material homogeneity and quality with limited data requirements [52].

1.3.2 Controlled release profiles

Mathematical modeling of release kinetics of controlled release of polymers utilizes different models to maximize drug delivery systems. Polysaccharides, owing to their biocompatibility and biodegradability, are widely used in nanoscale synthesis to encapsulate drugs and release, where mathematical models predict matrix behavior, minimizing experimental trial and error [53]. For chitosan/alginate membranes, kinetic models such as zero-order, first-order, Higuchi, Korsmeyer–Peppas, and Weibull functions examine release profiles, where the Weibull model fits some forms of membranes optimally [54]. Chitosan particulate carriers illustrate the pivotal role played by mathematical models in describing particle structure and release kinetics relationship [55]. Coupling physiologically based pharmacokinetic (PBPK) models with release kinetics permits simulation of systemic exposure over an extended period of time and release parameter optimization [56]. Design of experiments (DOE) and physiologically based biopharmaceutics modeling (PBBM) in extended-release formulation development predict absorption and formulate design optimization [57]. Programmable self-regulated molecular buffers provide an innovative strategy for maintaining target drug concentrations [58]. Coordination between drug release and targeting, as illustrated with lipid-like polymer nanoparticles, underlines the critical need to balance release behavior with targeting capability [59].

Stimuli-responsive polymer systems have opened new doors for achieving accurate and efficient drug delivery. These systems are triggered by internal and external stimuli, which widely probe to maximize the aimed therapeutic effect and controlled drug release. In gene therapy, stimulus-sensitive nanocarriers release therapeutic nucleic acids, triggered by responding to physiological conditions such as pH changes and redox gradients, within tumor microenvironments [60]. Nanomedicine platforms with distinctive properties allow for more effective drug delivery in cancer treatment through a response to microenvironmental stimuli such as enzyme activity and pH levels [61, 62]. The mechanism of drug release is increased with integration to AI by evaluating biological conditions and predicting the best release patterns. Through modulation based on patient-specific data, AI models are able to maximize drug delivery to obtain better accuracy and outcomes. Smart hydrogels and nanomaterials that respond to biological and environmental stimuli are investigated for their potential uses such as antimicrobial coatings and demand-driven drug delivery [63, 64]. Such systems have the potential to deliver therapeutic drugs that are triggered by particular stimuli, reducing systemic side effects and optimizing the therapeutic outcomes [65, 66]. AI is also able to facilitate on-time monitoring and adaptive drug release control, which allows for personalization of the treatment regimens. Hence, the prompt-responsive material synergy of AI is setting the stage for innovative healthcare remedies with higher specificity, lower side effects, and greater adaptability to patient demands [67, 68]. Integration represents a revolutionary new approach in controlled drug delivery, with a new line of more intelligent, efficient therapeutic regimens. Machine learning algorithms can improve accuracy in real-time monitoring systems of regulated polymer release by analyzing complex data and forecasting system behavior. Integration of machine learning with spectroscopic methods, i.e., deep ultraviolet Raman spectroscopy, allows accurate identification and quantification of chemical pollutants in environmental monitoring, which can be applied to polymer release monitoring by discriminating specific spectral features of polymer release or degradation processes [69]. Machine learning models, particularly with deep neural networks, can adjust parameters of coarse-grained models of polymers, enhancing the accuracy of predictions about polymer behavior under different conditions such as temperature fluctuation and stress relaxation [70]. Application of recurrent neural networks in predictive control structure modeling can improve real-time control of complex systems by forecasting control-relevant parameters, crucial for maintenance of target release rate in polymer systems [71]. Advanced feedforward neural networks, with optimization methods, have improved prediction accuracy for complex systems, such as wastewater treatment, by identifying nonlinear relationships and avoiding overfitting, which can be applied to polymer release monitoring [72]. Convolutional neural networks have been promising in analyzing high-dimensional data, such as fluorescence spectra, to forecast chemical byproducts, which can be applied to monitoring specific release of polymer constituents [73]. Real-time monitoring systems for regulated polymer release are challenging, mainly due to limitations, primarily because of complexities in accurate forecasting and regulation of release profiles in dynamic environments. One such vital limitation is compatibility between pharmaceutical agents and carrier materials, restricting the loading capacity and overall efficiency of drug delivery systems. This is countered with sophisticated techniques, such as the sequential solidification technique, which improve drug loading and release control by encapsulating drug nanoparticles in a polymeric matrix, as shown by high drug-loaded microspheres for spinal cord injury therapeutic use [74]. Dynamic characteristics of polymer interactions in systems that fluctuate pose a consistent challenge to standard monitoring methods, which tend to require static geometries of the sample. AI-based spectroscopic methods offer a desirable strategy by enabling rapid, real-time evaluation of structural alterations, with no limitations posed by standard strategies [75]. The capacity of AI in handling large amounts of data as well as in identifying low-dimensional patterns within complicated systems can be utilized to enhance flow control in polymer release systems so that there can be more adaptive and accurate regulation mechanisms [71]. AI also helps in enabling the development of feedback-controlled adaptive biomaterials, e.g., hydrogels that release anti-inflammatory drugs, upon activation by certain biological stimuli, to enhance the safety and efficacy of polymer-based drug delivery systems [76].

Polymeric systems of controlled release and AI hold immense potential in sustaining the stability and longevity, which is truly evident by a number of empirical analyses. Polylactic/glycolic acid (PLGA) has been utilized to prepare microspheres for payload delivery in single, time-delayed doses in a controlled fashion to provide patient compliance and therapeutic efficacies [77]. Hence, involvement of AI can optimize release profiles by anticipating and responding to release kinetics with real-time inputs, which can render therapy more effectively. Sustained drug release and mathematical modeling facilitate the prediction of release dynamics, which can be optimized through AI software to reduce trial and error by experimentation [32]. The potential of AI can be utilized for nanofibers that facilitate photothermally controlled release to regulate external stimuli, such as light, to release drugs at the most appropriate time intervals [78]. Supramolecular micelles for antioxidant encapsulation, such as fullerene, illustrate the potential of AI to manipulate environmental factors such as temperature to regulate the release mechanism [79]. Drug delivery to tumor site has been enabled by polymeric micelles and with the presence of AI, the process can be further improved by fine-tuning stimuli-responsive processes to release drugs [80]. Encapsulation of bacteriophages in PLGA/alginate composites to enhance in vivo viability illustrates the ability of AI to monitor and control release to maintain therapeutic concentrations for long periods of time [81]. Inflammation-reactive hydrogels for drug release, in response to particular biological triggers, can be designed with AI, which could identify inflammation and modulate release [76]. Merging polymer blends and AI in controlled release is a promising area of research for maximizing medical use of drug delivery systems in aspects of stability, durability, and efficacy.

1.4 Personalized and disease-specific treatment approaches

1.4.1 Customizing nanocarriers for patient profiles

Individualization of nanocarriers from individualized genetic and molecular profiles of patients requires a list of vital factors, as discussed by various applications and methodologies in relevant scholarly literature. First, employment of microRNAs (miRNAs), encapsulated in nanocarriers, against cancer stemness, for example, melanoma, necessitates the choice of miRNAs possessing the ability to modulate epigenetic signaling cascades. This necessitates nanocarriers to provide maximum entrapment efficiency, stability, and specificity to allow the delivery of such miRNAs, and potentially with chemotherapeutic drugs, to enhance therapeutic effects [82]. Moreover, mechanical properties of nanocarriers, e.g., silica nanocapsules, can be designed to govern the delivery of encapsulated drugs, in response to external mechanical stimulation, which is essential to targeted drug delivery and release to the target site [83]. In ocular diseases, lipid-DNA nanoparticles demonstrate the importance of nanocarriers to possess better biodistribution and sustained release, offsetting such limitations as rapid diffusion of drugs and poor tissue retention [84]. Lastly, the development of stimuli-responsive nanocarriers for gene therapy, responsive to physiological or external stimuli, demonstrates the critical importance of smart delivery systems able to target the therapeutic agents to the target site and overcome biological barriers [60].

Personalization of nanocarriers to the individual patient profiles in their biological and metabolic environments requires advanced technologies and complete awareness of the individual needs of each patient. Three-dimensional printing technologies, investigated in personalized medicine, allow the fabrication of customized nanomedicines to conform to the individual physiological profiles of patients, such as metabolic rates and biological milieus [85]. Personalization is of critical significance in illnesses such as ulcerative colitis, where sensitive nanosystems can direct pharmacological agents to the area of inflammation, producing maximum therapeutic responses and minimum systemic side effects [86]. Nanocarrier modulation of the innate immune response can be optimized to modulate infection and inflammatory processes in patients with varied immune profiles [87]. In the treatment of chronic wound healing, nanocarriers may be formulated to disassemble biofilms and release therapeutic molecules at wound sites, considering the patient's microbial niche and immune system [88]. Control-programmable molecular buffers that regulate on-demand drug concentration levels exemplify how nanocarriers can be custom designed to meet the distinct pharmacokinetic and pharmacodynamic needs of individual patients [89]. Employing nanocomposites for localized drug delivery, as illustrated by the targeted use of infliximab for inflammatory bowel disease, showcases the potential of nanocarriers to be precisely tailored to decrease systemic exposure while improving locoregional drug administration, in accordance with the patient's distinct disease phase and biological surroundings [90].

Several critical immunological parameters must be taken into account for the customization of nanocarriers, in alignment with patient profiles. Understanding the immunomodulatory effects of nanoparticles is essential, since they can influence drug biodistribution and toxicity, potentially triggering either favorable or unfavorable immune reactions, dictated by the formulation and administration technique. The immune system's role in the detection and subsequent clearance of nanoparticles can optimize therapeutic efficacy; however, this necessitates comprehensive toxicological evaluations to preclude potential harm [91]. Moreover, nanoparticles could play a crucial role in promoting antigen-specific T-cell tolerance within autoimmune diseases, indicating a beneficial method for reintroducing immune tolerance through focused interaction with designated antigen-presenting cells [92]. Immune response modulation in nanocarrier design must be taken into consideration, as in biomimetic nanoparticles for cancer immunotherapy, which can enhance the immunosuppressive tumor microenvironment [93]. Physicochemical characteristics of nucleic acid nanoparticles (NANPs) and delivery devices thereof play an important role in immune response modulation, with predictive resources such as AI-cell playing an important role in guiding NANP design for the best immunological benefit [94]. Dose and route of administration of nanoparticles control antibody response against lipid nanoparticles (LNPs) applied for mRNA delivery, with every route inducing a distinct immune response [95]. Nanoparticle surface modification has the potential to enhance their immune-activating profile, modulating protein loading and delivery to target tissues and being safe [96]. AI-driven models have the potential to maximize precision in nanocarrier personalization by incorporating molecular profiling data to stratify patients and forecast drug response, allowing for more personalized treatment for autoimmune and other multifactorial diseases [97].

1.4.2 Applications in precision medicine

The recent developments in precision medicine have discovered notable biomarkers that increase the efficacy of AI-augmented polymer drug delivery systems (DDS) for cancer treatment. They are drug response predictors and putative therapeutic targets, which can be used to design personalized treatment regimens. In hypopharyngeal cancer, four biomarkers – AGR2, PDE4D, NINJ2, and CDC25B – are associated with cellular sensitivity to chemotherapeutic drugs such as docetaxel, cisplatin, and 5-fluorouracil. These biomarkers are drug response predictors and therapeutic targets, enabling precision medicine strategies to chemotherapy in hypopharyngeal cancer patients [98]. In drug-resistant cancer, aggregation-induced emission (AIE)-active polymers are promising by regulating redox homeostasis, overcoming barriers to enhance drug accumulation and penetration, thus enhancing therapeutic efficacy [99]. In hepatocellular carcinoma (HCC), proteomic profiling and AI technologies have found new biomarkers like AFP, GPC3, and DKK1, involved in dysregulated pathways like Wnt/β-catenin and PI3K/Akt, providing leads for early diagnosis, prognosis, and targeted therapy [100]. Dysregulation of the CDK4/6 cell cycle regulatory pathway is a significant feature of pediatric and young adult sarcomas. CDK4/6 inhibition has been a promising therapy, though resistance is a concern, leading to the discovery of new combination therapies to enhance efficacy [101]. The integration of AI in bioinformatics has markedly progressed the knowledge surrounding biomarkers tied to CDK4/6 inhibitors, uncovering genes like ERBB2, AKT1, and PIK3CA as fresh targets in breast cancer therapy [102]. The fusion of quantitative molecular proteomics, alongside immunology, has highlighted immune checkpoint biomarkers, specifically PD-1/PD-L1, as significant predictors for immunotherapy responses. Moreover, deep learning neural network models have recognized genes such as CCR7 and LGALS1 as critical predictors of immune checkpoint therapy response, thereby enhancing patient stratification [103, 104],.

Ionizable drug delivery frameworks (IDDSs) have revealed considerable potential in alleviating toxicity and boosting nucleic acid delivery. These systems exhibit pH sensitivity, which is particularly advantageous within the acidic milieu of endosomes. Exposure to the reduced pH profile found in endosomes triggers IDDSs to undergo a structural shift that enhances the release of nucleic acids into the cytoplasm, which in turn boosts the success of gene delivery [103, 105]. One significant case of a positive application exists in the management of Duchenne muscular dystrophy (DMD), an intense genetic illness defined by advancing muscle loss. Established strategies for the delivery of antisense oligonucleotides (AONs) face challenges arising from their vulnerability and inadequate tissue absorption. To address these challenges, hydrogel-forming polymeric carriers have been meticulously engineered. These carriers establish a stable environment for AONs, safeguarding the molecules from degradation and enhancing their uptake within muscle tissue. Consequently, patients afflicted with DMD have reported improved therapeutic outcomes, encompassing enhanced muscle functionality and a deceleration in disease progression [106]. Yet another progress in drug formulations that are polymer-derived involves the implementation of ionizable LNPs (iLNPs), which have been adapted for the conveyance of mRNA COVID-19 vaccines. The function of iLNPs to encapsulate, stabilize, and proficiently convey therapeutic agents, including nucleic acids, to designated cells has surfaced as a point of great significance. Their proficiency in stabilizing mRNA against degradation and facilitating cellular uptake has expedited the rapid development of COVID-19 vaccines. Additionally, it has highlighted the prospects of iLNPs for different purposes in gene therapy, targeting the resolution of genetic ailments and cancerous growths [107]. Studies have considered DNA nanostructures as multifunctional delivery systems for therapeutic drugs in cancer-related treatments. These tiny structures are uniquely tailored with aptamers, enabling a focused approach to delivering cancer-fighting drugs to tumor cells, thus lowering off-target reactions and boosting the effectiveness of anticancer therapies. For example, the application of DNA nanostructures for the delivery of doxorubicin to cancer cells has resulted in improved tumor suppression while concurrently mitigating side effects [108]. Glycopolymers have also been researched for targeted gene delivery, particularly in HCC therapy. Glycopolymers incorporate sugar units that can interact with specific receptors on the liver cell surface. Through this interaction, therapeutic genes can be delivered with high specificity and transfection efficiency and thus render them as promising vectors for HCC gene therapy. In such a case, off-target avoidance and liver cell targeting need to be employed to enable effective gene therapy [109].

Polymeric drug delivery systems are a breakthrough approach for the treatment of infectious diseases, leading to maximum therapeutic effectiveness and no antimicrobial resistance (AMR). Polymeric characterization is utilized to ensure maximum bioavailability of antibiotics and site-specific delivery, with the best drug concentration at the site of infection. Fluoroquinolones, encapsulated in polymer-based nanocarriers, have achieved maximum antimicrobial activity against Staphylococcus aureus pathogens through the maintenance of sustained high local concentrations of drugs, reducing opportunities for resistance development [110]. Site-specific delivery maximizes efficacy and minimizes systemic exposure, which is vital in minimizing resistance-mediated selective pressure. AI maximizes the potential of these systems through drug combination optimization and dosing regimen optimization. Platforms like IDentif.AI can potentially identify synergistic drug combinations, useful for combating multidrug-resistant infections such as nontuberculous mycobacteria infections [111]. AI can predict effective antibiotic combinations, minimizing trial-and-error experimentation and maximizing patient outcomes. Machine learning facilitates treatment optimization by patient history through therapy selection based on reduced probabilities of resistance [112]. Personalized therapy maximizes therapy efficacy while minimizing risk of contribution to AMR. Novel polymeric systems like temperature-responsive hydrogels and hyaluronic-acid-based drug delivery systems attain antimicrobial precision of therapy through drug release, in accordance with specific physiological conditions. This has been successful in combating intra-abdominal infections and dermal infections and minimized systemic side effects and maximized efficacy [113, 114]. Sustaining drug concentrations over a long period at infection sites minimizes requirements for high systemic dosages of drugs, reducing the pressure to develop resistance [115]. AI maximizes the dosing regimens, especially for multifactorial conditions like sepsis. Algorithms such as the deep deterministic policy gradient help in calculating the optimal dose of antibiotics in real time and deliver precise volumes of medication to combat infection without overexposure. This level of precision optimizes survival and reduces risk of resistance by avoiding suboptimal or excessive dosing [116].

Computational modeling that is AI-based allows the design and optimization of polymeric carriers for the effective delivery of therapeutic agents with minimal side effects. One of the approaches includes peroxisome proliferator-activated receptor gamma (PPAR-γ) agonist-loaded nanocarriers for inflammatory eye disorders. These formulations enhance drug retention at the target site that suppresses inflammation without inducing cytotoxicity or compromising corneal integrity. Conventional approaches are unable to overcome rapid drug clearance and poor tissue penetration; hence it is a major advancement in ocular drug delivery [117]. Polymeric nanoparticles responsive to reactive oxygen species (ROS) have been designed for the treatment of inflammatory skin disorders, which utilize disease-specific oxidative stress to trigger controlled drug release in inflamed tissues, providing localized therapy with minimal systemic exposure [118]. AI-optimized nanocarriers are demonstrating improved drug delivery specificity in the treatment of rheumatoid arthritis. Dexamethasone-loaded flexible liposome hydrogels have been designed for transdermal delivery, which provides targeted drug release at inflamed joints that minimizes systemic corticosteroid exposure by avoiding immune suppression and metabolic side effects of long-term steroid treatment [119]. Poly(ε-caprolactone) nanoparticles have been designed for sustained intra-articular delivery of leflunomide for a disease-modifying anti-rheumatic drug (DMARD). By targeting drug action in inflamed joints, these nanoparticles suppress inflammatory markers without inducing systemic toxicity, overcoming a limitation of conventional systemic DMARD administration [120]. Advances in polymeric drug delivery include smart colonic delivery systems for inflammatory bowel diseases (IBD). These systems utilize pH-sensitive polymer formulations that ensure drug release selectively in the colon, improving therapeutic efficacy and minimizing dosing frequency. Such selectivity maintains drug stability in the gut and minimizes inflammation in other parts of the gastrointestinal tract [121]. AI is also driving bio-based polymeric material design at an ever-increasing pace. Bayesian optimization-based machine learning algorithms efficiently design high-performance, environmentally friendly polymers. These processes accelerate material design, minimize material cost, and maximize mechanical as well as biocompatibility functionalities of the polymer, enabling improved drug delivery platforms [122].

The drug delivery strategies for neurodegenerative diseases by improving bioavailability, targeting specificity, and penetration through biological barriers such as the blood-brain barrier (BBB) are transformed by the application of polymer formulations. These formulations employ polymeric nanocarriers to provide controlled and sustained release of the drug, which results in enhanced therapeutic effects and minimized systemic side effects. Strategies involving AI have played a key role in the optimization of these formulations for enhanced penetration through the BBB and controlled release of the drug at the diseased neuronal sites [123, 124]. Biodegradable polymeric carriers provide sustained and prolonged release of therapeutic agents. These carriers increase the half-life of the drug and induce nerve regeneration and functional recovery, which is essential in diseases with neuronal damage such as Parkinson's and Alzheimer's diseases. Research shows that polymeric drug delivery systems can improve neuroprotection by stabilizing and prolonging the activity of neurotrophic factors and anti-inflammatory drugs and preventing neuronal degeneration [125]. Machine learning algorithms optimize these polymer-based systems by revealing molecular transport mechanisms that can be optimized to enhance drug permeability and ion selectivity through polymeric membranes. It also helps in the selection of effective polymer compositions to penetrate the BBB while maintaining drug stability and activity in the central nervous system [126]. Optimization, which involves AI, helps in reduced trial-and-error experiments and effective neurotherapeutic formulations, which also are formulated at a very faster rate. Interpenetrating polymer networks (IPNs) render drug delivery by responding to external stimuli such as pH and temperature. The networks provide site-specific and sustained release of the drug that provides precise modulation of therapeutic action based on pathophysiological conditions in the brain. IPNs have also been found to target dysregulated disease pathways of neurodegenerative diseases that include chronic inflammation and oxidative stress, effectively. They help in facilitating co-delivery of drugs, improving biodistribution and synergistic therapeutic effects, which leads to more comprehensive neuroprotective effects [127]. Nanocarrier systems such as cell-derived nanovesicles and exosomes are currently the cutting-edge platforms for targeted brain drug delivery. These carriers are naturally biocompatible and can penetrate the BBB, making them effective vehicles to deliver therapeutic agents to diseased neurons [128]. Predictive modelling, in assistance with AI, optimizes their potential in precision medicine that allows treatments to be designed according to patient-specific needs. Drug design involving AI has made it possible to design polymer formulations with both synthetic and phytochemical compounds that target multiple pathogenic mechanisms at once. These formulations enhance neuroprotection by stabilizing antioxidants, modulating neuro inflammation, and enhancing neuronal survival pathways. By combining AI with polymer science, scientists can screen hundreds of thousands of potential formulations quickly and optimize those with the highest therapeutic potential [124]. Optimization through AI continues to optimize the design and application of polymer-based drug delivery systems and hence the future of neurotherapeutic interventions will be marked by continued advances in precision medicine, promising more effective and personalized interventions for disabling neurodegenerative diseases [129]. A summary of some applications of AI in precision medicine is presented in Table 1.1.

Table 1.1: An overview of applications of AI in precision medicine.

Application area	Key advancements	Impact	References
Cancer therapy	Identification of biomarkers (AGR2, PDE4D, NINJ2, and CDC25B) for drug sensitivity in hypopharyngeal cancer	Enhances chemotherapy precision and drug targeting	[98]
	AIE-active polymers for drug-resistant cancers	Improves drug penetration and overcomes resistance	[99]
	AI-driven proteomic profiling in HCC (AFP, GPC3, and DKK1)	Enables early diagnosis and targeted therapy	[100]
	CDK4/6 pathway dysregulation in sarcomas	Drives novel combination therapies for resistant cases	[101]
	AI-driven biomarker identification for CDK4/6 inhibitors (ERBB2, AKT1, and PIK3CA)	Expands treatment options for breast cancer	[102]
	AI-enhanced immune checkpoint therapy predictions (PD-1/PD-L1, CCR7, and LGALS1)	Enables precise patient stratification for immunotherapy	[103], [104]
Gene therapy	Ionizable drug delivery systems (IDDSs) for nucleic acid delivery	Enhances drug escape from endosomes	[105]
	Hydrogel-based polymeric carriers for Duchenne muscular dystrophy (DMD)	Improves antisense oligonucleotide (AON) stability and tissue uptake	[106]
	Ionizable lipid nanoparticles (iLNPs) for mRNA delivery (COVID-19 vaccines)	Demonstrates successful application in gene therapy	[107]
	DNA nanostructures functionalized with aptamers	Enables precise chemotherapeutic drug targeting	[108]
	Glycopolymers for hepatocellular carcinoma (HCC) gene therapy	Ensures high transfection efficiency and targeted delivery	[109]
Infectious diseases and AMR	Polymer nanocarriers for antibiotic delivery	Enhances antimicrobial efficacy against resistant bacteria	[110]
	AI-driven optimization of synergistic antibiotic combinations (IDentif.AI)	Identifies effective treatments for multidrug-resistant infections	[111]

Table 1.1 (continued)

Application area	Key advancements	Impact	References
	Machine learning for personalized antibiotic dosing	Reduces resistance risk and enhances patient outcomes	[112]
	Temperature-responsive hydrogels and hyaluronic acid platforms	Provides localized drug release in infections	[113], [114]
	AI-driven dosing models (deep deterministic policy gradient)	Optimizes antibiotic dosing in sepsis	[116]
Autoimmune and Inflammatory diseases	PPAR-γ agonist-loaded nanocarriers for inflammatory eye diseases	Improves drug retention without toxicity	[117]
	ROS-responsive polymeric nanoparticles for skin diseases	Enables disease-specific drug activation	[118]
	AI-optimized liposome hydrogels for rheumatoid arthritis (dexamethasone)	Reduces systemic toxicity and improves drug targeting	[119]
	Poly(ε-caprolactone) nanoparticles for intra-articular leflunomide delivery	Enhances localized drug action in rheumatoid arthritis	[120]
	Smart colonic delivery systems (pH-sensitive polymers) for IBD	Ensures targeted release and prolonged efficacy	[121]
	AI-guided bio-based polymer development (Bayesian optimization)	Reduces costs and improves sustainability of polymer formulations	[122]
Neurodegenerative diseases	AI-optimized polymeric carriers for BBB penetration	Enhances targeted drug delivery to neuronal sites	[123], [124]
	Biodegradable polymeric carriers for nerve regeneration	Supports prolonged neurotrophic factor release	[125]
	AI-driven molecular transport models	Improves polymeric permeability and drug stability	[126]
	Interpenetrating polymer networks (IPNs) for controlled drug release	Enables pH- and temperature-responsive drug targeting	[127]
	Cell-derived nanovesicles and exosomes for brain-targeted drug delivery	Facilitates BBB penetration and minimizes side effects	[128]
	AI-driven drug formulation optimization	Enables rapid evaluation of effective polymer–drug combinations	[129]

1.5 Conclusion

The synergy of AI and polymeric nanotechnology is a revolutionary strategy for drug discovery and targeted delivery by overcoming the limitations faced by conventional therapeutics. AI data analytics, predictive modeling, and algorithms facilitate the simple fabrication of polymeric nanocarriers. This technology offers solutions to the challenges of lack of bioavailability, off-targeting, non-patient-specific treatment tailoring, and most importantly, enables precise control of dosage. AI optimizes the design of biocompatible polymers so that the properties of the polymer can be tailored to match the needs of particular diseases and the individuals affected by them. This customization brings about greater encapsulation efficiency, improved drug release kinetics, precise targeting of diseased tissues, and in general, enhanced therapeutic effects with fewer side effects. The use of AI in the discovery of medicines via polymer nanotechnology overcomes the issues involved in long prototype and refinement procedures with polymeric nanocarrier systems. This substantially reduces the development time required and promotes the development of precision medicine such that all approaches to treatment are tailored to each person's individual genetic, molecular, and physiologic make-up. Possible applications range from cancer treatments to infectious disease therapies, providing more efficient treatment opportunities. In summary, the union of AI with polymeric nanotechnology is a major breakthrough in medicine. With the integration of sophisticated computational approaches and novel material sciences, the method holds high levels of precise, patient-specific therapeutic interventions. As AI improves, its conjunction with polymer nanotechnology will spearhead the future of pharmaceuticals, transforming the diagnosis, treatment, and control of disease to better the patients' quality of life across the world.

References

[1] Paul D, Sanap G, Shenoy S, Kalyane D, Kalia K and Tekade RK. Artificial intelligence in drug discovery and development. Drug Delivery and Translational Research, Jan 2021;26(1):80–93. doi: 10.1016/j.drudis.2020.10.010.

[2] Vora LK, Gholap AD, Jetha K, Thakur RRS, Solanki HK and Chavda VP. Artificial intelligence in pharmaceutical technology and drug delivery design. Pharmaceutics, Jul 2023;15(7):1916. doi: 10.3390/pharmaceutics15071916.

[3] Jena GK, Patra CN, Jammula S, Rana R and Chand S. Artificial intelligence and machine learning implemented drug delivery systems: A paradigm shift in the pharmaceutical industry. Journal of bio-X Research, Jan 2024;7. doi: 10.34133/jbioxresearch.0016.

[4] Ali KA, Mohin S, Mondal P, Goswami S, Ghosh S and Choudhuri S. Influence of artificial intelligence in modern pharmaceutical formulation and drug development. Future Journal of Pharmaceutical Sciences, Mar 2024;10(1):53. doi: 10.1186/s43094-024-00625-1.

[5] Serrano DR, et al. Artificial intelligence (AI) applications in drug discovery and drug delivery: Revolutionizing personalized medicine. Pharmaceutics, Oct 2024;16(10):1328. doi: 10.3390/pharmaceutics16101328.

[6] Giro R, et al. AI powered, automated discovery of polymer membranes for carbon capture. Npj Computational Materials, Jul 2023;9(1):133. doi: 10.1038/s41524-023-01088-3.
[7] Li D, Zhang X, Chen X and Li W. Research progress and prospects for polymeric nanovesicles in anticancer drug delivery. Frontiers in Bioengineering and Biotechnology, Feb 2022;10. doi: 10.3389/fbioe.2022.850366.
[8] Fatima M, Sheikh A, Abourehab MAS and Kesharwani P. Advancements in polymeric nanocarriers to mediate targeted therapy against triple-negative breast cancer. Pharmaceutics, Nov 2022;14(11):2432. doi: 10.3390/pharmaceutics14112432.
[9] Chen S, et al. Redox homeostasis modulation using theranostic AIE nanoparticles results in positive-feedback drug accumulation and enhanced drug penetration to combat drug-resistant cancer. Mater Today Bio, Dec 2022;16:100396. doi: 10.1016/j.mtbio.2022.100396.
[10] Palivan CG, et al. Advancing artificial cells with functional compartmentalized polymeric systems – In Honor of Wolfgang Meier. Biomacromolecules, Sep 2024;25(9):5454–5467. doi: 10.1021/acs.biomac.4c00769.
[11] Martinez-Borrajo R, Diaz-Rodriguez P and Landin M. Engineering mannose-functionalized nanostructured lipid carriers by sequential design using hybrid artificial intelligence tools. Drug Delivery and Translational Research, Jan 2025;15(1):343–354. doi: 10.1007/s13346-024-01603-z.
[12] Gautham SMB and Patra TK. Deep learning potential of mean force between polymer grafted nanoparticles. Soft Matter, 2022;18(41):7909–7916. doi: 10.1039/D2SM00945E.
[13] Parajuli D. MXenes-polymer nanocomposites for biomedical applications: Fundamentals and future perspectives. Frontiers in Chemistry, May 2024;12. doi: 10.3389/fchem.2024.1400375.
[14] de Almeida Campos LA, de Souza JB, de Queiroz Macêdo HLR, Borges JC, de Oliveira DN and Cavalcanti IMF. Synthesis of polymeric nanoparticles by double emulsion and pH-driven: Encapsulation of antibiotics and natural products for combating Escherichia coli infections. Applied Microbiology and Biotechnology, Dec 2024;108(1):351. doi: 10.1007/s00253-024-13114-5.
[15] Kou L, et al. High sensitivity and wide response range artificial synapse based on polyimide with embedded graphene quantum dots. Scientific Reports, May 2023;13(1):8194. doi: 10.1038/s41598-023-35183-8.
[16] Bong JH, et al. Graphene oxide–DNA/graphene oxide–PDDA sandwiched membranes with neuromorphic function. Nanoscale horizons, 2024;9(5):863–872. doi: 10.1039/D3NH00570D.
[17] Zhang Y, Li J, Lin S, Zhao J, Xiong Y and Wei D-Q. An end-to-end method for predicting compound-protein interactions based on simplified homogeneous graph convolutional network and pre-trained language model. Journal of Cheminformatics, Jun 2024;16(1):67. doi: 10.1186/s13321-024-00862-9.
[18] Wang X, Liu J, Zhang C and Wang S. SSGraphCPI: A novel model for predicting compound-protein interactions based on deep learning. International Journal of Molecular Sciences, Mar 2022;23(7):3780. doi: 10.3390/ijms23073780.
[19] Ji B-Y, You Z-H, Jiang H-J, Guo Z-H and Zheng K. Prediction of drug-target interactions from multi-molecular network based on LINE network representation method. Journal of Translational Medicine, Dec 2020;18(1):347. doi: 10.1186/s12967-020-02490-x.
[20] Watanabe N, Ohnuki Y and Sakakibara Y. Deep learning integration of molecular and interactome data for protein–compound interaction prediction. Journal of Cheminformatics, Dec 2021;13(1):36. doi: 10.1186/s13321-021-00513-3.
[21] Kabylda A, Vassilev-Galindo V, Chmiela S, Poltavsky I and Tkatchenko A. Efficient interatomic descriptors for accurate machine learning force fields of extended molecules. Nature Communications, Jun 2023;14(1):3562. doi: 10.1038/s41467-023-39214-w.
[22] Akkus E, Tayfuroglu O, Yildiz M and Kocak A. Accurate binding free energy method from end-state MD simulations. Journal of Chemical Information & Modeling, Sep 2022;62(17):4095–4106. doi: 10.1021/acs.jcim.2c00601.

[23] Bryant P, Pozzati G and Elofsson A. Improved prediction of protein-protein interactions using AlphaFold2. Nature Communications, Mar 2022;13(1):1265. doi: 10.1038/s41467-022-28865-w.

[24] Boz E and Stein M. Accurate receptor-ligand binding free energies from fast QM Conformational chemical space sampling. International Journal of Molecular Sciences, Mar 2021;22(6):3078. doi: 10.3390/ijms22063078.

[25] Muroga S, Miki Y and Hata K. A comprehensive and versatile multimodal deep-learning approach for predicting diverse properties of advanced materials. Advanced Science, Augs 2023;10(24). doi: 10.1002/advs.202302508.

[26] Lockner Y and Hopmann C. Induced network-based transfer learning in injection molding for process modelling and optimization with artificial neural networks. The International Journal of Advanced Manufacturing Technology, Feb 2021;112(11–12):3501–3513. doi: 10.1007/s00170-020-06511-3.

[27] Micheletti C, Hauke P and Faccioli P. Polymer physics by quantum computing. Apr. 2021; doi: 10.1103/PhysRevLett.127.080501.

[28] Jin L, Cao W, Wang P, Song N and Ding P. Interconnected Mxene/graphene network constructed by soft template for multi-performance improvement of polymer composites. Nano-micro Letters, Dec 2022;14(1):133. doi: 10.1007/s40820-022-00877-7.

[29] Rajpal S and Mizaikoff B. An *in silico* predictive method to select multi-monomer combinations for peptide imprinting. Journal of Materials Chemical B, 2022;10(35):6618–6626. doi: 10.1039/D2TB00418F.

[30] Jeršovaitė J, Šarachovaitė U, Matulaitienė I, Niaura G, Baltriukienė D and Malinauskas M. Biocompatibility enhancement via post-processing of microporous scaffolds made by optical 3D printer. Frontiers in Bioengineering and Biotechnology, Apr 2023;11. doi: 10.3389/fbioe.2023.1167753.

[31] Claus J, Brietzke A, Lehnert C, Oschatz S, Grabow N and Kragl U. Swelling characteristics and biocompatibility of ionic liquid based hydrogels for biomedical applications. PLoS One, Apr 2020;15(4):e0231421. doi: 10.1371/journal.pone.0231421.

[32] Bayer IS. Controlled drug release from nanoengineered polysaccharides. Pharmaceutics, Apr 2023;15(5):1364. doi: 10.3390/pharmaceutics15051364.

[33] Bingham NM, Un Nisa Q, Gupta P, Young NP, Velliou E and Roth PJ. Biocompatibility and physiological thiolytic degradability of radically made thioester-functional copolymers: opportunities for drug release. Biomacromolecules, May 2022;23(5):2031–2039. doi: 10.1021/acs.biomac.2c00039.

[34] Kłosiński KK, et al. Biocompatibility and mechanical properties of carboxymethyl chitosan hydrogels. Polymers (Basel), Dec 2022;15(1):144. doi: 10.3390/polym15010144.

[35] Valencia-Llano CH, et al. Histological evaluation of cassava starch/chicken gelatin membranes. Polymers (Basel), Sep 2022;14(18):3849. doi: 10.3390/polym14183849.

[36] Chen X and Li J. Bioinspired by cell membranes: Functional polymeric materials for biomedical applications. Materials Chemistry Frontiers, 2020;4(3):750–774. doi: 10.1039/C9QM00717B.

[37] Pelosi C, Constantinescu I, Son HH, Tinè MR, Kizhakkedathu JN and Wurm FR. Blood Compatibility of Hydrophilic Polyphosphoesters. ACS Applied Bio MaterialsACS Applied Bio Materials, Mar 2022;5(3):1151–1158. doi: 10.1021/acsabm.1c01210.

[38] Plohl O, et al. Screen-printing of chitosan and cationised cellulose nanofibril coatings for integration into functional face masks with potential antiviral activity. International Journal of Biological Macromolecules, May 2023;236:123951. doi: 10.1016/j.ijbiomac.2023.123951.

[39] Bonacci M, et al. Towards high-throughput many-body perturbation theory: Efficient algorithms and automated workflows. Npj Computational Materials, May 2023;9(1):74. doi: 10.1038/s41524-023-01027-2.

[40] Lyu W, Deng Z and Zhang G. High-throughput screening for epigenetic compounds that induce human β-Defensin 1 synthesis. Antibiotics, Jan 2023;12(2):186. doi: 10.3390/antibiotics12020186.

[41] Vollenbroek JC, et al. Droplet microreactor for high-throughput fluorescence-based measurements of single catalyst particle acidity. Microsystems & Nanoengineering, Mar 2023;9(1):39. doi: 10.1038/s41378-023-00495-2.

[42] Sleebs BE, et al. Development and application of a high-throughput screening assay for identification of small molecule inhibitors of the P. falciparum reticulocyte binding-like homologue 5 protein. International Journal for Parasitology – Drugs and Drug Resistance, Dec 2020;14:188–200. doi: 10.1016/j.ijpddr.2020.10.008.

[43] Li Q, et al. Microencapsulation with different starch-based polymers for improving oxidative stability of cold-pressed hickory (Carya cathayensis Sarg.) oil. Foods, Feb 2023;12(5):953. doi: 10.3390/foods12050953.

[44] Hauptstein N, et al. Polymer selection impacts the pharmaceutical profile of site specifically conjugated Interferon-α2a. Feb. 22, 2022; doi: 10.26434/chemrxiv-2022-j85br.

[45] Browne E, Worku ZA and Healy AM. Physicochemical properties of poly-vinyl polymers and their influence on ketoprofen amorphous solid dispersion performance: A polymer selection case study. pharmaceutics, May 2020;12(5):433. doi: 10.3390/pharmaceutics12050433.

[46] Pham P, Oliver S, Nguyen DT and Boyer C. Effect of cationic groups on the selectivity of ternary antimicrobial polymers. Macromolecular Rapid Communications, Nov 2022;43(21). doi: 10.1002/marc.202200377.

[47] Hettich BF, Bader JJ and Leroux J. Encapsulation of hydrophilic compounds in small extracellular vesicles: Loading capacity and impact on vesicle functions. Advanced Healthcare Materials, Mar 2022;11(5). doi: 10.1002/adhm.202100047.

[48] Barnett JW, et al. Designing exceptional gas-separation polymer membranes using machine learning. Science Advances, May 2020;6(20). doi: 10.1126/sciadv.aaz4301.

[49] Davidopoulou C and Ouranidis A. Pharma 4.0-artificially intelligent digital twins for solidified nanosuspensions. Pharmaceutics, Oct 2022;14(10):2113. doi: 10.3390/pharmaceutics14102113.

[50] Rahman MA, Rahman MM and Ashraf A. Automatic dispersion, defect, curing, and thermal characteristics determination of polymer composites using micro-scale infrared thermography and machine learning algorithm. Scientific Reports, Feb 2023;13(1):2787. doi: 10.1038/s41598-023-29270-z.

[51] Tsubaki M and Mizoguchi T. Quantum deep descriptor: Physically informed transfer learning from small molecules to polymers. Journal of Chemical Theory and Computation, Dec 2021;17(12):7814–7821. doi: 10.1021/acs.jctc.1c00568.

[52] Lampe C, Kouroudis I, Harth M, Martin S, Gagliardi A and Urban AS. Rapid data-efficient optimization of perovskite nanocrystal syntheses through machine learning algorithm fusion. Advanced Materials, Apr 2023;35(16). doi: 10.1002/adma.202208772.

[53] Bayer IS. Controlled drug release from nanoengineered polysaccharides. Pharmaceutics, Apr 2023;15(5):1364. doi: 10.3390/pharmaceutics15051364.

[54] Concha L, Pires ALR, Moraes AM, Mas-Hernández E, Berres S and Hernandez-Montelongo J. Cost function analysis applied to different kinetic release models of arrabidaea chica verlot extract from chitosan/alginate membranes. Polymers (Basel), Mar 2022;14(6):1109. doi: 10.3390/polym14061109.

[55] Weng J, Durand A and Desobry S. Chitosan-based particulate carriers: Structure, production and corresponding controlled release. Pharmaceutics, May 2023;15(5):1455. doi: 10.3390/pharmaceutics15051455.

[56] Kim S, et al. Model-based analysis of in vivo release data of levonorgestrel implants: Projecting long-term systemic exposure. Pharmaceutics, May 2023;15(5):1393. doi: 10.3390/pharmaceutics15051393.

[57] Issa MG, De Souza NV, Jou BWC, Duque MD and Ferraz HG. Development of extended-release minitablets containing metoprolol supported by design of experiments and physiologically based biopharmaceutics modeling. Pharmaceutics, Apr 2022;14(5):892. doi: 10.3390/pharmaceutics14050892.

[58] Desrosiers A, et al. Programmable self-regulated molecular buffers for precise sustained drug delivery. Nature Communications, Nov 2022;13(1):6504. doi: 10.1038/s41467-022-33491-7.

[59] Kovshova T, et al. Exploring the Interplay between drug release and targeting of lipid-like polymer nanoparticles loaded with doxorubicin. Molecules, Feb 2021;26(4):831. doi: 10.3390/molecules26040831.

[60] Lin M and Qi X. Advances and challenges of stimuli-responsive nucleic acids delivery system in gene therapy. Pharmaceutics, May 2023;15(5):1450. doi: 10.3390/pharmaceutics15051450.

[61] Cong X, Chen J and Xu R. Recent progress in bio-responsive drug delivery systems for tumor therapy. Frontiers in Bioengineering and Biotechnology, Jun 2022;10. doi: 10.3389/fbioe.2022.916952.

[62] Xia D, Xie J and Hu Y. Editorial: Microenvironmental stimuli-responsive nanomedicine for biomedical application. Frontiers in Bioengineering and Biotechnology, May 2023;11. doi: 10.3389/fbioe.2023.1206895.

[63] Tallet L, Gribova V, Ploux L, Vrana NE and Lavalle P. New smart antimicrobial hydrogels, nanomaterials, and coatings: Earlier action, more specific, better dosing?. Advanced Healthcare Materials, Jan 2021;10(1). doi: 10.1002/adhm.202001199.

[64] Sun X, Liu L, Zou H, Yao C, Yan Z and Ye B. Intelligent drug delivery microparticles with visual stimuli-responsive structural color changes. International Journal of Nanomedicine, Jul 2020;15:4959–4967. doi: 10.2147/IJN.S249009.

[65] Zhang M, Hu W, Cai C, Wu Y, Li J and Dong S. Advanced application of stimuli-responsive drug delivery system for inflammatory arthritis treatment. Mater Today Bio, Mar 2022;14:100223. doi: 10.1016/j.mtbio.2022.100223.

[66] Chen M, Lan H, Jin K and Chen Y. Responsive nanosystems for targeted therapy of ulcerative colitis: Current practices and future perspectives. Drug Delivery, Dec 2023;30(1). doi: 10.1080/10717544.2023.2219427.

[67] Hughes KA, Misra B, Maghareh M and Bobbala S. Use of stimulatory responsive soft nanoparticles for intracellular drug delivery. Nano Research, May 2023;16(5):6974–6990. doi: 10.1007/s12274-022-5267-5.

[68] Longo R, Gorrasi G and Guadagno L. Electromagnetically stimuli-responsive nanoparticles-based systems for biomedical applications: Recent advances and future perspectives. Nanomaterials, Mar 2021;11(4):848. doi: 10.3390/nano11040848.

[69] Post C, et al. Application of laser-induced, deep UV Raman spectroscopy and artificial intelligence in real-time environmental monitoring – solutions and first results. Sensors, Jun 2021;21(11):3911. doi: 10.3390/s21113911.

[70] Shireen Z, et al. A machine learning enabled hybrid optimization framework for efficient coarse-graining of a model polymer. NPJ Computational Materials, Nov 2022;8(1):224. doi: 10.1038/s41524-022-00914-4.

[71] Bieker K, Peitz S, Brunton SL, Kutz JN and Dellnitz M. Deep model predictive flow control with limited sensor data and online learning. Theoretical & Computational Fluid Dynamics, Aug 2020;34(4):577–591. doi: 10.1007/s00162-020-00520-4.

[72] Xie Y, et al. Enhancing real-time prediction of effluent water quality of wastewater treatment plant based on improved feedforward neural network coupled with optimization algorithm. Water (Basel), Mar 2022;14(7):1053. doi: 10.3390/w14071053.

[73] Peleato NM. Application of convolutional neural networks for prediction of disinfection by-products. Science Reports, Jan 2022;12(1):612. doi: 10.1038/s41598-021-03881-w.

[74] Li W, et al. High drug-loaded microspheres enabled by controlled in-droplet precipitation promote functional recovery after spinal cord injury. Nature Communications, Mar 2022;13(1):1262. doi: 10.1038/s41467-022-28787-7.

[75] Ye S, Zhang G and Jiang J. AI-based spectroscopic monitoring of real-time interactions between SARS-CoV-2 and human ACE2. Proceedings of the National Academy of Sciences, Jun 2021;118(26). doi: 10.1073/pnas.2025879118.

[76] Helmecke T, et al. Inflammation-controlled anti-inflammatory hydrogels. Advanced Science, Mar 2023;10(7). doi: 10.1002/advs.202206412.

[77] Langer R, Jaklenec A, Traverso G and Hartman D. From molecule to patient: A biotech perspective. Clinical Pharmacology & Therapeutics, Jan 2020;107(1):65–67. doi: 10.1002/cpt.1676.

[78] Gutberlet B, Preis E, Roschenko V and Bakowsky U. Photothermally controlled drug release of poly (d,l-lactide) nanofibers loaded with indocyanine green and curcumin for efficient antimicrobial photodynamic therapy. Pharmaceutics, Jan 2023;15(2):327. doi: 10.3390/pharmaceutics15020327.

[79] Huang C-W, Chang -Y-Y, Cheng -C-C, Hung M-T and Mohamed MG. Self-assembled supramolecular micelles based on multiple hydrogen bonding motifs for the encapsulation and release of fullerene. Polymers (Basel), Nov 2022;14(22):4923. doi: 10.3390/polym14224923.

[80] Fatfat Z, Fatfat M and Gali-Muhtasib H. Micelles as potential drug delivery systems for colorectal cancer treatment. World Journal of Gastroenterology, Jul 2022;28(25):2867–2880. doi: 10.3748/wjg.v28.i25.2867.

[81] Kim S-G, et al. Prolongation of fate of bacteriophages In Vivo by Polylactic-Co-Glycolic-Acid/Alginate-Composite Encapsulation. Antibiotics, Sep 2022;11(9):1264. doi: 10.3390/antibiotics11091264.

[82] Liu J, et al. Perspectives on the nanocarriers with miRNAs for targeting melanoma stemness through epigenetic regulation. International Federation of Pigment Cell Societies, May 2023;36(3–4):268–287. doi: 10.1111/pcmr.13081.

[83] Uebel F, Thérien-Aubin H and Landfester K. Tailoring the mechanoresponsive release from silica nanocapsules. Nanoscale, 2021;13(36):15415–15421. doi: 10.1039/D1NR04697G.

[84] Schnichels S, et al. Lipid-DNA Nanoparticles as drug-delivery vehicles for the treatment of retinal diseases. Pharmaceutics, Feb 2023;15(2):532. doi: 10.3390/pharmaceutics15020532.

[85] Serrano DR, et al. 3D printing technologies in personalized medicine, nanomedicines, and biopharmaceuticals. Pharmaceutics, Jan 2023;15(2):313. doi: 10.3390/pharmaceutics15020313.

[86] Chen M, Lan H, Jin K and Chen Y. Responsive nanosystems for targeted therapy of ulcerative colitis: Current practices and future perspectives. Drug Delivery, Dec 2023;30(1). doi: 10.1080/10717544.2023.2219427.

[87] Ko C-N, Zang S, Zhou Y, Zhong Z and Yang C. Nanocarriers for effective delivery: Modulation of innate immunity for the management of infections and the associated complications. Journal of Nanobiotechnology, Aug 2022;20(1):380. doi: 10.1186/s12951-022-01582-8.

[88] Mallick S, et al. Engineered nanotechnology: An effective therapeutic platform for the chronic cutaneous wound. Nanomaterials, Feb 2022;12(5):778. doi: 10.3390/nano12050778.

[89] Desrosiers A, et al. Programmable self-regulated molecular buffers for precise sustained drug delivery. Nature Communications, Nov 2022;13(1):6504. doi: 10.1038/s41467-022-33491-7.

[90] Kim JM, et al. Nanocomposites-based targeted oral drug delivery systems with infliximab in a murine colitis model. Jun. 05, 2020; doi: 10.21203/rs.3.rs-32688/v1.

[91] Dobrovolskaia MA. Lessons learned from immunological characterization of nanomaterials at the nanotechnology characterization laboratory. Frontiers in Immunology, Oct 2022;13. doi: 10.3389/fimmu.2022.984252.

[92] Benne N, Ter Braake D, Stoppelenburg AJ and Broere F. Nanoparticles for inducing antigen-specific T Cell tolerance in autoimmune diseases. Frontiers in Immunology, Mar 2022;13. doi: 10.3389/fimmu.2022.864403.

[93] Yu H, Wu M, Chen S, Song M and Yue Y. Biomimetic nanoparticles for tumor immunotherapy. Frontiers in Bioengineering and Biotechnology, Nov 2022;10. doi: 10.3389/fbioe.2022.989881.
[94] Chandler M, et al. Artificial immune cell, *AI-cell*, a new tool to predict interferon production by peripheral blood monocytes in response to nucleic acid nanoparticles. Small, Nov 2022;18(46). doi: 10.1002/smll.202204941.
[95] Münter R, Christensen E, Andresen TL and Larsen JB. Studying how administration route and dose regulates antibody generation against LNPs for mRNA delivery with single-particle resolution. Molecular Therapy Methods & Clinical Development, Jun 2023;29:450–459. doi: 10.1016/j.omtm.2023.05.008.
[96] Punz B, et al. Surface functionalization of silica nanoparticles: Strategies to optimize the immune-activating profile of carrier platforms. Pharmaceutics, May 2022;14(5):1103. doi: 10.3390/pharmaceutics14051103.
[97] Moingeon P. Artificial intelligence-driven drug development against autoimmune diseases. Trends in Pharmacological Sciences, Jul 2023;44(7):411–424. doi: 10.1016/j.tips.2023.04.005.
[98] Kawata-Shimamura Y, et al. Biomarker discovery for practice of precision medicine in hypopharyngeal cancer: A theranostic study on response prediction of the key therapeutic agents. BMC Cancer, Dec 2022;22(1):779. doi: 10.1186/s12885-022-09853-1.
[99] Chen S, et al. Redox homeostasis modulation using theranostic AIE nanoparticles results in positive-feedback drug accumulation and enhanced drug penetration to combat drug-resistant cancer. Mater Today Bio, Dec 2022;16:100396. doi: 10.1016/j.mtbio.2022.100396.
[100] Moldogazieva N, Mokhosoev I, Zavadskiy S and Terentiev A. Proteomic profiling and artificial intelligence for hepatocellular carcinoma translational medicine. Biomedicines, Feb 2021;9(2):159. doi: 10.3390/biomedicines9020159.
[101] Barghi F, et al. Precision medicine highlights dysregulation of the CDK4/6 cell cycle regulatory pathway in pediatric, adolescents and young adult sarcomas. Cancers (Basel), Jul 2022;14(15):3611. doi: 10.3390/cancers14153611.
[102] Wager K, et al. 76P Using bioinformatics and artificial intelligence (AI) to map the cyclin-dependent kinase 4/6 inhibitor (CDK4/6i) translational biomarker landscape. ESMO Open, May 2023;8(1):101299. doi: 10.1016/j.esmoop.2023.101299.
[103] Miles J, Ward SG and Larijani B. The fusion of quantitative molecular proteomics and immune-oncology : A step towards precision medicine in cancer therapeutics. FEBS Letters, Nov 2022;596(21):2721–2735. doi: 10.1002/1873-3468.14480.
[104] Kang Y, Vijay S and Gujral TS. Deep neural network modeling identifies biomarkers of response to immune-checkpoint therapy. iScience, May 2022;25(5):104228. doi: 10.1016/j.isci.2022.104228.
[105] Zhang Y-Q, et al. Ionizable drug delivery systems for efficient and selective gene therapy. Military Medical Research, Feb 2023;10(1):9. doi: 10.1186/s40779-023-00445-z.
[106] Cohen SA, Bar-Am O, Fuoco C, Saar G, Gargioli C and Seliktar D. In vivo restoration of dystrophin expression in mdx mice using intra-muscular and intra-arterial injections of hydrogel microsphere carriers of exon skipping antisense oligonucleotides. Cell Death and Disease, Sep 2022;13(9):779. doi: 10.1038/s41419-022-05166-0.
[107] González-Rioja R, Salazar VA, Bastús NG and Puntes V. The development of highly dense highly protected surfactant ionizable lipid RNA loaded nanoparticles. Frontiers in Immunology, Feb 2023;14. doi: 10.3389/fimmu.2023.1129296.
[108] Wang N, et al. Self-assembly of DNA nanostructure containing cell-specific aptamer as a precise drug delivery system for cancer therapy in non-small cell lung cancer. Journal of Nanobiotechnology, Nov 2022;20(1):486. doi: 10.1186/s12951-022-01701-5.
[109] Santo D, Cordeiro RA, Mendonça PV, Serra AC, Coelho JFJ and Faneca H. Glycopolymers mediate suicide gene therapy in ASGPR-expressing hepatocellular carcinoma cells in tandem with docetaxel. Biomacromolecules, Mar 2023;24(3):1274–1286. doi: 10.1021/acs.biomac.2c01329.

[110] Le H, Dé E, Cerf DL and Karakasyan C. Using targeted nano-antibiotics to improve antibiotic efficacy against staphylococcus aureus infections. Antibiotics, Jun 2023;12(6):1066. doi: 10.3390/antibiotics12061066.
[111] Mukherjee D, et al. Addressing antimicrobial resistance with the IDentif.AI platform: Rapidly optimizing clinically actionable combination therapy regimens against nontuberculous mycobacteria. Theranostics, 2022;12(16):6848–6864. doi: 10.7150/thno.73078.
[112] Stracy M, et al. Minimizing treatment-induced emergence of antibiotic resistance in bacterial infections. Science (1979), Feb 2022;375(6583):889–894. doi: 10.1126/science.abg9868.
[113] Heffernan JM, McLaren AC and Overstreet DJ. Local antimicrobial delivery from temperature-responsive hydrogels reduces incidence of intra-abdominal infection in rats. Comp Immunol Microbiol Infect Dis, Jul 2022;86:101823. doi: 10.1016/j.cimid.2022.101823.
[114] Cesaro A, et al. Novel retro-inverso peptide antibiotic efficiently released by a responsive hydrogel-based system. Biomedicines, Jun 2022;10(6):1301. doi: 10.3390/biomedicines10061301.
[115] Bourguignon T, Godinez-Leon JA and Gref R. Nanosized drug delivery systems to fight tuberculosis. Pharmaceutics, Jan 2023;15(2):393. doi: 10.3390/pharmaceutics15020393.
[116] Lin T, et al. A dosing strategy model of deep deterministic policy gradient algorithm for sepsis patients. BMC Medical Informatics and Decision Making, May 2023;23(1):81. doi: 10.1186/s12911-023-02175-7.
[117] Miralles E, et al. Assessment of efficacy and safety using PPAR-γ agonist-loaded nanocarriers for inflammatory eye diseases. International Journal of Molecular Sciences, Sep. 2022;23(19):11184. doi: 10.3390/ijms231911184.
[118] Noddeland HK, et al. Reactive oxygen species-responsive polymer nanoparticles to improve the treatment of inflammatory skin diseases. ACS Omega, Jul 2022;7(29):25055–25065. doi: 10.1021/acsomega.2c01071.
[119] Zhao Y-P, et al. Flexible nano-liposomes-based transdermal hydrogel for targeted delivery of dexamethasone for rheumatoid arthritis therapy. Drug Delivery, Dec 2022;29(1):2269–2282. doi: 10.1080/10717544.2022.2096718.
[120] Singh E, et al. Poly ε-Caprolactone nanoparticles for sustained intra-articular immune modulation in adjuvant-induced arthritis rodent model. Pharmaceutics, Feb 2022;14(3):519. doi: 10.3390/pharmaceutics14030519.
[121] Bayan MF, Marji SM, Salem MS, Begum MY, Chidambaram K and Chandrasekaran B. Development of Polymeric-based formulation as potential smart colonic drug delivery system. Polymers (Basel), Sep 2022;14(17):3697. doi: 10.3390/polym14173697.
[122] Albuquerque RQ, Rothenhäusler F and Ruckdäschel H. Designing formulations of bio-based, multicomponent epoxy resin systems via machine learning. MRS Bullettin, Jan 2024;49(1):59–70. doi: 10.1557/s43577-023-00504-9.
[123] Fernandes AI and Jozala AF. Polymers Enhancing bioavailability in drug delivery. Pharmaceutics, Oct 2022;14(10):2199. doi: 10.3390/pharmaceutics14102199.
[124] Fakhri S, Abdian S, Zarneshan SN, Moradi SZ, Farzaei MH and Abdollahi M. Nanoparticles in combating neuronal dysregulated signaling pathways: recent approaches to the nanoformulations of phytochemicals and synthetic drugs against neurodegenerative diseases. International Journal of Nanomedicine, Jan 2022;17:299–331. doi: 10.2147/IJN.S347187.
[125] Bianchini M, Micera S and Riva ER. Recent advances in polymeric drug delivery systems for Peripheral Nerve Regeneration. Pharmaceutics, Feb 2023;15(2):640. doi: 10.3390/pharmaceutics15020640.
[126] Ritt CL, Liu M, Pham TA, Epsztein R, Kulik HJ and Elimelech M. Machine learning reveals key ion selectivity mechanisms in polymeric membranes with subnanometer pores. Science Advances, Jan 2022;8(2). doi: 10.1126/sciadv.abl5771.

[127] Aldaais EA. A comprehensive review on the COVID-19 vaccine and drug delivery applications of interpenetrating polymer networks. Drug Delivery and Translational Research, Mar 2023;13(3):738–756. doi: 10.1007/s13346-022-01254-y.
[128] Liang Y, et al. Cell-derived nanovesicle-mediated drug delivery to the brain: Principles and strategies for vesicle engineering. Molecular Therapy, May 2023;31(5):1207–1224. doi: 10.1016/j.ymthe.2022.10.008.
[129] Siafaka PI, et al. Protein and gene delivery systems for neurodegenerative disorders: Where do we stand today?. Pharmaceutics, Nov 2022;14(11):2425. doi: 10.3390/pharmaceutics14112425.

Leandro Rodrigues Oviedo, Maurício Dalla Costa Rodrigues da Silva, Pâmela Cristine Ladwig Muraro, Raisi Natalia Lenz Baldez, Tiago Moreno Volkmer, and William Leonardo da Silva*

Chapter 2 Machine learning applications in material science: unveiling innovative solutions

Abstract: Since the twentieth century, polymers have transformed modern society due to their versatility, durability, flexibility, and cost-effectiveness, making them suitable for industrial applications, such as packaging, electronics, automotive, medical devices, and consumer goods. Precise control over polymer synthesis is crucial to ensure consistency and tailored properties for specific applications. The prediction of polymeric properties, such as flame resistance, mechanical stability, and thermal stability, is crucial for complying with industry standards and promoting safety and technological innovation. However, optimizing synthesis and investigating properties at lab scale is time-consuming and costly, which emphasizes the need for computational methods to overcome these drawbacks and fasten commercialization. Artificial intelligence algorithms have been promising in materials science due to their good performance in materials synthesis optimization. Thus, this chapter provides an overview of the main machine learning (ML) and deep learning algorithms used to optimize polymer synthesis, with a focus on the tuning of polymeric properties and investigating durability. Moreover, supervised and unsupervised learning algorithms are discussed as well as their main advantages/challenges in polymer science. Thereafter, the main assays for polymer characterization are also discussed. These assays include X-ray diffraction, Raman, and Fourier-transform infrared spectroscopy, which can be associated with ML algorithms (e.g., convolutional neural networks, recurrent neural

Acknowledgment: Authors would also like to thank the Federal University of Health Sciences of Porto Alegre (UFCSPA), Technological University of Uruguay (UTEC), Federal University of Pelotas (UFPEL) and Franciscan University (UFN) for their support.

***Corresponding author: William Leonardo da Silva**, Applied Nanomaterials Research Group (GPNAp), Franciscan University (UFN), Santa Maria, Rio Grande do Sul, Brazil,
e-mail: williamleonardo_silva@hotmail.com
Leandro Rodrigues Oviedo, Applied Nanomaterials Research Group (GPNAp) Franciscan University (UFN), Santa Maria, Rio Grande do Sul, Brazil
Maurício Dalla Costa Rodrigues da Silva, Federal University of Health Sciences of Porto Alegre (UFCSPA), Porto Alegre, Rio Grande do Sul, Brazil
Pâmela Cristine Ladwig Muraro, Raisi Natalia Lenz Baldez, Northern Regional Technological Institute Technological University of Uruguay (UTEC), Rivera, Uruguay
Tiago Moreno Volkmer, Federal University of Pelotas (UFPEL), Pelotas, Rio Grande do Sul, Brazil

https://doi.org/10.1515/9783111681870-002

networks, especially long short-term memory), reinforcement learning, and computer vision. These approaches are considered promising for the design of optimized polymers to predict their degradation rates/recycling efficiency and can be helpful in classification tasks (non-compostable and compostable polymers).

Keywords: Artificial intelligence (AI), machine learning (ML), deep learning (DL), polymer, artificial neural network (ANN)

2.1 Introduction to machine learning paper in polymer science

Polymers, including microplastics, are ubiquitous in the environment [1]. The term "polymer" refers to a wide variety of materials that are constituted of extended molecular chains formed by the repetition of monomeric units [2].

Polymeric materials are present in almost every aspect of our lives, ranging from common synthetic plastics (e.g., polystyrene) to natural biopolymers (e.g., DNA and proteins). Polymers exhibit mechanical strength, flexibility, high-temperature resistance, and electrical conductivity, which are determined by their unique molecular composition and structure [3, 4]. Its remarkable properties (mechanical, chemical, physical, and biological) provide a wide range of applications in areas such as biomedicine, chemistry, and materials science [5].

Due to their characteristics, polymers are widely used in various areas of industry, including packaging, manufacturing, textiles, automotive, medical, and electronics. An example of a polymer widely and historically used in industry is polypropylene. It has been used as a material with dielectric properties in high-energy-density capacitors, due to its high self-compensation in cases of failure and low production costs [6].

The synthesis/development of new polymers, combined with the existence of a wide range of polymers, requires a comprehensive understanding and documentation of the physical and mechanical properties associated with these materials [6]. Using traditional experimental methods, it is time-consuming and costly to search for polymers that satisfy specific properties for different applications. However, their exploration can be aided and increased by computationally and data-driven design algorithms.

Machine learning, a branch of artificial intelligence (AI), is a powerful tool for prediction of material properties, synthesis optimization, and pattern recognition within experimental data, especially when working with large datasets [7]. These algorithms can be used to understand and construct the intricate relationships between input and output results [8].

In recent years, ML and AI have become increasingly popular in materials science [9]. The techniques provide viable solutions for prognosticating the properties of polymers and have demonstrated significant impacts in several areas [10]. This technique allows, for example, to establish a direct correspondence between the characteristics of fibers and the properties of composites, among other diverse applications, including catalyst design, design and creation of new materials, and drug research [11].

It is worth highlighting that neural network-based algorithms often require large datasets for training/learning process. For this reason, the learning process can be computationally intensive or, in other words, expensive. When dealing with limited datasets, learning algorithms such as regression trees, Gaussian process regression (GPR), support vector regression (SVR), or kernel ridge regression are more suitable for certain applications [12].

In materials science, the direct use of polymers to train ML models is not practicable at present. Therefore, it is essential to represent polymer structures in a way that is easily comprehensible by the machine (computer) [13]. In this field, empirical methods and simulations were used to develop and meet the demand for new materials [14]. This approach requires significant instruments and often involves getting certain advantages and skills.

ML has enabled the creation of new chemical potentials for polymers, the development of membrane materials, the prediction of polymer properties with defined sequences, the design of bioplastics, and the optimization of 3D printing.

Despite these advances, the field still presents challenges, unique to polymer science. Different from other materials, the structure of polymers is stochastic rather than fixed, which makes the representation of polymers in ML algorithms challenging [15]. Defining all atoms in complex polymers is not feasible due to the high computational demand of input representations needed for the use of the technique. An option is to, instead of directly using all sequenced atoms in a polymer chain as a source of feature representations, consider alternatives such as the use of chemical compounds or functional groups that may be more valuable and practical for representing polymers [16].

The field of polymer science faces major challenges when it comes to using ML in materials prediction. Databases of polymeric materials are not easily accessible to the public. Many of the essential measurements require instruments manufactured by companies with limited open interfaces/models. This makes it difficult to create databases and nearly impossible to integrate these devices into high-performance systems [17].

In short, too much research has been developed to better represent the polymeric structure and to incorporate expert knowledge about polymers into ML algorithms. The future of the technique is promising for polymeric materials, but further research still needs to be encouraged in this area.

2.2 Prediction and optimization of physicochemical properties

Comprehension of the physical and chemical properties of polymeric materials is essential for their effective application in various fields. Polymer properties are highly dependent on the structure of the polymer chains, molecular weight, and physical structure, affecting chemical, mechanical and thermal properties [18]. The relationship between structure and properties helps in designing polymers for specific applications, ensuring their optimal performance and durability [19]. Predicting polymer properties is therefore crucial, not only for advancing material science but also for enabling the design of next-generation functional materials [20].

Traditional approaches for polymer property estimation rely heavily on experimental measurements and computational simulations such as molecular dynamics (MD), for example, LAMMPS [21], GROMACS [22], NAMD [23], and DL_POLY [24], which are frequently employed to estimate a wide range of polymeric material properties such as crystallization, polymer drug uptake, structural properties, reverse osmosis membranes [21–24], while Monte Carlo based methods such as DL_MONTE [25] and Cassandra [26] are used for gas transport and dilation resistance of separation membranes [25, 26]. At the quantum level, packages like Quantum ESPRESSO [27] and VASP [28] are commonly used for DFT and first principles approach.

Even with all the advancements and consistency of experimental or simulation-based methods, they often require extensive property measurements, which are both time-consuming and expensive [29, 30]. Hence, there is a pressing need for innovative methodologies to accelerate polymers' design process without compromising accuracy [31]. To address this gap, Machine Learning (ML) has emerged as an informatic innovative tool to analyze, predict, and optimize polymer properties with remarkable accuracy [32]. However, the complex and periodic nature of polymer structures presents difficulties in data representation [33], while the lack of standardized, high-quality datasets limits model generalization [34]. Another challenge is that many ML models function as black-box optimization tools, raising concerns about interpretability and experimental validation. Therefore, overcoming these challenges requires collaborative efforts in data curation, model explainability, and integration with traditional simulation methods [35, 36].

2.2.1 Polymer featurization

Since machine learning algorithms require numerical data, it is essential to convert these chemical groups into machine-readable representations that can be used to build predictive models, which can be used to predict and/or optimize polymeric properties [37].

One of the most common strategies for transforming categorical data into a mathematical description suitable for ML models is One-hot Encoding (OHE) [38], in which the polymer sequence is transformed into a binary vector, allowing efficient processing of categorical features by ML algorithms [38, 39]. OHE can also be applied to SMILES strings, which are text-based representations of molecular structures. These strings are often used to describe small molecules or repeating units in polymers. By converting SMILES into vector formats, such as binary images, they become more accessible for machine learning models to work with [40–42].

Property coloring offers another approach by representing polymers as three-dimensional arrays, with each dimension corresponding to specific chemical properties of the monomeric units. These arrays are then processed through filters, generating complex, image-like structures that are well-suited for feature extraction in machine learning models [43]. Another advanced technique, known as the chemical tree, leverages atomic-level descriptions to define the chemical composition of polymer subunits. This method organizes polymers into hierarchical tree structures that encapsulate both monomer-level and atomic-level information [44, 45].

Autoencoding represents another powerful featurization technique. This neural network-based approach converts discrete molecular representations, such as SMILES strings, into continuous variables within a latent space. The encoder component performs dimensionality reduction, while the decoder reconstructs the original molecular structure from the compressed form, providing compact, yet meaningful features for ML tasks [46–48]. Lastly, motif-based fingerprinting extracts molecular features by examining the number and connectivity of specific atoms within a molecular structure. By identifying recurring chemical motifs and leveraging principles of chemical bonding, this method generates rich, descriptive representations of polymers [49, 50].

2.2.2 ML for properties prediction

Machine learning algorithms are typically divided into the following categories: supervised, semisupervised, unsupervised, and reinforcement learning [51].

2.2.2.1 Unsupervised machine learning models

Unsupervised models have emerged as essential tools in polymer research, offering novel approaches for identifying and classifying polymers without relying on labeled datasets [52]. It is primarily used for exploration, clustering, dimensionality reduction, and anomaly detection. Methods such as principal component analysis, diffusion maps, self-organizing maps, and autoencoder networks have been successfully used to distinguish different polymer states and configuration states, and enabled the detection and quantification of complex structural features such as voids and grains [53–55].

2.2.2.2 Supervised machine learning models

In supervised ML models, the data used for training and testing are labeled with property values. After the featurization process to ensure that the data are in a format suitable for the algorithms, the dataset is typically divided into training data (usually 70–80% of dataset) and test data subsets. During the training phase, several regression models such as random forest, gradient boosting, XGBoost, bagging regressor, AdaBoost can be employed to learn from the inputs and, as an output, predict the physical and chemical properties of polymeric materials [32]. To improve predictive accuracy and prevent issues like bias, underfitting, or overfitting, each model relies on a specific set of hyperparameters, adjustable settings defined before training starts. These hyperparameters can be tuned either manually or through automated methods such as grid search and random search. These tools systematically or randomly explore different combinations to identify the configuration that yields the best model performance [54]. To assess model performance, commonly used evaluation metrics include the determination coefficient (R^2), mean absolute error (MAE), mean percentage error, mean squared error (MSE), and root MSE. Once training is complete, the best metric values for each feature across all models are recorded and analyzed using graphical tools, allowing for a clear comparison and identification of the most effective machine learning model [55].

2.2.2.3 Semisupervised machine learning models

This approach integrates both supervised and unsupervised learning techniques to train predictive models. The input data may originate from a single dataset or from multiple datasets [56]. Semisupervised regression presents a promising strategy for predicting labels on unlabeled data, while also estimating label confidence – information that can be used to expand the training set and improve model accuracy [57].

2.2.2.4 Reinforcement machine learning models

Reinforcement learning is a class of algorithms that uses machines and software agents to automatically determine optimal behavior within a specific context to improve overall efficiency [51]. These algorithms rely on iterative interactions between agents and their environments, aiming to maximize a defined reward signal. Techniques such as Markov decision processes and active learning exemplify this approach [58]. Active learning is valuable in scenarios with limited labeled data. Rather than relying solely on available examples, the model actively selects the most informative new instances, known as queries, to suggest the "next-best experiment." These selected examples are then incorporated into the training set, and the model is retrained. This cycle continues, allowing the model to improve more rapidly by concentrating on the most valuable data points [58, 59].

2.2.3 Machine learning studies examples

To enable comparison across studies utilizing ML to predict the mechanical and physicochemical properties of polymers, Table 2.1 provides a summary of key information extracted from the selected literature, including the models, data sources, predicted properties, and R^2 values.

In the first study, Ding et al. [60] used ML algorithms to predict elastic modulus (E), tensile strength (σ_{max}), and elongation at break (ε_r) of polyurethane elastomers. The dataset was compiled through data mining from academic literature and refined using recursive data elimination, which helped to remove noise and inconsistencies to improve data quality. Among the five tested models, XGB achieved the highest predictive accuracy across all three target properties [60].

Another study employed a deep neural network (DNN), improved by Bayesian optimization, to predict E, σ_{max}, and Poisson's ratio (ν) from nanoindentation load–depth curves of polycarbonate (PC) and polymethyl methacrylate. The dataset, generated via finite element simulations, was used to train the DNN architecture comprising two hidden layers with exponential linear unit activation function, and optimizers such as adaptive moment estimation function (Adam) and stochastic gradient descent. Bayesian optimization was applied to minimize the MAE during training, resulting in a model with R^2 values of 0.99 or higher for all predicted properties. Experimental validation showed that the prediction results were close to the literature results [61].

Similarly, the influence of nano-TiO_2 and cellulose nanocrystals on fracture strength, yield strength, and Young's modulus in high-density polyethylene nanobiocomposites was investigated through a multifaceted methodological framework. This included the application of analysis of variance to determine statistically significant factors, response surface methodology for constructing predictive statistical models, and an artificial neural network coupled with genetic algorithm (ANN-GA) hybrid model for optimization purposes. The experimental dataset was derived from specimens fabricated via injection molding and subsequently subjected to mechanical testing protocols. The ANN model demonstrated high predictive accuracy for the target mechanical properties, while the ANN-GA approach effectively mitigated overfitting risks and facilitated model-driven optimization of formulation parameters [62].

In addition to mechanical analysis, thermal property prediction has become a growing area of interest in polymer research using ML. Malashin et al. [32] applied 11 ML models to predict polymer properties, including glass transition temperature (T_g), thermal decomposition temperature (T_ad), melting temperature (T_m), and crystallization kinetics (k). The dataset was extracted from the PolYinfo database [63], and was used to train the algorithms according to the tested property. Before training, molecular features were derived from SMILES representations. Among the tested models, random forest, XGBoost, and gradient boosting showed the strongest performance, particularly with larger datasets, while bagging, K neighbor regressor, and SVR were more effective for smaller sample sizes.

Table 2.1: Compilation of examples ML papers to predict polymer's properties.

Predicted properties	Polymer	Dataset source	ML models	Best model and determination coefficient (R^2)	Reference
Elastic modulus (E), tensile strength (σ_{max}), and elongation at break (ε_r)	Polyurethane elastomers (PUEs)	Data mining from academic literature	Xtreme gradient boosting (XGB), random forest (RF), support vector regression (SVR), neural network (NN), and k-nearest neighbors (KNN)	XGB: E: $R^2 = 0.89$, σ_t: $R^2 = 0.91$, ε_r: $R^2 = 0.85$	[45]
Elastic modulus (E), tensile strength (σ_{max}), and Poisson's ratio (ν)	Polycarbonate (PC) and polymethyl methacrylate (PMMA)	Finite element simulations	Deep neural network (DNN) improved via Bayesian optimization	E: $R^2 \geq 0.99$, σ_t: $R^2 \geq 0.99$, ν: $R^2 \geq 0.99$	[46]
Elastic modulus (E), fracture strength (σ_f), and yield strength (σ_y)	High-density polyethylene (HDPE) nanobiocomposites	Experimental	Artificial neural networks (ANN) coupled with genetic algorithm (GA) hybrid model	E: $R^2 = 96.30\%$, σ_y: $R^2 = 97.49\%$, σ_f: $R^2 = 98.80\%$	[47]
Glass transition temperature (T_g), thermal decomposition temperature (T_ad), and melting temperature (T_m), crystallization kinetics (k)	18,311 unique polymers	Online polymer database (PoLyInfo)	Random forest, gradient boosting, XGBoost, bagging regressor, AdaBoost, lasso regression, elastic net, support vector regressor (SVR), decision tree (DT) regressor, K-neighbor regressor, and linear regression	Random forest: T_g: $R^2 = 0.88$ (8,092)* T_ad: $R^2 = 0.73$ (6,325)* T_m: $R^2 = 0.71$ (3,844)* XGBoost: k: $R^2 = 0.97$ (59)*	[3]

Glass transition temperature (T_g)	Poly(meth)acrylates	Data mining from academic literature	Recursive neural networks (RvNN)	$R^2 = 0.9351$	[34]
Glass transition temperature (T_g)	More than 200 polymers	Data mining from academic literature	Convolutional neural networks (CNN) combined with fully connected (FC) layers	$R^2 \approx 0.91$	[30]

*Number o⁻ samples in the dataset.

Researchers predicted the T_g of poly(meth)acrylates, including polymers with phenyl rings, using RvNN. Two molecular representation approaches were explored: group representation, treating cyclic groups as single units, and cycle breaking representation using standard formats like Unique SMILES. The dataset containing 217 training samples and 54 test samples yielded an R^2 of 0.9351 using the recursive cascade correlation recurrent neural network (RNN) with InChI representation [45].

Finally, researchers introduced a deep learning approach that leverages CNN with fully connected layers to predict the glass transition temperature (T_g) directly from monomeric chemical structures. In this method, SMILES strings were transformed into binary images and passed through a CNN architecture comprising two convolutional layers with rectified linear unit (ReLU) activation function, max-pooling, and dropout rates between 20% and 30%. The model was evaluated on both a restricted dataset (~100 polymers) and a larger dataset (>200 polymers), achieving prediction errors below 6% [42].

In summary, the reviewed studies consistently demonstrate that machine learning-based approaches can significantly lower experimental costs while improving the accuracy of physicochemical property predictions.

2.2.4 Language-based and predictive platforms

Limitations in polymer data previously delayed their application, but advancements in data standardization and augmentation now enable breakthroughs. Language models excel at addressing multimodal and multidomain challenges, making them ideal for understanding the complex relationships between polymer compositions as also their properties.

PolyNC, developed by researchers is an advanced natural/chemical language model designed to predict polymeric properties with high accuracy and efficiency. By using advanced data augmentation and tokenization strategies and by integrating natural and SMILES chemical languages, it operates as an end-to-end solution for multitask learning, reducing data preprocessing and enhancing generalization for polymer discovery and property prediction [36].

Another language-based model is TransPolymer. Instead of relying solely on conventional machine learning techniques, it employs masked language model to pretrain approximately 5 million unlabeled polymers, followed by fine-tuning. The predictive performance of the model was tested against 10 conventional machine learning algorithms across multiple properties and outperformed the traditional algorithms [63].

Polymer Genome (available at polymergenome.org) is a practical, easy-to-use platform that uses supervised machine learning models to predict polymer properties from chemical structure and composition. A dataset containing both computationally derived and experimentally measured polymer properties is used. To improve property predictions, the platform uses a hierarchical fingerprinting scheme that captures structural features from atomic-level connectivity to morphological characteristics, using GPR, ensuring accurate predictions with uncertainty quantification. The online platform enables users to input polymer structures via SMILES strings or an interactive molecular sketch tool, allowing for rapid on-demand property predictions [64].

2.3 Molecular structure design and automated polymer synthesis using machine learning

For many years, the study of structure–property relationships in macromolecular materials, including block copolymers, polymer blends, and polymer nanocomposites, has been central to polymer science and engineering. Research has shown that, beyond choosing the polymer chemistry and architecture, the final material properties are determined by the assembled structures, which are often hierarchical and span multiple scales. Due to the recent breakthroughs in polymer synthesis/processing technologies, the range of achievable equilibrated and non-equilibrated structures has expanded significantly. Identifying the optimal structures that produce the desired properties, while minimizing costs and efforts for industrial-scale production, is a key driver for high-throughput experimentation. The rising demand for high-throughput experiments also creates a greater need for rapid, and ideally automated, analysis of synthesized materials and their characterization, highlighting the importance of developing appropriate ML workflows [65].

The molecular architecture of polymers is fundamentally based on macromolecules that are either synthesized through chemical processes or derived from natural sources. A critical feature of both natural and synthetic polymers is the capacity of carbon atoms to form extensive molecular chains. Polymers are constructed from fundamental repeating units known as monomers. In synthetic polymerization processes, monomers are covalently bonded to form molecular chains through mechanisms such as addition polymerization or condensation polymerization. These macromolecules can exhibit either linear or branched configurations. Typically, a single macromolecule consists of several hundred monomers linked together [66].

An essential determinant of a polymer's physical properties is the spatial arrangement of its molecular chains, which is helpful in the material's processability and suitability for molding. This arrangement is divided into two distinct categories: physical and chemical configurations. Chemical arrangements are characterized by irreversible cross-linking of molecular chains, where covalent bonds form a permanent three-dimensional network. In contrast, physical arrangements are reversible, involving merely entangled chains without the formation of permanent chemical bonds [67].

In the context of new material design, various techniques assist researchers in pushing the limits of polymer design. One such method is the use of robust design of experiments (DoE) approaches, which are essential for methodically exploring the design space and achieving the desired properties of materials. This process often requires decisions to be made across several areas, such as material properties, operational conditions, and system configurations. Conventional design methodologies, which often rely on experience-based approaches, present numerous challenges that can hinder the optimization and reliability of the material design process [68].

The reduction in the number of experiments (and associated costs) and the exploratory nature of DoE have had significant impact in the field of architecture. However, DoE typically relies on a range of conditions within which the mathematical models are valid and provide satisfactory predictions. Conditions outside the selected range may reduce the accuracy of the mathematical models, causing the experimental response to differ significantly from the predicted response [69].

ML regression algorithms, in conjunction with feature engineering and extensive datasets, are instrumental in predicting material properties for advanced fabrication and practical applications. By leveraging available data for training, researchers can markedly reduce the time and costs traditionally required for experimental procedures. These algorithms are helpful in the optimization and discovery of functional materials across a range of disciplines, including thermoelectric, photovoltaics, catalysis, and optics [70].

A notable limitation of existing algorithms is the insufficient availability of data, which hinders the development of AI/ML-based algorithms for predicting the behavior of polymer materials. Machine learning algorithms rely on large datasets for training and improving their predictive accuracy. In the absence of comprehensive data, these models are unable to generate reliable predictions of polymer behavior across varied scenarios, thereby complicating the development of robust, ML-driven materials for polymer development [71]. However, where conventional physical and chemical models exhibit limitations, ML models are increasingly addressing these gaps with remarkable speed and precision.

The integration of these approaches has been referred to in multiple ways (such as self-driving laboratory, autonomous platform, or closed-loop experiments), but the common result is a faster pace of discovery and a deeper understanding of highly refined materials [72].

2.4 Polymer characterization through computer vision and spectrum analysis

Polymers are increasingly being used in various areas of knowledge. Therefore, before using these materials, it is necessary to understand their characteristics and applications. For this, several analyses are carried out in order to identify polymeric properties (e.g., hardness, impact resistance, and tensile strength). The main spectral analyses and computer vision are described below.

2.4.1 Fourier transform infrared spectroscopy (FTIR)

In this technique, an infrared light beam interacts with the polymer, inducing vibrations in the molecule's atoms. These vibrations are linked to specific frequencies of chemical bonds, reflecting the molecule's energy levels. They are influenced by the molecular structure, the shape of the molecule's potential energy surface, atomic masses, and possibly, vibrational interactions [73]. One of the main advantages of using Attenuated Total Reflectance spectroscopy (ATR) is that it allows samples to be analyzed with very little or no preparation.

2.4.2 Raman spectroscopy

Raman spectroscopy is a vibrational technique that can obtain information about the structure of molecules and the nature of chemical bonds [74]. Raman scattering signals are obtained due to the excitation of the normal vibrational modes of molecules by monochromatic excitation radiation, which undergoes inelastic scattering when demonstrating changes in the polarizability of molecules [75].

2.4.3 Light scattering

Light scattering techniques are classified into elastic or dynamic and static modes. Both are employed in the characterization of polymers [76]. The almost elastic light scattering allows information about the hydrodynamic radius to be obtained, but it does not directly provide the average molar mass of the polymer [77]. For this reason, its application in polymer characterization is limited when the average molar mass is an essential parameter. However, this technique is useful to evaluate the degree of branching of polymers formed from the same monomers, since more branched polymers have a smaller hydrodynamic radius in a suitable solvent, i.e., a solvent that interacts well with polymeric chains [78].

Static laser light scattering in the so-called Debye system allows the measurement of the average molar mass (weighted by the mass) of polymers, ranging from a few tens to millions of Daltons [79].

2.4.4 Computer vision

The use of computational models based on ML is a valuable tool for optimizing processes, which makes them faster and more reliable in predicting polymeric properties. In this way, it is possible to evaluate the efficiency of different compounds in specific applications, extracting data from chemoinformatics libraries without the need to synthesize them in the laboratory [80].

To successfully apply predictive ML models in problem solving, it is critical to employ proper preprocessing and structuring techniques for datasets [81]. When combined with high-quality data sources, these approaches ensure robust and efficient models [82]. In broad terms, ML techniques are typically categorized into supervised learning and unsupervised learning, with the former being most useful in polymer science [83].

For example, one of the applications of computer vision in polymer science involves the training of overlapping and embedded chemical substructures through the unsupervised CORPUS method, a chemoinformatics tool that preserves information about the properties of the compounds contained in a training database [84].

Furthermore, this method allows efficient grouping of high-quality data based on chemical structures. In this view, chemical descriptors represent different levels of characterization of molecular structure and can be classified based on their dimensions. This classification facilitates the interpretation of molecular properties, since the 40 dimensions of the descriptors encompass a wide variety of properties, characteristics, and physicochemical phenomena. These dimensions can be organized into categories ranging from one to six dimensions.

2.5 Modeling and simulation of polymer degradation and recycling processes

The increasing environmental concerns regarding polymeric waste have prompted significant research into the degradation and recycling of polymers. Thus, ML algorithms are a powerful tool for the prediction of polymeric behavior in both regular and extreme conditions, which helps the optimization of recycling processes [85]. In this sense, this topic explores the application of ML models, particularly RNN, such as LSTM (long short-term memory) networks and reinforcement learning (RL), in modeling polymer degradation and enhancing recyclability. Furthermore, Figure 2.1 shows the amount of scientific research carried out on this topic, where the search was performed at scopus.com using the descriptors "polymer degradation" OR "polymer recycling" AND "RNN" OR "reinforcement learning."

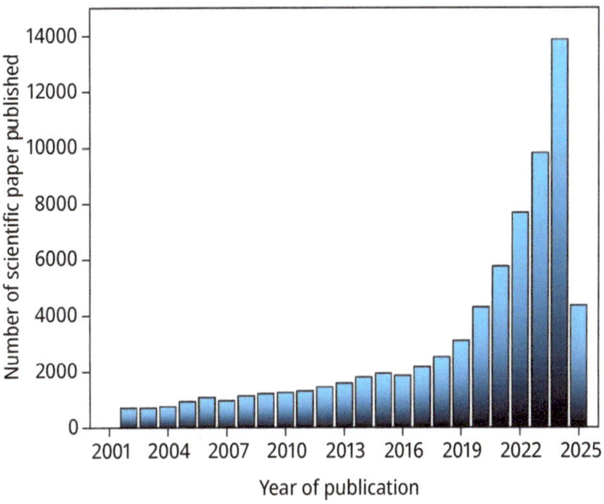

Figure 2.1: Number of published scientific papers dealing with polymer science assisted by ML models from January 2002 to February 2025.

According to Figure 2.1, there is an observable increase in the number of published scientific research articles from 2002 to 2025. All these studies focus on polymer science, particularly the synthesis and application of stable polymers, with the assistance of machine learning (ML) algorithms. ML is primarily employed to investigate polymer degradation and recycling, highlighting the growing interest of researchers in this promising field [86–89]. Consequently, the classification of ML models for polymer degradation and recycling has emerged as a hot topic. In this view, the following sections are concerned with the concepts of RNN and Reinforcement Learning and with the discussion of the application of these ML algorithms in the polymer degradation/recycling studies.

2.5.1 Applied models: RNNs and reinforcement learning

Machine learning algorithms are valuable and helpful tools for understanding complex interactions governing polymer degradation. For example, ML algorithms can effectively identify the variables (features) that result in higher influence on polymer degradation, and use the variables to perform simulations of the polymer behavior under various scenarios and conditions [90]. Moreover, this approach enables predictions of specific variables (e.g., degradation rates, weight loss, transition temperatures), which can help in the production of optimized materials [91].

In parallel, RNN, particularly LSTM, can be used for modelling the degradation processes of polymers over time by capturing time dependency within the data [92]. This approach is valuable to analyze the change in polymeric properties with respect to time, being helpful in the synthesis of biodegradable polymers. Additionally, RL-based algorithms (e.g., Q-learning and deep Q-learning) can provide enhanced predictions since they are especially designed to interpret and reproduce complex systems [93]. Thus, it is useful to simulate recycling processes and optimize them. For instance, RL can distinguish between compostable and non-compostable polymers [94]. In this way, RL significantly improves the sorting task commonly used in recycling facilities [95].

2.5.2 RNNs: the role of LSMT in polymer science

RNN is a type of ANN. However, the main difference between conventional ANN and RNN is the capability of the latter to retain information during the learning process (i.e., forward and backpropagation steps) [96, 97]. Thus, RNN is extremely useful in time series and the investigation of dynamic systems such as financial forecast tasks and polymer behavior over time [98]. Gated recurrent units (GRUs) and LSTM networks are the main RNN algorithms used in industrial applications [99–103]. It is worth pointing out that in polymer science and for systems with high complexity, LSMT tends to be most effective when there are long-term patterns to be learned, such as progressive degradation over months or years, and hence will be deeply discussed in this book.

LSTM networks overcome some of these challenges by the utilization of memory cells and some mechanisms that deal with the flow of information [104]. Thus, LSTM architecture can retain significant information over time and selectively forget irrelevant details. In polymer science, LSTMs have proven particularly beneficial for investigating degradation and recycling processes [105]. Thus, Figure 2.2(a and b) illustrates the typical architecture and functioning of an RNN with LSMT layers and some of the responses expected in polymer science applications.

According to Figure 2.2, RNN-LSTM, or simply LSTM, involve the feeding of data to the input layer (Five variables: pH, time, hardness, toughness, and modulus of elasticity), processing in hidden layers (one or more LSTM layers), and the generation of a response based on pattern learning from data.

The LSTM layers are mainly concerned with pattern detection that varies with time, whereas dense layers deal only with pattern recognition without considering the time dependency of input data. Thus, LSTM deals with pattern recognition and learning from time-dependent data, being well-suited for analyzing the temporal evolution of material properties such as stability, degradation, and thermal resistance, which also enables more accurate predictions and material design optimizations

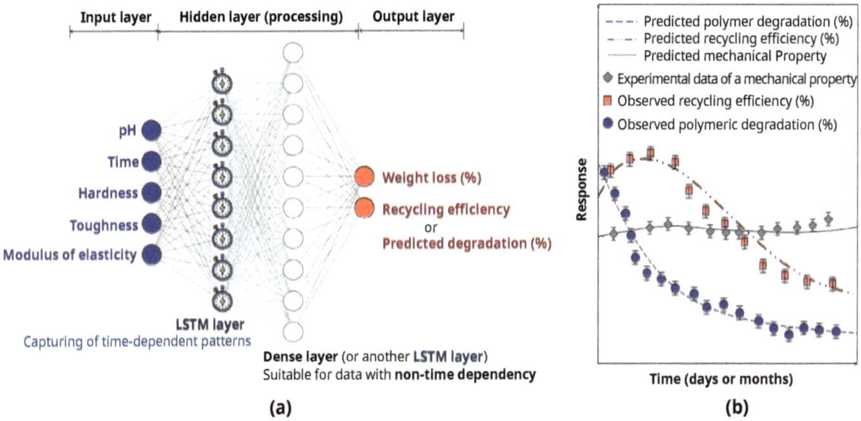

Figure 2.2: (a) Architecture and functioning principle of RNN-LSTM algorithms and (b) example of some predictions carried out by the algorithm.

[106]. Thus, LSTM is used in various applications within polymer science such as predicting degradation rates, monitoring polymerization processes, optimizing recycling, and modeling mechanical performance [107]. These capabilities help design durable materials, improve product quality, enhance recycling efficiency, and tailor polymers for specific applications [108]. Moreover, LSTM facilitates real-time monitoring of polymerization processes, enabling researchers to make adjustments that improve product quality and properties [109]. Additionally, it can simulate the recyclability of various polymers by considering their degradation profiles as well as predict specific properties (tensile strength or elasticity based on historical data), which allows optimization of polymer recycling [110].

Therefore, RNN and LSTM algorithms are characterized as powerful AI tools for use in polymer science, especially for modelling and simulation of degradation and recycling processes on bench and industrial scale. Furthermore, Table 2.2 presents some scientific papers involving the use of these AI-based algorithms in polymer science.

According to Table 2.2, LSTM is extremely useful in polymer science, particularly in sorting tasks (classification of compostable and non-compostable polymers and to get insight into the biodegradability of these materials), degradation profile (based on the prediction of thermal/mechanical properties) and in the recycling process. Moreover, RNNs in these reported works have been presenting high accuracy in processing data with temporal dependency, being applicable in industrial situations.

2.5.3 RL applied to polymeric material science

RL is a subarea of ML, with a focus on training based on interactions between an environment and an agent, considering the use of an award to improve algorithm

Table 2.2: Scientific works utilizing LSTM in polymer science.

Application	LSTM configuration	Findings/commentary	Reference
Predicting polymer properties	Three LSTM layers, allied with dropout and feature extraction	Prediction of mechanical strength and thermal behavior of polymers with high accuracy	[111]
Monitoring polymerization	Bidirectional LSTM	Improved real-time monitoring capabilities, leading to enhanced product quality	[112]
Analyzing degradation rates	Ensemble of LSTM, GRU, and Elman recurrent neural network (ERNN)	Successfully modeled degradation behavior under various environmental conditions	[113]
Optimizing recycling processes	LSTM layers associated with genetic algorithms	Successfully used in recycling facilities (sorting tasks)	[114]
Mechanical performance prediction	1 LSTM layer associated with feature extraction and projection layers (long short-term memory projection – LSTMP)	Use of degradation history for the prediction of mechanical properties	[115]

performance as the learning process occurs [116]. Thus, RL makes decisions based on a given state of the system, receiving feedback in the form of a reward-and-learn pattern for optimizing the cumulative rewards over time [117]. By trial and error, the agent explores various situations in the dataset based on the outcomes obtained in previous experiences [118, 119]. In this sense, RL can identify scenarios where optimal situations are not known and should be discovered through experimentation [120]. However, to get good results in predictions, a large dataset (and computational resources) is required [121].

Based on these observations, RL can be used in optimization processes. For example, RL can assist in the material selection to find polymers that degrade more efficiently [122, 123]. Moreover, it can be used for the simulation of different synthesis conditions to obtain optimized materials [124, 125]. For further clarification, Figure 2.3 shows the example of how RL algorithms work, with a focus on their application in polymer science.

According to Figure 2.3, the RL algorithms are based on a cycle of actions done to get information about experimental observation. In the illustration, the agent (polymeric material used as a sample) suffers an action by the environment (mechanical resistance assay equipment) to provide an observation (response). Additionally, each action is associated with a reward (the investigation of the response of the polymeric material to the applied force) to classify whether this polymer can be either suitable or unsuitable for a specific application (e.g., in building, commercial materials, or as

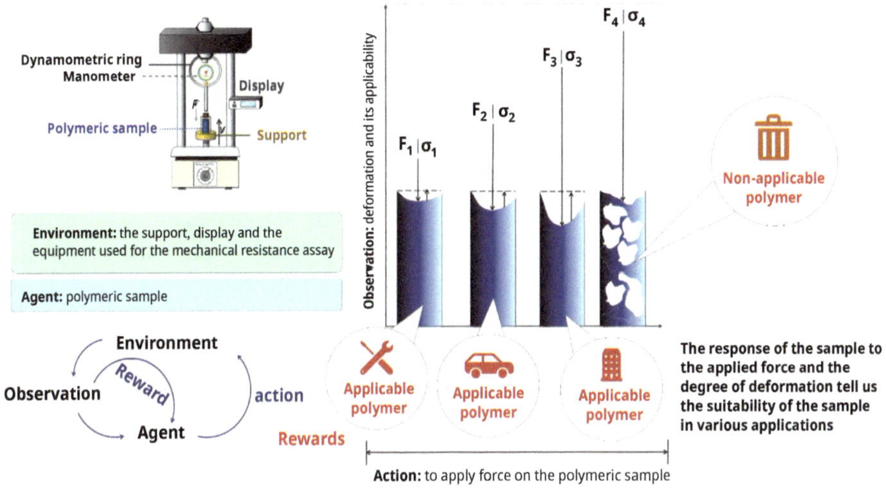

Figure 2.3: Working principle of reinforcement learning (RL) in studying polymer deformation and its suitability for industrial applications.

pieces of an automobile). Then, from repetition and sequential learning, the algorithm can make predictions of some polymeric properties and material classifications, that is, whether the polymer will be considered stable or recyclable. All these actions are performed by AI and operated by humans.

The use of RL in polymer discovery is supported by machine learning models that predict polymer properties, enabling the RL agents to make informed decisions based on predicted outcomes [126]. Thus, several studies have explored the application of RL in polymer degradation and recycling. To mention a few, two studies can be cited.

In the first study, the RL algorithm was used to synthesize polymers with enhanced thermal properties, such as thermal conductivity [127]. To carry out this type of study, dynamic simulations and different operational conditions, as well as high-quality and number of data, are required to get precise predictions. Additionally, in the second one, RL was used to perform cost analysis allied to polymer properties predictions [128]. Thus, with the aid of RL, it was possible to use agents capable of previewing scenarios (experimental conditions) that might result in undesirable outcomes for the polymers in industrial applications [129].

2.6 Conclusion

This chapter discussed the use of AI algorithms in polymer science over the years, bringing information on ML model categories and specific neural networks designed for time-series analysis. It was confirmed that the use of AI tools in polymer science is an emerging topic since it enables high applicability in real systems and in industrial-scale conditions. Thus, ML algorithms (e.g., clustering, RNN-LSTM, GRU, RF, XGB, DT, and SVM) have been continuously applied to predict polymer properties, as well as their degradation and recycling. Among these techniques, supervised AI algorithms have received particular attention in the field. Furthermore, characterization techniques (e.g., Raman spectroscopy, FTIR, dynamic light scattering, and computer vision) used to describe polymers can be valuable in constructing databases for AI algorithms applied to polymer science. The integration of AI with polymer science, along with the enhancement of key characterization techniques, can help reduce costs associated with the synthesis and commercialization of polymers, as well as decrease the time of bench-scale experiments, supporting optimization processes. As a perspective, the integration of AI with polymer science should be continuously encouraged, as it has the potential to enhance the efficiency of material design, accelerate the discovery of new polymers, and optimize experimental processes.

List of abbreviations

Adam	Adaptive moment estimation function
AI	Artificial intelligence
ANN	Artificial neural networks
ANN-GA	Artificial neural network–genetic algorithm hybrid model
ATR	Attenuated total reflectance spectroscopy
CNN	Convolutional neural networks
DFT	Density functional theory
DL	Deep learning
DNN	Deep neural network
DoE	Design of experiments
DT	Decision tree
ERNN	Elman recurrent neural networks
FC	Fully connected
FTIR	Fourier-transform infrared spectroscopy
GA	Genetic algorithm
GB	Gradient boosting
GPR	Gaussian process regression
GRU	Gated recurrent units
LSTM	Long short-term memory
LSTMP	Long short-term memory projection
MAE	Mean absolute error
MSE	Mean squared error
ML	Machine learning
NN	Neural networks
OHE	One-hot encoding
PC	Polycarbonate
PUEs	Polyurethane elastomers

ReLU	Rectified linear unit
RF	Random forest
RL	Reinforcement learning
RNN	Recurrent neural networks
RvNN	Recursive neural networks
SVR	Support vector regression
XGB	Xtreme gradient boosting

References

[1] Tian X, Beén F, Sun Y, Van Thienen P and Bäuerlein PS. Identification of polymers with a small data set of mid-infrared spectra: A comparison between machine learning and deep learning models. Environmental Science & Technology Letters, 2023;10(11):1030–1035. doi: 10.1021/acs.estlett.2c00949.

[2] Uddin MJ and Fan J. Interpretable machine learning framework to predict the glass transition temperature of polymers. Polymers (Basel), 2024;16(8):1049–1058. doi: 10.3390/polym16081049.

[3] Champa-Bujaico E, Díez-Pascual AM, Lomas Redondo A and Garcia-Diaz P. Optimization of mechanical properties of multiscale hybrid polymer nanocomposites: A combination of experimental and machine learning techniques. Composites Part B: Engineering, 2024;269:111099–111115. doi: 10.1016/j.compositesb.2023.111099.

[4] Jeon J, Rhee B and Gim J. Melt temperature estimation by machine learning model based on energy flow in injection molding. Polymers (Basel), 2022;14(24):5548–5555. doi: 10.3390/polym14245548.

[5] Chen G, Shen Z, Iyer A, Ghumman UF, Tang S, Bi J, et al. Machine-learning-assisted de novo design of organic molecules and polymers: Opportunities and challenges. Polymers (Basel), 12(1):163–208. 10.3390/polym12010163.

[6] Kern J, Chen L, Kim C and Ramprasad R. Design of polymers for energy storage capacitors using machine learning and evolutionary algorithms. Journal of Materials Science, 2021;56(35):19623–19635. doi: 10.1007/s10853-021-06520-x7.

[7] Daniel DT, Mitra S, Eichel R-A, Diddens D and Granwehr J. Machine learning isotropic g values of radical polymers. Journal of Chemical Theory and Computation, 2024;20(6):2592–2604. doi: 10.1021/acs.jctc.3c01252.

[8] Subeshan B, Atayo A and Asmatulu E. Machine learning applications for electrospun nanofibers: A review. Journal of Materials Science, 2024;59(31):14095–14140. doi: 10.1007/s10853-024-09994-7.

[9] Zuccarini C, Ramachandran K and Jayaseelan DD. Material discovery and modeling acceleration via machine learning. APL Mater, 2024;12(9):1–8. doi: 10.1063/5.0230677.

[10] Xu Y, Ge J and Ju C-W. Machine learning in energy chemistry: Introduction, challenges and perspectives. Energy Advances, 2023;2(7):896–921. doi: https://doi.org/10.1039/D3YA00057E.

[11] Zhao G, Xu T, Fu X, Zhao W, Wang L, Lin J, et al Machine-learning-assisted multiscale modeling strategy for predicting mechanical properties of carbon fiber reinforced polymers. Composites Science & Technology, 2024;248:110455–110465. doi: 10.1016/j.compscitech.2024.110455.

[12] Luka BS, Yuguda TK, Adnouni M, Zakka R, Abdulhamid IB and Gargea BG. Drying temperature-dependent profile of bioactive compounds and prediction of antioxidant capacity of cashew apple pomace using coupled Gaussian process regression and support vector regression (GPR–SVR) model. Heliyon, 2022;8(9):e10461–e10469. doi: 10.1016/j.heliyon.2022.e10461.

[13] Alshannaq AA, Tamimi MF and Abu Qamar MI. Machine learning-assisted prediction of durability behavior in pultruded fiber-reinforced polymeric (PFRP) composites. Results in Engineering, 2025;25:104198–104210. doi: 10.1016/j.rineng.2025.104198.

[14] Geng Y, Zhou J, Liu M, Hu Z, Zhu L, Wang L, et al A machine learning-enabled real-time temperature response system based on polymer-filler interactions for conductive network assembly. Chemical Engineering Journal, 2025;504:158778–158781. doi: 10.1016/j.cej.2024.158778.

[15] Martin TB and Audus DJ. Emerging trends in machine learning: A polymer perspective. ACS Polymer Au, 2023;3(3):239–258. doi: doi: 1021/acspolymersau.2c00053.

[16] Sattari K, Xie Y and Lin J. Data-driven algorithms for inverse design of polymers. Soft Matter, 2021;17 (33):7607–7622. doi: 10.1039/D1SM00725D.

[17] Chen L, Pilania G, Batra R, Huan TD, Kim C, Kuenneth C, et al. Polymer informatics: Current status and critical next steps. Materials Science and Engineering R: Reports, 2021;144:100595–100613. https://doi.org/10.1016/j.mser.2020.100595.

[18] Su W-F. Chemical and physical properties of polymers. In: 2013. p. 61–88. Available at: https://link.springer.com/10.1007/978-3-642-38730-2_4

[19] Aguilar-Vega M. Structure and mechanical properties of polymers. In: Handbook of Polymer Synthesis, Characterization, and Processing. 2013; (pp. 425–434). John Wiley & Sons, Ltd.

[20] Pai SM, Shah KA, Sunder S, Albuquerque RQ, Brütting C and Ruckdäschel H. Machine learning applied to the design and optimization of polymeric materials: A review. Next Materials, 2025;7:100449–100469. doi: 10.1016/j.nxmate.2024.100449.

[21] Luo C and Sommer JU. Coding coarse grained polymer model for LAMMPS and its application to polymer crystallization. Computer Physics Communications, 2009;180(8):1382–1391.

[22] Subashini M, Devarajan PV, Sonavane GS and Doble M. Molecular dynamics simulation of drug uptake by polymer. Journal of Molecular Modeling, 2011;17(5):1141–1147.

[23] Roberts BP, Krippner GY, Scanlon MJ and Chalmers DK. Molecular dynamics of variegated polyamide dendrimers. Macromolecules, 2009;42(7):2784–2794.

[24] Araki T, Cruz-Silva R, Tejima S, Takeuchi K, Hayashi T, Inukai S, et al Molecular dynamics study of carbon nanotubes/polyamide reverse osmosis membranes: polymerization, structure, and hydration. ACS Applied Materials and Interfaces, 2015;7(44):24566–24575.

[25] Dutta RC and Bhatia SK. Structure and gas transport at the polymer-zeolite interface: insights from molecular dynamics simulations. ACS Applied Materials and Interfaces, 2018;10(6):5992–6005.

[26] Anstine DM, Mendez NF and Colina CM. Sulfonyl PIM-1: A diverse separation membrane with dilation resistance. AIChE Journal, 2021;67(3):1–10.

[27] Marley FA, Asare J, Sekyi-Arthur D, Lukas T, Appiah ANS, Charway D, et al First-principles approach to finite element simulation of flexible photovoltaics. Energies, 2024;17(16):4064–4074. doi: 10.3390/en17164064.

[28] Kubaib A, Mohamed Imran P and Aathif Basha A. Applications of the Vienna Ab initio simulation package, DFT and molecular interaction studies for investigating the electrochemical stability and solvation performance of non-aqueous NaMF6 electrolytes for sodium-ion batteries. Computational and Theoretical Chemistry, 2022;1217:113934–1139344. doi: 10.1016/j.comptc.2022.113934.

[29] Shen C, Zhang Y, Han F and Xia K. Molecular topological deep learning for polymer property prediction. 2024;3:1–22. doi: 10.48550/arXiv.2410.04765.

[30] Xu C, Wang Y and Barati Farimani A. TransPolymer: A Transformer-based language model for polymer property predictions. NPJ Computational Materials, 2023;9(1):1–14.

[31] Gao L, Lin J, Wang L and Du L. Machine learning-assisted design of advanced polymeric Materials. Accounts of Materials Research, 2024;5(5):571–584.

[32] Malashin IP, Tynchenko VS, Nelyub VA, Borodulin AS and Gantimurov AP. Estimation and prediction of the polymers' physical characteristics using the machine learning models. Polymers (Basel), 16(1):115–142. http://dx.doi.org/10.3390/polym16010115.

[33] Antoniuk ER, Li P, Kailkhura B and Hiszpanski AM. Representing polymers as periodic graphs with learned descriptors for accurate polymer property predictions. Journal of Chemical Information & Modeling, 2022;62(22):5435–5445.

[34] Ma R, Liu Z, Zhang Q, Liu Z and Luo T. Evaluating polymer representations via quantifying structure-property relationships. Journal of Chemical Information & Modeling, 2019;59(7):3110–3119.
[35] Hu X, Wang D, Ying W and Fu Y. Reinforcement feature transformation for polymer property performance prediction. In: International Conference on Information and Knowledge Management, Proceedings. Association for Computing Machinery. 2024; (vol. 1, pp. 4538–4545).
[36] Qiu H, Liu L, Qiu X, Dai X, Ji X and Sun ZY. PolyNC: A natural and chemical language model for the prediction of unified polymer properties. Chemical Science, 2023;15(2):534–544.
[37] Patra TK. Data-driven methods for accelerating polymer design. ACS Polymer Au, 2022;2(1):8–26.
[38] Nguyen D, Tao L and Li Y. Integration of machine learning and coarse-grained molecular simulations for polymer materials: Physical understandings and molecular design. Frontiers in Chemistry, 2022;9:1–26.
[39] Chao TH, Rekhi S, Mittal J and Tabor DP. Data-driven models for predicting intrinsically disordered protein polymer physics directly from composition or sequence. Molecular Systems Design & Engineering, 2023;8(9):1146–1155.
[40] Babbar A, Ragunathan S, Mitra D, Dutta A and Patra TK. Explainability and extrapolation of machine learning models for predicting the glass transition temperature of polymers. Journal of Polymer Science: Polymer Letters Edition, 2024;62(6):1175–1186.
[41] Org C, Bjerrum EJ, Rastemo T, Irwin R, Kannas C and Genheden S. Software release note PySMILESUtils-enabling deep learning with the SMILES chemical language. 2021;10:1–8. doi: 10.33774/chemrxiv-2021-kzhbs.
[42] Miccio LA and Schwartz GA. From chemical structure to quantitative polymer properties prediction through convolutional neural networks. Polymer, 2020;193:122341–122348.
[43] Webb MA, Jackson NE, Gil PS and De Pablo Jj. Targeted sequence design within the coarse-grained polymer genome. Science Advances, 2020;6(43):6216–6236. doi: 10.1126/sciadv.abc6216.
[44] Bernazzani L, Duce C, Micheli A, Mollica V, Sperduti A, Starita A, et al Predicting physical-chemical properties of compounds from molecular structures by recursive neural networks. Journal of Chemical Information & Modeling, 2006;46(5):2030–2042. doi: 10.1021/ci060104e.
[45] Bertinetto C, Duce C, Micheli A, Solaro R, Starita A and Tiné MR. Prediction of the glass transition temperature of (meth)acrylic polymers containing phenyl groups by recursive neural network. Polymer, 2007;48(24):7121–7129.
[46] Gómez-Bombarelli R, Wei JN, Duvenaud D, Hernández-Lobato JM, Sánchez-Lengeling B, Sheberla D, et al Automatic chemical design using a data-driven continuous representation of molecules. ACS Central Science, 2018;4(2):268–276.
[47] Batra R, Dai H, Huan TD, Chen L, Kim C, Gutekunst WR, et al Polymers for extreme conditions designed using syntax-directed variational autoencoders. Chemistry of Materials, 2020;32 (24):10489–10500.
[48] Bhattacharya D and Patra TK. DPOLY: Deep learning of polymer phases and phase transition. Macromolecules, 2021;54(7):3065–3074.
[49] Huan TD, Mannodi-Kanakkithodi A and Ramprasad R. Accelerated materials property predictions and design using motif-based fingerprints. Physical Review. B, Condensed Matter Materials Physics, 2015;92(1):1–10.
[50] Mannodi-Kanakkithodi A, Pilania G, Huan TD, Lookman T and Ramprasad R. Machine learning strategy for accelerated design of polymer dielectrics. Scientific Reports, 2016;6:1–10.
[51] Sharker IH. Machine learning: Algorithms, real-world applications and research directions. SN Computer Science, 2021;2:60–87.
[52] Parker Q, Perera D, Li YW and Vogel T. Supervised and unsupervised machine learning of structural phases of polymers adsorbed to nanowires. Physical Review E, 2022;105(3):1–9. https://doi.org/10.1007/s42979-021-00592-x.

[53] Xu X, Wei Q, Li H, Wang Y, Chen Y and Jiang Y. Recognition of polymer configurations by unsupervised learning. Physical Review E, 2019;99(4):43307–43317. doi: 10.1103/physreve.99.043307.
[54] Yang L and Shami A. On hyperparameter optimization of machine learning algorithms: Theory and practice. Neurocomputing, 2020;415:295–316.
[55] Chan H, Cherukara M, Loeffler TD, Narayanan B and Sankaranarayanan SKRS. Machine learning enabled autonomous microstructural characterization in 3D samples. NPJ Computational Materials, 2020;6(1):1–9. doi: 10.1038/s41524-019-0267-z.
[56] Lu S and Jauaramn A. Machine learning for analyses and automation of structural characterization of polymer materials. Progress in Polymer Science, 2024;153:101828–101851. https://doi.org/10.1016/j.progpolymsci.2024.101828.
[57] Liang Y, Liu Z and Liu W. A co-training style semi-supervised artificial neural network modeling and its application in thermal conductivity prediction of polymeric composites filled with BN sheets. Energy AI, 2021;4:100052–100061.
[58] Sha W, Li Y, Tang S, Tian J, Zhao Y, Guo Y, Zhang W, Zhang X, Lu S, Cao Y-C and Cheng S. Machine learning in polymer informatics. InfoMat, 2021;3:353–361. https://doi.org/10.1002/inf2.12167.
[59] Kim C, Chandransekaran A, Jha A and Ramprasad R. Active-learning and materials design: The example of high glass transition temperature polymers. MRS Communications, 2019;9:860–866. https://doi.org/10.1557/mrc.2019.78.
[60] Ding F, Liu LY, Liu TL, Li YQ, Li JP and Sun ZY. Predicting the mechanical properties of polyurethane elastomers using machine learning. Chinese Journal of Polymer Science, 2023;41(3):422–431. https://doi.org/10.1007/s10118-022-2838-6.
[61] Park S, Marimuthu KP, Han G and Lee H. Deep learning based nanoindentation method for evaluating mechanical properties of polymers. International Journal of Mechanical Sciences, 2023;246:108162–108173. doi: 10.1016/j.ijmecsci.2023.108162.
[62] Mairpady A, Mourad AHI and Mozumder MS. Statistical and machine learning-driven optimization of mechanical properties in designing durable hdpe nanobiocomposites. Polymers (Basel), 2021;13 (18):3100–3127. https://doi.org/10.3390/polym13183100.
[63] Otsuka S, Kuwajima I, Hosoya J, Xu Y and Yamazaki M. PoLyInfo: Polymer database for polymeric materials design. Proc – 2011 Int Conf Emerg Intell Data Web Technol EIDWT, 2011;2011:22–29.
[64] Kim C, Chandrasekaran A, Huan TD, Das D and Ramprasad R. Polymer genome: A data-powered polymer informatics platform for property predictions. Journal of Physical Chemistry C, 2018;122 (31):17575–17585. http://dx.doi.org/10.1021/acs.jpcc.8b02913.
[65] Lu S and Jayaraman A. Machine learning for analyses and automation of structural characterization of polymer materials. Progress in Polymer Science, 2024;153:101828–101843. doi: 10.1016/j.progpolymsci.2024.101828.
[66] Worgull M. Hot Embossing: Theory of Microreplication. 2024; Amsterdam: Elsevier.
[67] Callister DWRD. Materials Science and Engineering. 2013; New York: Wiley-Blackwell.
[68] Pai SM, Shah KA, Sunder S, Albuquerque RQ, Brütting C and Ruckdäschel H. Machine learning applied to the design and optimization of polymeric materials: A review. Next Materials, 2025;7:100449–100469. doi: 10.1016/j.nxmate.2024.100449.
[69] Cenci F, Pankajakshan A, Facco P and Galvanin F. An exploratory model-based design of experiments approach to aid parameters identification and reduce model prediction uncertainty. Computers & Chemical Engineering, 2023;177:108353–108374. doi: 10.1016/j.compchemeng.2023.108353.
[70] Ramprasad R, Batra R, Pilania G, Mannodi-Kanakkithodi A and Kim C. Machine learning in materials informatics: Recent applications and prospects. NPJ Computational Materials, 2017;3(1):54–67. doi: 10.1038/s41524-017-0056-5.
[71] Jafari P, Zhang R, Huo S, Wang Q, Yong J, Hong M, et al Machine learning for expediting next-generation of fire-retardant polymer composites. Composites Communications , 2024;45:101806. https://doi.org/10.1016/j.coco.2023.101806.

[72] Beaucage PA, Sutherland DR and Martin TB. Automation and machine learning for accelerated polymer characterization and development: past, potential, and a path forward. Macromolecules, 2024;57(18):8661–8670. doi: 10.1021/acs.macromol.4c01410.
[73] Velandia Cabra JR. Identificación de polímeros por espectroscopía infrarroja. Rev Ontare, 2018;5:1–11. doi: 10.21158/23823399.v5.n0.2017.2005.
[74] Sala O. Fundamentals of Raman and Infrared Spectroscopy. 2008; (2nd ed). São Paulo: UNESP.
[75] Xu Z, He Z, Song Y, Fu X, Rommel M, Luo X, et al Topic review: Application of Raman spectroscopy characterization in micro/nano-machining. Micromachines, 2018;9(7):361–374. doi: 10.3390/mi9070361.
[76] Florenzano FH. Perspectivas atuais para a obtenção controlada de polímeros e sua caracterização. Polímeros, 2008;18(2):100–105. doi: 10.1590/S0104-14282008000200006.
[77] Berne BJ and Pecora R. Dynamic Light Scattering. 1976; New York: John Wiley and Sons.
[78] Rovere J, Correa CA, Grassi VG and Dal Pizzol MF. Caracterização morfológica do poliestireno de alto impacto (HIPS). Polímeros, 2008;18(1):12–19. doi: 10.1590/S0104-14282008000100007.
[79] Sperling LH. Introduction to Physical Polymer Science. 2005; New York: Wiley.
[80] Alves V, Braga R, Muratov E and Andrade C. Quimioinformática: Uma Introdução. Química Nova, 2017;41(2):202–212. doi: 10.21577/0100-4042.20170145.
[81] Gasteiger J and Funatsu K. Chemoinformatics – An important scientific discipline. Journal of Computer Chemistry, Japan, 2006;5(2):53–58. https://doi.org/10.2477/jccj.5.53.
[82] Wang W, Moreau NG, Yuan Y, Race PR and Pang W. Towards machine learning approaches for predicting the self-healing efficiency of materials. Computational Materials Science, 2019;168:180–187. https://doi.org/10.1016/j.commatsci.2019.05.050.
[83] Sha W, Li Y, Tang S, Tian J, Zhao Y, Guo Y, et al Machine learning in polymer informatics. InfoMat, 2021;3(4):353–361. https://doi.org/10.1002/inf2.12167.
[84] Baughman AK, Chuang W, Dixon KR, Benz Z and Basilico J. DeepQA Jeopardy! gamification: A machine-learning perspective. IEEE Trans Comput Intell AI Games, 2014;6(1):55–66.
[85] Struble DC, Lamb BG and Ma B. A prospective on machine learning challenges, progress, and potential in polymer science. MRS Communications, 2024;14(5):752–770. https://doi.org/10.1557/s43579-024-00587-8.
[86] Wang Y, Li S, Meng K, Zhang Y, Fang Z and Sun S. Exploring the ambient-temperature degradation reactions of pet through two-step machine learning and high-throughput experimentation. ACS Sustainable Chemistry & Engineering, 2024;12(14):5415–5426. https://doi.org/10.1021/acssuschemeng.3c06977.
[87] Otaru AJ, Alhulaybi ZA and Dubdub I. Machine learning backpropagation prediction and analysis of the thermal degradation of poly (Vinyl Alcohol). Polymers (Basel), 2024;16(3):437–449. doi: 10.3390/polym16030437.
[88] Zhang N, Xi Josie LZ and Ji PW. Utilizing machine learning algorithms in conjunction with terahertz spectra for enhanced plastic sorting efficiency. In 2024 49th International Conference on Infrared, Millimeter, and Terahertz Waves (Irmmw-thz). 2024; (pp. 1–2). IEEE.
[89] Ramos E, Lopes AG and Mendonça F. Application of machine learning in plastic waste detection and classification: A systematic review. Processes, 2024;12(8):1632–1641. doi: 10.3390/pr12081632.
[90] Safari H and Bavarian M. Enhancing polymer reaction engineering through the power of machine learning. In: 2024. p. 367–372. Available at: https://psecommunity.org/LAPSE:2024.1550
[91] Yuan W, Hibi Y, Tamura R, Sumita M, Nakamura Y, Naito M, et al Revealing factors influencing polymer degradation with rank-based machine learning. Patterns, 2023;4(12):100846–100857.
[92] Nimma D, Prakash N, Khan SA, Manchanda M, Poongavanam N and Infant Raj I. Optimizing battery recycling processes in electric vehicle life cycle assessment using a hybrid model of LSTM and deep reinforcement learning. In 2024 8th International Conference on I-SMAC (IoT in Social, Mobile, Analytics and Cloud) (I-SMAC). 2024; (pp. 1591–1596). IEEE. Available at: https://ieeexplore.ieee.org/document/10714684/.

[93] Malashin I, Martysyuk D, Tynchenko V, Gantimurov A, Semikolenov A, Nelyub V, et al Machine learning-based process optimization in biopolymer Manufacturing: A review. Polymers (Basel), 2024;16(23):3368–3379. doi: 10.3390/polym16233368.

[94] Fan Z. An exploration of reinforcement learning and deep reinforcement learning. Applied and Computational Engineering, 2024;73(1):154–159. doi: 10.54254/2755-2721/73/20240386.

[95] Lubongo C, Bin Daej MAA and Alexandridis P. Recent developments in technology for sorting plastic for recycling: The emergence of artificial intelligence and the rise of the robots. Recycling, 2024;9(4):59–108. doi: 10.3390/recycling9040059.

[96] He Y, Huang P, Hong W, Luo Q, Li L and Tsui K-L. In-Depth insights into the application of recurrent neural networks (RNNs) in traffic prediction: A comprehensive review. Algorithm, 2024;17(9):398–401. doi: 10.3390/a17090398.

[97] Ahlawat S. Recurrent Neural Networks. In: Reinforcement Learning for Finance. 2023; (pp. 177–232). Berkeley, CA: Apress. Available at: https://link.springer.com/10.1007/978-1-4842-8835-1_4.

[98] Hopfield JJ. Neural networks and physical systems with emergent collective computational abilities. Proceedings of the National Academy of Sciences, 1982;79(8):2554–2558. doi: 10.1073/pnas.79.8.2554.

[99] Sibruk L and Zakutynskyi I. Recurrent neural networks for time series forecasting. choosing the best architecture for passenger traffic data. Electronics and Control Systems, 2022;2(72):38–44. doi: 10.18372/1990-5548.72.16941.

[100] Bhuvaneswari S, Abirami A, Kurup D, Hardlin Sherin R and Harishma S. Predictive modelling of polymer properties from polymer genomic data using machine learning approaches. In: 2023 3rd International Conference on Innovative Mechanisms for Industry Applications (ICIMIA). 2023; (pp. 1069–1075). IEEE. Available at: https://ieeexplore.ieee.org/document/10425864/.

[101] Nazarova AL, Yang L, Liu K, Mishra A, Kalia RK, Nomura K, et al Dielectric polymer property prediction using recurrent neural networks with optimizations. Journal of Chemical Information & Modeling, 2021;61(5):2175–2186. doi: 10.1021/acs.jcim.0c01366.

[102] Agarwal H, Mahajan G, Shrotriya A and Shekhawat D. Predictive data analysis: leveraging RNN and LSTM techniques for time series dataset. Procedia Computer Science, 2024;235:979–989. https://doi.org/10.1016/j.procs.2024.04.093.

[103] Gupta BB, Gaurav A, Arya V, Bansal S, Attar RW, Alhomoud A, et al Earthworm optimization algorithm based cascade LSTM-GRU model for android malware detection. Cyber Security and Applications, 2025;3:100083–100096. https://doi.org/10.1016/j.csa.2024.100083.

[104] Okut H. Deep learning for subtyping and prediction of diseases: long-short term memory. In: Deep Learning Applications. IntechOpen; 2021. Available at: https://www.intechopen.com/books/deep-learning-applications/deep-learning-for-subtyping-and-prediction-of-diseases-long-short-term-memory

[105] Li D, Ru Y, Chen Z, Dong C, Dong Y and Liu J. Accelerating the design and development of polymeric materials via deep learning: Current status and future challenges. APL Machine Learning, 2023;1(2):21501–21521. https://doi.org/10.1063/5.0131067.

[106] Mohan RE, et al Investigating the use of deep learning, in materials research for predicting material properties, identifying new materials, and optimizing material selection for mechanical components. Dandao Xuebao/Journal Ballist, 2024;36(1):12–23. https://doi.org/10.52783/dxjb.v36.124.

[107] Noda K and Shibuta Y. Predicting long-term trends in physical properties from short-term molecular dynamics simulations using long short-term memory. Journal of Physics. Condensed Matter, 2024;36(38):385902–385914.

[108] Zhu X, Xu J, Fu Z, Damarla SK, Wang P and Hao K. Novel dynamic data-driven modeling based on feature enhancement with derivative memory LSTM for complex industrial process. Neurocomputing, 2025;626:129619–129639. doi: 10.1016/j.neucom.2025.129619.

[109] Cardoso R, da Costa CA, Figueiredo de RM, Zehetmeyer G and Schmith J. A method to predict the percentage of biodegradation in polymeric materials. Computers & Electrical Engineering, 2024;118:109473–109493. doi: 10.1016/j.compeleceng.2024.109473.
[110] Li D, Ru Y and Liu J. GATBoost: Mining graph attention networks-based important substructures of polymers for a better property prediction. Materials Today Communications, 2024;38:107577–107599. doi: 10.1016/j.mtcomm.2023.107577.
[111] Malashin I, Tynchenko V, Gantimurov A, Nelyub V and Borodulin A. Applications of long short-term memory (LSTM) networks in polymeric sciences: A review. Polymers (Basel), 2024;16(18):2607–2651. doi: 10.3390/polym16182607.
[112] Munir N, McMorrow R, Mulrennan K, Whitaker D, McLoone S, Kellomäki M, et al Interpretable machine learning methods for monitoring polymer degradation in extrusion of polylactic acid. Polymers (Basel), 2023;15(17):3566–3589. doi: 10.3390/polym15173566.
[113] Quradaa FH, Shahzad S and Almoqbily RS. A systematic literature review on the applications of recurrent neural networks in code clone research. Soldani J, organizador. PLoS One, 2024;19(2): e0296858–e0296898. doi: 10.1371/journal.pone.0296858.
[114] Cui X, Chipusu K, Ashraf MA, Riaz M, Xiahou J and Huang J. Symmetry-enhanced LSTM-Based recurrent neural network for oscillation minimization of overhead crane systems during material transportation. Symmetry (Basel), 2024;16(7):920–936. doi: 10.3390/sym16070920.
[115] Dridi S. Reinforcement Learning – A systematic literature review. 2024;1:1–19. doi: 10.31219/osf.io/svykj.
[116] Raihen NM and Tran J. Optimizing reinforcement learning in complex environments using neural networks. International Journal of Science and Research Archive (IJSRA), 2024;12(2):2047–2062. doi: 10.30574/ijsra.2024.12.2.1471.
[117] Costa RD and Hirata CM. Reinforcement learning applied to a situation awareness decision-making model. Information Science (Ny), 2025;121928–121957. https://doi.org/10.1016/j.ins.2025.121928.
[118] Shuford J. Deep reinforcement learning unleashing the power of AI in decision-making. The Journal of Artificial Intelligence General Science, 2024;1(1):1–29. https://doi.org/10.60087/jaigs.v1i1.36.
[119] Makkar P. Reinforcement learning: A comprehensive overview. International Journal of Innovative Research in Computer Science and Technology (IJIRCST), 2024;12(2):119–125. doi: 10.55524/ijircst.2024.12.2.21.
[120] Sandhu M, Saini A, Kaur G and Purnima. Reinforcement learning: Framework, Applications and challenges. International Journal of Engineering Science and Humanities, 2024;14(1):58–64. https://doi.org/10.62904/s3qf5660.
[121] Zhang W, Yu Y, Yuan Z, Tang P and Gao B. Data-driven pre-training framework for reinforcement learning of air-source heat pump (ASHP) systems based on historical data in office buildings: field validation. Energy and Build, 2025;2025:115436–115448. doi: 10.1016/j.enbuild.2025.115436.
[122] Teruel R, Alcalá N, Crespo C and Laspalas M. Machine learning model and input batch management tool for the composition of new recycled polypropylene plastic material with reduced variability in target properties. Journal of Cleaner Production, 2024;435:140390–140420. https://doi.org/10.1016/j.jclepro.2023.140390.
[123] Motadayen M, Devabharathi N and Agarwala S. Advancing sustainability: Biodegradable electronics and materials discovery through artificial intelligence. International Journal of AI for Materials and Design , 2024;1(2):1–20. doi: 10.36922/ijamd.3173.
[124] Chew AK, Afzal MAF, Chandrasekaran A, Kamps JH and Ramakrishnan V. Designing the next generation of polymers with machine learning and physics-based models. Machine Learning Science and Technology, 2024;5(4):45031–45049. doi: 10.1088/2632-2153/ad88d.
[125] MA R, Zhang H and Luo T. Explore high thermal conductivity amorphous polymers using reinforcement learning. Polymer Science, 2021;2:1–20. doi: 10.26434/chemrxiv-2021-6jftj-v2.

[126] Inayathullah S and Buddala R. Review of machine learning applications in additive manufacturing. Results in Engineering, 2025;25:103676–103683. doi: 10.1016/j.rineng.2024.103676.
[127] Wang S, Yue H and Yuan X. Accelerating polymer discovery with uncertainty-guided PGCNN: Explainable AI for predicting properties and mechanistic insights. Journal of Chemical Information & Modeling, 2024;64(14):5500–5509. doi: 10.1021/acs.jcim.4c00555.
[128] Ballard N, Farajzadehahary K, Hamzehlou S, Mori U and Asua JM. Reinforcement learning for the optimization and online control of emulsion polymerization reactors: Particle morphology. Computers & Chemical Engineering, 2024;187:108739–108752. doi: 10.1016/j.compchemeng.2024.108739.
[129] Patel AD, Schyns ZOG, Franklin TW and Shaver MP. Defining quality by quantifying degradation in the mechanical recycling of polyethylene. Nature Communications, 2024;15(1):8733–8741. doi: 10.1038/s41467-024-52856-8.

Shikha Kumari and Pratima Parashar Pandey*

Chapter 3
Machine learning for material simulation: revolutionizing polymer science

Abstract: This chapter explains the role of machine learning in elevating polymer research, emphasizing on prediction of properties, material simulations, and the discovery of novel polymer configurations. Polymers pose unique challenges in simulations and experiments due to the complexity in their molecular structures and their varying properties. The traditional methods of polymer design rely greatly on experimental techniques that are based on large scale trial-and-error processes. These techniques are mostly costly and time-consuming. Machine learning offers a promising solution by providing powerful tools that can predict the properties of polymers before they are synthesized, which can drastically reduce the time and resource requirements. By utilizing vast datasets of experimental results and molecular descriptors, machine learning models can predict polymer properties such as mechanical strength, conductivity, thermal stability, and biodegradability with high accuracy. These predictive models can be used to quickly access all promising candidates of polymers and select the efficient one, thereby reducing the requirement of time-consuming and uneconomical experiments. Not only prediction, but machine learning can also support the design process for new polymer materials. Generative models such as variational autoencoders and generative adversarial networks are now being utilized to propose new polymer architectures with requisite properties. These models are capable of generating novel polymer structures that may not have been considered through traditional design methods. Additionally, reinforcement learning algorithms are being applied to optimize polymerization processes. By these algorithms, researchers can tweak parameters such as temperature, solvent, and pressure, which can further accelerate the design process by minimizing the number of experimental trials needed.

Keywords: Polymer science, machine learning, material simulation

***Corresponding author: Pratima Parashar Pandey**, Department of Physics, School of Sciences, IILM University, Greater Noida, Uttar Pradesh, India, e-mail: pratimaparashara@rediffmail.com
Shikha Kumari, Department of Physics, School of Sciences, IILM University, Greater Noida, Uttar Pradesh, India

https://doi.org/10.1515/9783111681870-003

3.1 Introduction

Polymers are large molecules made by linking monomers together through covalent bonds. DNA and cellulose are naturally occurring polymers, while polyethylene and polystyrene are examples of manmade polymers. Polymer science is the study of polymers: that is, its synthesis, structure, properties, and applications [1, 2]. The unique characteristics of polymers such as flexibility, lightweight, strength and durability, and cost effectiveness have made them a vital component of modern technology and all fields of industries, including packaging [3], healthcare [4], electronics [5], automobile [6], and textile [7]. Despite numerous unique characteristics, the design and synthesis of polymers is very challenging that limits their performance. The prime limitations are in controlling the synthesis process, molecular structure, and properties of the polymer. Depending on the desired application of the polymer, its structure and properties like strength, flexibility, durability, etc. need to be tailored. Traditional practices require extensive trial and error until the desired polymer is synthesized, which can be time-consuming and costly [8]. Additionally, the polymerization process is also difficult to control; a small variation can cause significant change in the end product [9]. The stability and long-term performance of polymers requires prediction, which is also difficult to forecast using traditional methods. With increasing awareness toward environment and sustainability, the impact of certain synthetic polymers raises concern in relation to pollution [10, 11]. The traditional methods to recycle polymers degrade their characteristics, which limits their ability to be reused [12–14].

To address these challenges, technologies like machine learning (ML) and artificial intelligence (AI) are serving as essential tools. When applied to polymer science, they are enabling a fast and more accurate prediction of molecular structure, design, and synthesis of polymers. AI can process and analyze large databases and make predictions related to polymers more accurately, which can reduce time and cost considerably, in comparison to traditional trial-and-error-based methods [15, 16]. These technologies allow researchers to have more control over polymerization conditions like temperature, pressure, etc. Thus, researchers can predict right combinations to produce the desired polymers with the required properties. In addition, the technologies also optimize polymerization to improve the synthesis process. AI can also be useful in identifying recycling methods for polymer waste by examining the physical and chemical properties of the polymer [17]. To summarize, Al and ML are revolutionary tools to overcome the challenges of polymer science related to polymerization, synthesis, recycling, etc., making polymer science more efficient and sustainable. The rest of the chapter explains ML techniques and their applications to polymer science, and the challenges of AI and ML that limit their full utilization in the development of polymer technology. Lastly, how ML techniques can open new future directions in polymer science is discussed in detail.

3.2 Machine learning techniques in polymer science

AI is a field in computer science that is capable of performing tasks (e.g., reasoning, understanding, and problem solving) that require human intelligence. L is an interesting sub-branch of AI that analyzes and learns from available data and makes predictions and decisions, instead of following a strict set of instructions [18]. ML algorithms can learn using different approaches, which may include making decisions by directly studying the provided data or by analyzing hidden patterns within the data by taking feedback from the interacting environment [19] (see Figure 3.1 for better understanding of various ML algorithms).

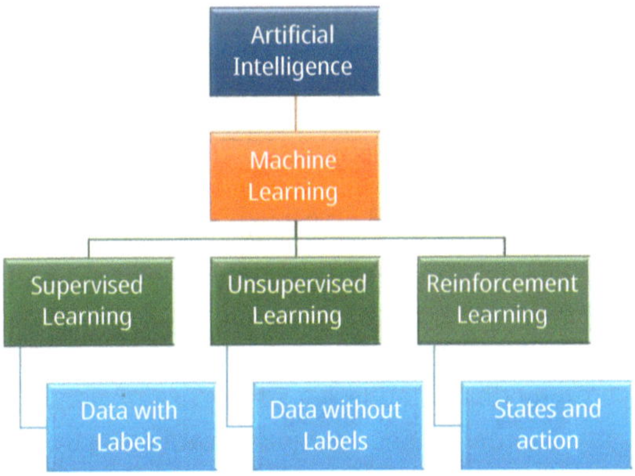

Figure 3.1: Schematic flow diagram showing machine learning algorithms.

Following are the key components of ML:

(a) Supervised learning – The model is prepared by using existing data, with known output known as labeled data.

(b) Unsupervised learning – The model is prepared using hidden patterns and structures within the data known as unlabeled data.

(c) Reinforcement learning (RL) – The model is prepared by receiving feedback from the interacting environment. The feedback can be in the form of benefits or punishments.

All these training methods can help build intelligent tools that can adapt and improve with time. AI and ML has become a vital component of our daily lives, with influence in transforming healthcare, customer experience, security enhancement, driving innovation, and facilitating research and technological development. The integration of ML with polymer science and engineering has accelerated the development of the branch, opening new research opportunities. ML has transformed this branch by making predictions, designing and optimizing new and more efficient polymers. The complete workflow of ML includes data processing, modeling, and validation (see Figure 3.2).

Figure 3.2: Schematic workflow of machine learning.

3.2.1 Supervised learning models

Supervised learning such as regression models, decision trees, and neural networks play a crucial role in predicting polymer properties. These models utilize the existing data on structure and properties of polymers to understand relationships and predicting new polymers with unknown properties. Regression techniques such as linear and polynomial regression models are also leveraged to predict the mechanical properties of polymers, such as tensile strength. Such models can relate chemical structures with macroscopic properties of the polymers. A recent study utilized a polynomial regression model to predict the mechanical properties of thermoplastic materials on the basis of their molecular structure [20]. Another supervised model used in polymer science is the decision tree. This algorithm breaks data into subsets, making them easier to understand and interpret. Numerous studies have employed decision trees to predict structural properties of polymers. One such study involves prediction of adhesive properties of polymer blends [21]. Artificial neural networks are another predictive model that is effective in determining biodegradability, thermal stability, and mechanical properties of polymers [22, 23]. The Polymer Genome is one such framework that offers effective tools that helps in this development [24]. The Polymer Genome is a computational tool that uses ML to analyze vast datasets of polymer molecular descriptors and experimental results to predict polymer properties such as mechanical strength, thermal stability, etc.

3.2.2 Deep learning models

Deep learning model is a powerful tool that can analyze complex data. It can understand complex nonlinear relationships in molecular structure and can handle high-dimensional molecular datasets. Convolutional neural networks (CNNs) and recurrent neural networks (RNNs) are two such deep learning models that can identify complex molecular structure and predict behavior and properties of polymer for much more efficient material design. CNNs are efficient in processing grid-like data, making them suitable for understanding spatial hierarchies in molecular structures [25], while RNNs can analyze sequential data, making them suitable for understanding temporal dynamics in materials, making it suitable for understanding fracture propagation [26]. Thus, by using these two techniques, it is possible to predict potential energy surfaces and energy-conserving force fields, short-term local dynamics and long-term statistical properties of microstructures, and fracture prediction in brittle materials [27].

3.2.3 Generative models

Generative models are capable of generating novel polymer structures. These models can learn from existing data and discover new polymer structures with desired characteristics like conductivity, thermal stability, and biodegradability. Variational autoencoders (VAEs) and generative adversarial networks (GANs) are two such generative models that enable material designing with requisite properties, and have shown great advancements in computational material science. Both these models have several characteristics such as inverse design capabilities, where these models are capable of inverting the design process, which is practically impossible in traditional methods. These models can also enable exploration of the vast chemical spaces to uncover novel polymers that could not be identified using conventional approaches. These models can also learn from existing data to predict properties of novel polymers and provide efficient methods to speed up the discovery and optimization of materials.

VAEs can learn a low-dimensional latent space to replicate complex data distributions. These data distributions are then helpful in generating new data points. VAEs can analyze the existing polymer datasets and get trained by learning underlying patterns to generate novel polymer structures. VAEs, when combined with Gaussian process regression, can be used to design polymers with optimized properties for conditions like high temperatures and electric fields, which meet the required performance goals [28]. The GANs comprise two neural networks; a generator and a discriminator. Generator creates artificial data while discriminator distinguishes between authentic and artificial data. This adversarial process trains GANs to produce highly realistic data. GANs are also being used along with molecular dynamics simulations to create novel vitrimers. Vitrimers are a class of polymers that have dynamic bond exchange properties [29, 30]. In recent studies, GANs were combined with graph-based representations of polymers and this combination was capable of designing new polymer architectures that exhibit glass transition temperatures [29].

3.2.4 Reinforcement learning models

RL is a powerful tool that can optimize polymerization processes and speedup material design process by reducing the number of trials required in traditional experimental techniques. RL algorithms can be useful in adjusting variables such as temperature, pressure, and solvent characteristics to achieve requisite polymer properties. RL works on a feedback mechanism, where the algorithm interacts with the surrounding environment, adjusts the parameters, and receives feedback based on the outcomes [31]. This iterative loop continues till optimal polymerization conditions are achieved. Various studies where RL is trained to automate material synthesis and discovery are available in literature. Beeler et al. exploited a deep RL system to plan and execute chemical synthesis for creating novel materials [32]. There are studies where an RL-based approach is utilized for the solvent switch process in the pharmaceutical industry [33]. In this study, a digital twin is employed, which is a virtual representation of a physical process. This helps in fine-tuning the process parameters to achieve the desired results that enhance efficiency and reduce environmental impacts. All the different types of ML models that have been discussed above are summarized in Table 3.1.

Table 3.1: Summary of different machine learning techniques utilized in polymer science.

	Model	Algorithms
1	Supervised learning	Regression techniques
		Decision tree
		Artificial neural networks
2	Deep learning	Convolutional neural networks (CNNs)
		Recurrent neural networks (RNNs)
3	Generative	Variational autoencoders (VAEs)
		Generative adversarial networks (GANs)
4	Reinforcement learning	Feedback models

3.3 Applications of machine learning in polymer research

In Section 3.2, the integration of ML and polymer research has been discussed in detail. By analyzing vast datasets available, the designing, synthesis, characterization, and optimization of polymers can be accelerated more efficiently. This section covers the applications of ML in polymer research:

1. Design and property prediction: By analyzing existing datasets, ML can predict polymer properties like mechanical behavior, such as tensile strength, elasticity, and thermal and chemical properties. This information can help in designing new materials with the requisite properties that can considerably save time and efforts required in experimental trials [34]. This generally uses supervised learning and deep learning techniques.

2. Optimization of synthesis: Polymer synthesis requires optimization of reaction conditions such as temperature, pressure, and catalysts, which is done by trial-and-error methods. These methods are time-consuming and costly. By analyzing experimental data, ML can identify optimal conditions for polymer synthesis [35]. This can significantly improve polymer quality and save lots of time and money.
3. Characterization: The deep learning models in ML can be used to analyze spectroscopy data such as FTIR and NMR. By using these models, the morphology of the polymer like crystal structure, pore size distribution, and roughness can be assessed, and thus the characterization of the desired polymer can be improved [36].
4. Lifetime prediction: ML models can analyze polymer degradation data to predict the lifetime of polymers. This information is important for polymers that have applications in automotive, aerospace, and packaging. These models predict the long-term behavior of polymers under various environmental conditions, which leads to more durable polymers [37, 38].
5. Blends and composites: High-performance materials are usually created by combining multiple polymers together in a blend or composite. The combined properties of the polymers can create a highly efficient material with superior qualities, such as greater thermal resistance or high conductivity. High-performance plastics are examples of such a material. ML can be used here to predict the interactions among the various polymers and optimize the properties of polymer blend and composites [39].
6. Polymer nanocomposites: Polymer nanocomposites are formed by the inclusion of nanoparticles on a polymer matrix. This is done to incorporate the characteristics of nanomaterials to enhance the properties of nanomaterials. ML can be used to predict the properties of these polymer nanocomposites and design enhanced nanocomposites [35, 40, 41].
7. Environmental impact and recycling of polymers: To align with the global sustainability goals, it is necessary to shift toward sustainable polymers that are eco friendly [42, 43]. ML can help in this by predicting and classifying polymers based on their recyclability and optimizing the recycling process. The algorithm analyses the spectral data for this purpose.

3.4 Challenges and limitations

ML offers novel advancements in the development of polymer science, however there are several challenges that obstruct its full utilization. ML algorithms rely on high-quality and large datasets for better training and prediction. In polymer systems, due to the various molecular structures and properties, the datasets are complex and difficult to obtain. For example, new polymer families under development have limited data. Also, the sources of this data are diverse, including experimental data, data from research papers, etc. Such data lack proper format and uniformity required for ML models [44, 45]. Also, the model that is trained on a specific polymer dataset may not perform well when applied on another polymer dataset. Thus, the ability of the model to work on varied, new and unseen data remains a challenge.

Another major limitation of ML is that often tools like deep learning are visualized as black-box. The underlying mechanism that drives the material behavior is important in polymer science. However, the insights of how these models predict the polymer are not provided, making it difficult to verify the findings and thus, lacks trust [46, 47].

Applying ML to optimize the entire development process of polymer requires integration of models with the traditional experimental approaches [48]. This will ensure that the predictions are relevant and precise. Since experimental verification is based on trial and error, it is time-consuming and costly. Integrating it with ML for the training of the model, based on experimental feedback, can thus be a slow and iterative process. The hybrid models that combine ML and experimental data can be promising, but the success will depend on the quality of experimental data and the ability to integrate the data into ML models [49].

Combined efforts of computational researchers and experimentalists can help in overcoming these challenges in order to improve efficiency and drive innovation in polymer science.

3.5 Future directions and potential impact

In this section, we will discuss how the integration of ML in polymer science can impact its application in various industries and highlight key future directions in polymer science. ML techniques can open new opportunities in exploring material design and performance enhancement, promoting sustainability.

3.5.1 Advances in polymer design

ML models can be used to overcome challenges involved in polymer design. Hybrid model is a combination of deep learning, RL, genetic algorithm, and graph-based models. Deep Learning techniques are ideal in handling large unstructured datasets and predicting polymer properties [50], while RL can help in optimizing pathways to polymer synthesis. Graph-based methods can improve accuracy in both prediction and synthesis [51, 52]. The ability of hybrid models to combine multiple techniques together makes them more robust and versatile. Thus, they can handle the complexities and challenges involved in polymer design much better, providing more accurate and efficient design. Hybrid models can thus have a great potential in advancing polymer science, as they can help discover new materials with unknown properties and can also design polymers with requisite properties for specific applications.

3.5.2 Automated discovery of materials

AI has the potential of creating fully automated platforms that can predict, synthesize, and test polymers autonomously. AI-driven prediction models can be integrated with robotics or simulation systems to create such platforms that can do real-world testing. These systems will be capable of rapid assessment and prototyping of polymers that can accelerate the development of new polymers, which will be high performing, biocompatible, and sustainable. This can be beneficial for various industries, including medical science [53, 54].

3.5.3 Cross-disciplinary collaboration

AI can be fully utilized in polymer science when experts from various fields collaborate together to build an extensive far-reaching AI system that has deep scientific knowledge. Material scientists can contribute in understanding the output of the models, as they have expertise in polymer science. They can understand the structure, property, and behavior of polymers. Data scientists can focus on model selection, validity, and data processing, ensuring that the data is processed and analyzed appropriately, while chemists and engineers can provide insights of synthesis and polymerization processes for creating AI designs that mimic real-world labs. Thus, a strong collaboration between material scientists, data scientists, chemists, and engineers is required for the future of ML in the polymer industry [55].

3.5.4 Industrial implications

Integration of ML tools into polymer science can benefit multiple industries [56, 57]. The manufacturing industry can synthesize polymers at reduced cost that are more eco-friendly and sustainable, and produce less waste. The biomedicine industry can exploit AI tools to create polymers that are less toxic and biocompatible. These polymers can be used in tissue engineering, implants, and drug delivery. The electronics industry can leverage AI to design next-generation devices that are more flexible, with improved electrical, thermal, chemical and mechanical properties. All these technological advancements can break through existing bottlenecks and challenges in polymer science, revolutionizing manufacturing, biomedicine, electronics, packaging, aerospace, automobile, and many more industries.

3.6 Conclusion

ML plays a promising role in predicting the properties of polymers such as thermal stability, mechanical strength, conductivity, and biodegradability before their synthesis. Generative models such as VAEs and GANs are exploited to design specific polymer structures for particular application. This can save time, resources, and cost incurred during the experimental process. Also, these algorithms help in the optimization of the synthesized polymer by tweaking various parameters such as temperature, solvent, and pressure to help in making a tailored, desired polymer structure. Table 3.1 provides all the ML techniques used in polymer science. These different models are employed for various aspects of polymers. Regression techniques can relate chemical structures with macroscopic properties of the polymers. CNNs and RNNs are two deep learning models that can predict behavior and properties of polymers for much more efficient material design. Generative models can learn from existing data and invent new polymer structures with requisite characteristics like conductivity, thermal stability, and biodegradability desired for the particular application. RL is used in optimization of polymerization processes by reducing the time consumed in traditional experimental techniques. Further, it helps in characterization of polymer structure, durability of polymers under various environment conditions, and optimization of polymer nanocomposites properties. Using ML in polymer science achieves new heights despite many limitations in its implications due to unavailability of data in the proper format. Also, optimization needs to integrate various models into one, which is a slow and iterative process. But, the current scenario looks at AI for advancement in polymer design, collaboration in different disciplines, and industrial application along with an automated platform for new polymer structure in future.

List of abbreviations

AI	Artificial intelligence
ML	Machine learning
VAEs	Variational autoencoders
GANs	Generative adversarial networks
CNNs	Convolutional neural networks
RNNs	Recurrent neural networks
RL	Reinforcement learning
FTIR	Fourier-transform infrared spectroscopy
NMR	Nuclear magnetic resonance

References

[1] Ebewele RO. Polymer Science and Technology. 2000 Mar 23; Boca Raton: CRC press.

[2] Gedde UW and Hedenqvist MS. Fundamental Polymer Science. 2019 Dec 20; Cham: Springer.

[3] Ellis LD, Rorrer NA, Sullivan KP, Otto M, McGeehan JE, Román-Leshkov Y, Wierckx N and Beckham GT. Chemical and biological catalysis for plastics recycling and upcycling. Nature Catalysis, 2021 Jul;4 (7):539–556.

[4] Grey EL. Polymeric Microparticles and Stimuli-Responsive Hydrogels Targeting Macrophages. (Doctoral dissertation University of Colorado at Boulder).

[5] Abdellaoui H, Raji M, Essabir H, Bouhfid R and Qaiss AE. Nanofibrillated cellulose-based nanocomposites. Bio-based Polymers and Nanocomposites: Preparation, Processing, Properties & Performance, 2019;67–86.

[6] Hürkamp A, Dér A, Gellrich S, Ossowski T, Lorenz R, Behrens BA, Herrmann C, Dröder K and Thiede S. Integrated computational product and production engineering for multi-material lightweight structures. The International Journal of Advanced Manufacturing Technology, 2020 Oct;110:2551–2571.

[7] Parida SK, Satpathy A, Dalai A and Mishra S. Introduction of polymers and polymer composites: Basic fundamentals. In: Polymer Composites: Fundamentals and Applications. 2024 May 4; (pp. 1–37). Singapore: Springer Nature Singapore.

[8] Gupta KM Engineering materials: research, applications and advances.

[9] Su WF. Principles of Polymer Design and Synthesis. 2013 Oct 9; Berlin, Heidelberg: Springer Berlin Heidelberg.

[10] Kumar R, Verma A, Shome A, Sinha R, Sinha S, Jha PK, Kumar R, Kumar P, Shubham Das S and Sharma P. Impacts of plastic pollution on ecosystem services, sustainable development goals, and need to focus on circular economy and policy interventions. Sustainability, 2021 Sep 6;13(17):9963.

[11] Andrady AL and Koongolla B. Degradation and fragmentation of microplastics. Plastics and the Ocean: Origin, Characterization, Fate, and Impacts, 2022 Apr;20:227–268.

[12] Sadhukhan J and Sekar K. Economic conditions to circularize clinical plastics. Energies, 2022 Nov 27;15(23):8974.

[13] Hamad K, Kaseem M and Deri F. Recycling of waste from polymer materials: An overview of the recent works. Polymer Degradation and Stability, 2013 Dec 1;98(12):2801–2812.

[14] Stein RS. Polymer recycling: Opportunities and limitations. Proceedings of the National Academy of Sciences, 1992 Feb 1;89(3):835–838.

[15] Winner Olabiyi EF and Gracee P. The role of AI and machine learning. 2024.

[16] Shandilya SK, Datta A, Kartik Y and Nagar A. Role of artificial intelligence and machine learning. In: Digital Resilience: Navigating Disruption and Safeguarding Data Privacy. 2024 Jan 2; (pp. 313–399). Cham: Springer Nature Switzerland.

[17] Dey, et al Dey A, Dhumal CV, Sengupta P, Kumar A, Pramanik NK and Alam T. Challenges and possible solutions to mitigate the problems of single-use plastics used for packaging food items: A review. Journal of Food Science and Technology, 2021 2021 Sep;58(9):3251–3269.

[18] Ertel W. Introduction to Artificial Intelligence. 2024; Boca Raton: Springer Nature.

[19] Singh Y, Bhatia PK and Sangwan O. A review of studies on machine learning techniques. International Journal of Computer Science and Security, 2007 Jun;1(1):70–84.

[20] Khosravi A, Fereidoon A, Khorasani MM and Saeb MR. Experimental and theoretical mechanical behavior of compatibilized polylactic acid/polyolefin elastomer blends for potential packaging applications. Iranian Polymer Journal, 2022 May;31(5):651–663.

[21] Zhang Y and Karimi S. Strength prediction of nanoparticle-reinforced adhesive and hybrid joints under unaged and hygrothermal conditions using machine learning and experimental methods. Journal of Composite Materials, 2024;59(7)00219983241300322.

[22] Wang F, Semirumi DT, He A, Pan Z and Alizadeh A. Physical, mechanical characterization, and artificial neural network modeling of biodegradable composite scaffold for biomedical applications. Engineering Applications of Artificial Intelligence, 2024 Oct 1;136:108889.
[23] Alhulaybi ZA and Otaru AJ. Machine learning analysis of enhanced biodegradable phoenix dactylifera L./HDPE composite thermograms. Polymers, 2024 May 27;16(11):1515.
[24] Doan Tran H, Kim C, Chen L, Chandrasekaran A, Batra R, Venkatram S, Kamal D, Lightstone JP, Gurnani R, Shetty P and Ramprasad M. Machine-learning predictions of polymer properties with Polymer Genome. Journal of Applied Physics, 2020 Nov 7;128(17):171104.
[25] Schütt KT, Sauceda HE, Kindermans PJ, Tkatchenko A and Müller KR. Schnet–a deep learning architecture for molecules and materials. The Journal of Chemical Physics, 2018 Jun 28;148 (24):171104.
[26] Yang K, Cao Y, Zhang Y, Tang M, Aberg D, Sadigh B and Zhou F. Self-supervised learning and prediction of microstructure evolution with recurrent neural networks. arXiv Preprint arXiv, 2020 Aug 17;2008:07658.
[27] Schwarzer M, Rogan B, Ruan Y, Song Z, Lee DY, Percus AG, Chau VT, Moore BA, Rougier E, Viswanathan HS and Srinivasan G. Learning to fail: Predicting fracture evolution in brittle material models using recurrent graph convolutional neural networks. Computational Materials Science, 2019 May 1;162:322–332.
[28] Batra R, Dai H, Huan TD, Chen L, Kim C, Gutekunst WR, Song L and Ramprasad R. Polymers for extreme conditions designed using syntax-directed variational autoencoders. Chemistry of Materials, 2020 Dec 14;32(24):10489–10500.
[29] Zhang S, He X, Xia X, Xiao P, Wu Q, Zheng F and Lu Q. Machine-learning-enabled framework in engineering plastics discovery: A case study of designing polyimides with desired glass-transition temperature. ACS Applied Materials & Interfaces, 2023 Jul 25;15(31):37893–37902.
[30] Liu Z, Huo Y, Chen Q, Zhan S, Li Q, Zhao Q, Cui L and Liu J. Predicting the glass transition temperature of polymer based on generative adversarial networks and automated machine learning. Materials Genome Engineering Advances, 2024 Dec;2(4):e78.
[31] Dadashi D, Kaedi M, Dadashi P and Sinha Ray S. Prediction of the appropriate temperature and pressure for polymer dissolution using machine learning models. Molecular Informatics, 2025 Feb;44(2):e202400193.
[32] Beeler C, Subramanian SG, Sprague K, Baula M, Chatti N, Dawit A, Li X, Paquin N, Shahen M, Yang Z and Bellinger C. ChemGymRL: A customizable interactive framework for reinforcement learning for digital chemistry. Digital Discovery, 2024;3(4):742–758.
[33] Elmaz F, Di Caprio U, Wu M, Wouters Y, Van Der Vorst G, Vandervoort N, Anwar A, Leblebici ME, Hellinckx P and Mercelis S. Reinforcement learning-based approach for optimizing solvent-switch processes. Computers & Chemical Engineering, 2023 Aug 1;176:108310.
[34] Voisin T, Grapes MD, Li TT, Santala MK, Zhang Y, Ligda JP, Lorenzo NJ, Schuster BE, Campbell GH and Weihs TP. In situ TEM observations of high-strain-rate deformation and fracture in pure copper. Materials Today, 2020 Mar 1;33:10–16.
[35] Champa-Bujaico E, García-Díaz P and Díez-Pascual AM. Machine learning for property prediction and optimization of polymeric nanocomposites: A state-of-the-art. International Journal of Molecular Sciences, 2022 Sep 14;23(18):10712.
[36] Alrfou K, Zhao T and Kordijazi A. Deep learning methods for microstructural image analysis: The state-of-the-art and future perspectives. Integrating Materials and Manufacturing Innovation, 2024 Sep;13(3):703–731.
[37] Fu Z, Liu W, Huang C and Mei T. A review of performance prediction based on machine learning in materials science. Nanomaterials, 2022 Aug 26;12(17):2957.
[38] Goyal S, Mondal S, Mohanty S, Katari V, Sharma H and Sahu KK. AI-and ML-based models for predicting remaining useful life (RUL) of nanocomposites and reinforced laminated structures. In:

Fracture Behavior of Nanocomposites and Reinforced Laminate Structures. 2024 Oct 15; (pp. 385–425). Cham: Springer Nature Switzerland.
[39] Malashin IP, Tynchenko VS, Nelyub VA, Borodulin AS and Gantimurov AP. Estimation and prediction of the polymers' physical characteristics using the machine learning models. Polymers, 2023 Dec 29;16(1):115.
[40] Saxena A, Mehta A, Vasudev H, Prashar G and Jasim NY. A short review on machine learning for the purpose of optimizing and predicting the properties of polymeric nanocomposites. Materials Today: Proceedings, 2023 Aug 1.
[41] Dananjaya V, Marimuthu S, Yang R, Grace AN and Abeykoon C. Synthesis, properties, applications, 3D printing and machine learning of graphene quantum dots in polymer nanocomposites. Progress in Materials Science, 2024 Mar;12:101282.
[42] Tran H, Gurnani R, Kim C, Pilania G, Kwon HK, Lively RP and Ramprasad R. Design of functional and sustainable polymers assisted by artificial intelligence. Nature Reviews Materials, 2024 Aug;19:1–21.
[43] Batra R, Song L and Ramprasad R. Emerging materials intelligence ecosystems propelled by machine learning. Nature Reviews Materials, 2021 Aug;6(8):655–678.
[44] Kusne AG, Mueller T and Ramprasad R. Machine learning in materials science: Recent progress and emerging applications, Reviews in computational chemistry 29, 186–273 (2016).
[45] Ward L, Agrawal A, Choudhary A and Wolverton C. A general-purpose machine learning framework for predicting properties of inorganic materials. NPJ Computational Materials, 2016 Aug 26;2(1):1–7.
[46] Oviedo F, Ferres JL, Buonassisi T and Butler KT. Interpretable and explainable machine learning for materials science and chemistry. Accounts of Materials Research, 2022 Jun 3;3(6):597–607.
[47] Ribeiro MT, Singh S and Guestrin C. "Why should i trust you?" Explaining the predictions of any classifier. In: Proceedings of the 22nd ACM SIGKDD International Conference on Knowledge Discovery and Data Mining. 2016 Aug 13; (pp. 1135–1144).
[48] Amrihesari M, Kern J, Present H, Moreno Briceno S, Ramprasad R and Brettmann B. Machine learning models for predicting polymer solubility in solvents across concentrations and temperatures. The Journal of Physical Chemistry B, 2024 Dec 12;128(51):12786–12797.
[49] Barbosa RD, Lima CC, Oliveira FF, Câmara GB, Viseras C, Moura TF, Souto EB, Severino P, Raffin FN and Fernandes MA. New Machine Learning Approach for the Optimization of Nano-Hybrid Formulations. Nanomanufacturing, 2022 Jul 18;2(3):82–97.
[50] Andraju N, Curtzwiler GW, Ji Y, Kozliak E and Ranganathan P. Machine-learning-based predictions of polymer and postconsumer recycled polymer properties: A comprehensive review. ACS Applied Materials & Interfaces, 2022 Sep 14;14(38):42771–42790.
[51] Patra TK. Data-driven methods for accelerating polymer design. ACS Polymers Au, 2021 Dec 28;2(1):8–26.
[52] Li H, Collins CR, Ribelli TG, Matyjaszewski K, Gordon GJ, Kowalewski T and Yaron DJ. Tuning the molecular weight distribution from atom transfer radical polymerization using deep reinforcement learning. Molecular Systems Design & Engineering, 2018;3(3):496–508.
[53] Upadhya R, Kosuri S, Tamasi M, Meyer TA, Atta S, Webb MA and Gormley AJ. Automation and data-driven design of polymer therapeutics. Advanced Drug Delivery Reviews, 2021 Apr 1;171:1–28.
[54] Yue T, He J, Tao L and Li Y. High-throughput screening and prediction of high modulus of resilience polymers using explainable machine learning. Journal of Chemical Theory and Computation, 2023 Jun 20;19(14):4641–4653.
[55] Struble DC, Lamb BG and Ma B. A prospective on machine learning challenges, progress, and potential in polymer science. MRS Communications, 2024 Oct;14(5):752–770.
[56] Corzo D, Tostado-Blázquez G and Baran D. Flexible electronics: Status, challenges and opportunities. Frontiers in Electronics, 2020 Sep 30;1:594003.
[57] Tran H, Gurnani R, Kim C, Pilania G, Kwon HK, Lively RP and Ramprasad R. Design of functional and sustainable polymers assisted by artificial intelligence. Nature Reviews Materials, 2024 Aug;19:1–21.

Neha Minocha* and Parijat Pandey

Chapter 4 The autonomous revolution: role of AI in self-driving labs for materials sciences

Abstract: Currently, the use of autonomous technologies within materials science studies is transforming the entire discipline. High-quality large datasets generated through autonomous experimentation would facilitate the advancement of artificial intelligence (AI) in materials science and chemistry, improving machine learning (ML) and deep learning (DL) models, and augmenting the decision-making efficacy of data-driven algorithms. The so-called self-driving laboratories (SDLs), as an innovative analytical platform, are leading this change, using AI, robots, and high-throughput testing to accelerate discovery and innovation. An SDL is an advanced experimental platform featuring various hardware modules that systematically execute a sequence of processes, as determined and organized by the ML algorithm in a closed-loop configuration to attain a specified goal. These systems make use of ML analyses to scrutinize vast databases, improve the conditions of experiments and support decision-making, where the input from a human is marginal. SDLs help to free the researcher's time, who can then focus on creating a hypothesis and work within complex problems, while repetitive tasks and overlooked patterns can be solved algorithmically. This chapter discusses the various roles of AI in shaping SDLs, with a focus on how it affects predictive modeling and learning, adoption and optimization of experimentation, and enhances the reusability of materials research. The chapter describes the design of SDLs for catalytic materials discovery, energy storage materials optimization, and the development of complex polymers. The chapter also covers challenges like algorithmic bias and poor data quality, and the need to cooperate with an interdisciplinary team to properly unlock the full potential of SDLs. The use of AI-based SDLs is not just a breakthrough in the current research and progression in materials science, but the promise for a new paradigm shift, in which the existing structures are replaced with faster, innovative, and groundbreaking findings. It is thus an autonomous revolution of a complementarity of creativity between humans and the intelligence of machines to make a rather good step in materialism and use.

Keywords: Self-driving laboratories (SDLs), artificial intelligence (AI), autonomous systems, AI-accelerated innovation

---

***Corresponding author: Neha Minocha**, School of Medical and Allied Sciences, K. R. Mangalam University, Gurgaon 122103, Haryana, India, e-mail: neha.minocha.bmu@gmail.com, orcid.org/0000-0002-0870-4972
Parijat Pandey, Department of Pharmaceutical Sciences, Gurugram University, Gurugram 122018, Haryana, India

https://doi.org/10.1515/9783111681870-004

4.1 Introduction

4.1.1 Intersection of AI, polymer science, and nanotechnology

The complexity of the interdependence between experimentation and theoretical modeling has been a hallmark of the domain of materials science in the past and continues to be in the present. New technologies such as artificial intelligence (AI) and its automation have created the foundation for the adaptation of self-driving laboratories (SDLs) [1]. The race to create the next best technological breakthrough has become not a matter of human tenacity only, but also of the utilization of AI. These complete autonomous systems integrate machine learning (ML) with robotic systems with high-throughput testing that significantly facilitates research and development in the field of materials science [2]. Combining theoretical computations with ML speeds up discovery due to the use of known material properties and computed possibilities to construct virtually indefinite lists of new materials. Until recently, the human scientist was termed the link between theoretical and experimental study. In synthesizing in the laboratory, ML can support the formation of hypotheses, computational model training, setting of realistic parameters, and the creation of new hypotheses, as well as the conceptualization and operationalization of future innovation, but human interaction is relevant. It is this change that is essentially stimulated by the potential of AI to determine patterns from large sets of data, as well as make compelling, accurate predictions [3]. The integration of AI, polymer science, and nanotechnology represent a brand-new horizon in materials research; one that has no parallel in terms of prospects. Polymers and nanoparticles are currently important subjects in modern science and technologies involving flexible electronics, biocompatible materials, energy storage, and enhanced coatings. These materials are complex and given that factors of the parameter space governing their synthesis and behavior are numerous, this is a huge challenge for traditional approaches [4, 5]. Based on the ability of AI to identify and predict challenging relationships in massive databases, this approach has become an innovative driver for the development of SDLs for polymer and nanoscale science research. SDLs combine high computational potential with highly accurate measurement of automated experiments to transform the discovery and optimization processes in polymer science and nanotechnology [6]. All of these SDLs employ ML methods to identify the association between material property, processing condition, and performance characteristic, and allow the systematic and self-driven exploration of these mutually connected elements. Robotic capabilities for high-throughput synthesis and advanced characterization techniques for feedback significantly enhance the research productivity of SDLs, while providing capabilities not accessible by other methods [7]. The introduction of AI into polymer science and nanotechnology holds a good promise toward overcoming modern challenges. In the field of polymer research, specifically enhanced mechanical, thermal, and optical performance, and applications such as sustainable food packaging, biomedical devices, and solar cell devices (Figure 4.1), these systems can improve the manufacturing of nanostructures with specific surface and electrical properties or catalytic activities, which could be addressed in energy storage, environmental treatments and sensing technologies. The combination of AI in these fields not only accelerates the development of innovation but also opens up the opportunity to create new classes of materials [8].

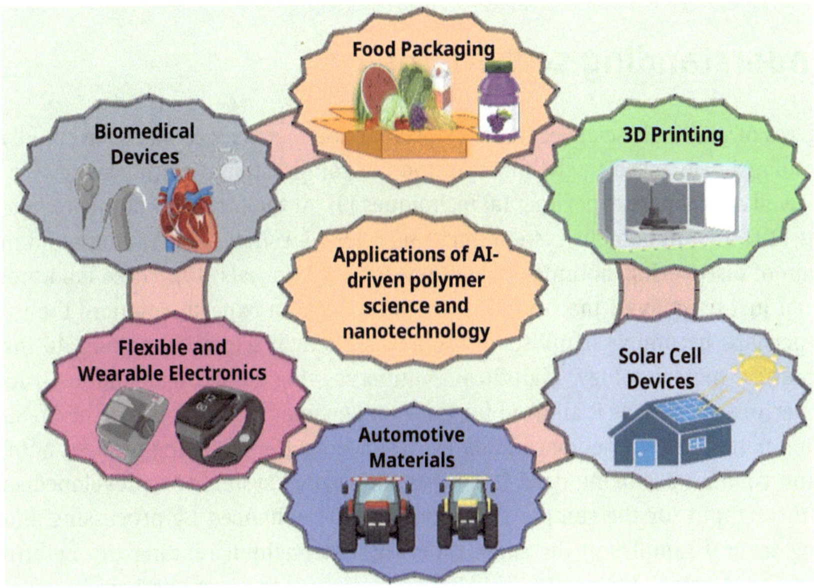

Figure 4.1: Diagram illustrating applications of AI-driven polymer science and nanotechnology in various fields, including food packaging, 3D printing, solar cell devices, automotive materials, wearable electronics, and biomedical devices.

4.1.2 Aim and scope of the chapter

This chapter outlines how integrated SDLs advanced by AI work at the interface of polymer science and nanotechnology. That is why the aim is to provide a comprehensive review of how these autonomous systems are revolutionizing the institutions and processes relating to material discovery and optimization using AI and automation. The chapter discusses the usage of SDLs in elaborating new generation polymeric materials, with specific properties for definite purposes, such as biodegradable plastics, conductive polymers, and self-healing polymeric materials using synthetic strategies and characterization methods of nanomaterials, including quantum dots, carbon nanostructures, and metal-organic frameworks based on AI methodologies. The incorporation of SDLs to highlight how polymer science and nanotechnology areas benefited from the interaction between AI and material sciences remains the focal point. This chapter performs a comprehensive review of modern developments and the existing applications of AI, thus garnering inspiration for further technological advances at this critical juncture that is relevant to the direction of AI integration in the creative domain, which will help foster an ideal future in which advanced AI and human ingenuity interfuse with one another to pioneer a revolution in the search for functional materials.

4.2 Understanding self-driving labs

The progress of materials science, engineering, and polymer science has been rapidly and continuously advanced by the incorporation of computational methods and techniques as well as complex experimental techniques [9]. At the forefront of this revolution is the SDL, an autonomous system that integrates AI, robotics, and automation in the domain of discovering, optimizing, and developing materials [10]. These laboratories are not just progeny of the traditional high-throughput experimentation; the robots are perhaps becoming signposts of the new paradigm scientific method. In the pharmaceutical biotechnology, high-throughput screening gained tremendous traction and performed well, as it allowed for the rapid evaluation of thousands of chemicals to find potential therapeutic candidates. To improve the screening process, automation and robotics are being used [11]. Custom-designed systems are developed so that the throughput for the sample evaluation can be enhanced by processing and evaluating several samples at the same time. High-throughput reading can be efficiently processed using automation. This has been motivated by researchers in materials science who have used high-throughput screening methodologies in experiments to quickly find new materials with attractive properties. Rapid synthesis, enabled by high-throughput synthesis (HTS), has been able to rapidly generate extensive arrays of material, enabling efficient exploration of a large materials landscape. HTS typically employs automated [12, 13], parallel processing techniques, which facilitate the concurrent synthesis of many materials under a variety of conditions. As a result of automating experiments for materials discovery, costs are lowered per experiment, human error and repetitive chores are minimized, and large amounts of data are obtained over a broad range of composition and processing parameters. It is advisable to incorporate this methodology across all laboratory components: human-in-the-loop process for synthesis, characterization, and optimization [14].

4.2.1 Self-driving labs: role in scientific research

Fully automated research systems with self-driving capability are intelligent laboratories that enable the creation of experiments, data analysis, and design improvement for new experiments, completely independent of any human intervention. SDLs signify essentially a feedback loop, and the study's experimental process is guided solely by AI algorithms. These systems operate with the help of ML models, having as base the historical and actual values of ideal experiment parameters [15]. After these criteria have been set, the robotic systems run the proper tests, and the results are incorporated back into the system. In materials science, SDLs serve as an appropriate means to characterize and map the vast and complex area of compositions and process variables. In the synthesis of new polymers or nanomaterials [16], SDLs can choose candidates from combinatorial libraries of chemical precursors, reaction conditions, and structural configurations in synthesizing polymers or nanomaterials of a specific feature. Besides, they help in identifying the materials, which otherwise may remain out of view, while using a trial-and-error approach. Organizations use SDLs to generate and test hypotheses on intellectual assets and their utilization. This system can develop a new scientific hypothesis based on the analysis of experimental results and the patterns recognized from experiments [17]. The fusion of autonomous experimentation with the institutionalization of insights represents a major advancement, where SDLs fortify or improve human cognition and imagination rather than eclipse them [18].

4.2.2 Core technologies in materials development

The transformative capacity of SDLs is supported by three fundamental technologies:

1. **AI:** AI plays the role of the cognitive core for SDLs and supports modeling and optimization as well as decision-making capabilities [19]. Free-form material properties, as well as complex links between these properties and the subsequent processing parameters, are well described by ML techniques, including deep learning and reinforcement learning [20, 21]. Bayesian optimization is used to control the experiment definitions to achieve the exploration of new parameter spaces, while at the same time leveraging successful areas that have already been proven to perform well [22]. However, AI improves complex dataset evaluation of high dimensionality by highly specialized characterization methods such as spectroscopy, microscopy, and diffraction processes [23]. AI increases preprocessing tasks of pattern and anomaly identification, making real-time work of the SDLs effective [24, 25].
2. **Robotics:** Robotics provides the hardware support for SDLs and ensures that the experiments are carried out to a fine level of precision [26]. When it comes to automated pipetting, sample handling and delicate synthesis, and characterization tasks, robotic systems are designed to perform the tasks consecutively and with a very small error margin. Superior robots offer adjustable feedback mechanisms that can be fine-tuned on the fly during testing, meaning that they enhance the system's flexibility and accuracy [27, 28].
3. **Automation:** Instrumentation is the base of SDLs, which provides the means for incorporating experimental procedures naturally [29, 30]. This ensures that all steps in the research process – design, synthesis, characterization, and analysis – are

interconnected. Automated systems include sensors and monitoring instruments that provide real-time feedback, and therefore SDLs can presume the change in the experiment configuration. Furthermore, automation also increases the reproducibility of SDLs in a way that allows for the management of many experiments at once, thereby increasing the scope of investigation [31, 32]. Various benefits of core technologies in SDLs are mentioned in Figure 4.2.

Core Technologies in SDLs		
Artificial Intelligence	**Robotics**	**Automation**
Real-time Optimization	Precision and Accuracy	Increased Reproducibility
Predictive Capabilities	High Throughput Processing	Real-time Monitoring
Enhanced Data Analysis	Flexibility in Experimentation	Scalability
Improved Efficiency	Safety Enhancement	Time and Cost Efficiency
Automation of Complex Tasks	Integration with AI	Seamless Workflow Integration

Figure 4.2: Core technologies driving self-driving laboratories (SDLs): AI for data-driven decision-making, robotics for precision execution, and automation for seamless workflow integration.

4.2.3 Importance of self-driving labs

ML and automated laboratories are changing materials discovery, and they carry with them a few problems. The linear molecular and materials discovery tradition has traditionally started with theoretical screening (including computational), followed by synthesis and characterization, typically with little feedback [10, 13]. ML techniques applied in automated laboratories are growing this method, which is now becoming a more dynamic and flexible approach. The main obstacle in this area is the traversal of large, often intractable, high-dimensional search spaces associated with compositions, processing settings, and molecular structures. The dimensions, the attributes of the parameter space, and the functions intended for exploration characterize these spaces [33]. These spaces are systematically endowed with dimensionality, completeness, and differentiability, bases for classifying problems that are amenable to standard ML methods and those for which more novel approaches are required. In addition, the dimension of practical chemical optimization is to access certain places inside the search area [15]. SDLs represent a great advancement in the field of material as using AI, robotics, and automation, redefine the boundaries of research. SDLs enable efficient, precise, and scalable experimentation, and quickly engage key scientific–technological challenges and set the stage for the emergence of a new interdisciplinary paradigm . This combination of the advanced technologies not only brings benefits to research efficiency but also creates a fundamental groundwork for introducing extraordinary improvements in the material's creation [34, 35]. Core components of SDLs in material science are tabulated in Table 4.1.

Table 4.1: Components of self-driving laboratories.

Component	Role in self-driving labs	Examples in materials science	Challenges	References
Machine learning algorithms	Analyze data and predict material properties	Predicting polymer thermal stability or nanomaterial conductivity	Requires large, high-quality datasets	[12]
Robotic systems	Automate synthesis and experimental workflows	Automated mixing, heating, and testing of polymers	High initial setup and maintenance costs	[13]
High-throughput screening	Rapid testing of material combinations	Screening catalyst materials for energy storage applications	Limited by equipment precision	[11]
Data integration frameworks	Combine data from multiple experiments	Integrating microscopy, spectroscopy, and simulation data	Managing heterogeneous datasets	[36]
Simulation and modeling tools	Predict material behavior under different conditions	Molecular dynamics simulations for polymer behavior prediction.	High computational demands	[15]
Advanced sensing technologies	Provide real-time data during experiments	In situ monitoring of nanomaterial fabrication processes	Ensuring accuracy under varying conditions	[33]

The advent of SDLs in materials science tackles some significant difficulties and creates new opportunities for innovation, like:

i. **Expediting discovery**: Traditional investigation of materials is often a time-consuming activity. SDLs decrease the discovery cycle time drastically since they enable and encourage the automation of multiple iterations. This acceleration is especially useful for addressing the urgent global challenges, for instance, the creation of sustainable energy sources or the efficient delivery of complex medications [37].

ii. **Expanding exploration**: The parameter space in materials science is vast, ranging from various compositions and types of alloy to the conditions of their processing and the structure of the finished product. SDLs make it possible to systematically search these complex areas and obtain materials with additional features that are unlikely to be identified using conventional approaches [13].
iii. **Augmenting reproducibility**: Reproducibility is a key component of sound scientific work, to the extent that works that are not reproducible should not be published. SDLs favor replication and the steadiness of results through consolidating several experiments to minimize human influence or standardizing execution procedures [11].
iv. **Cost and resource efficiency**: Based on the experimental results, autonomous systems reduce reliance on manual labor while avoiding wasting resources to identify highly promising regions [36]. This efficiency makes SDLs very attractive, especially in environments where there are fewer resources are available.
v. **Fostering interdisciplinary collaboration**: It is for this reason that many scientific disciplines interact through SDLs – chemists, physicists, engineers and computer scientists. AI integration into materials science produces a synthesis environment where specialists confront complex issues [36].

4.3 Artificial intelligence in polymer science and nanotechnology

As AI is incorporated into nanotechnology, a new domain called AI nanotechnology has emerged. As a result of this integration, progress in nanotechnology has been made on novel material and device creation, enhancing characterization and analytical methods, and making manufacturing processes more efficient [38]. Future developments in information sciences based on these technological mixtures will depend on innovative computer architectures and data representations, hybrid technologies combining animals with the support of nanodevices, microengineering, and other related fields. A future paradigm for material science, AI has a role in the discovery, analysis, and design of nanomaterials. AI-driven science has replaced the expensive and laborious empirical error method, which is the way of doing things in materials research, with AI. The automated identification and production of functional nanomaterials with desirable characteristics rely on AI. The advancement of novel paradigms and forecasting materials is accelerated by AI [39]. It allows effective exploration of optimal synthesis parameters to realize materials with increased characteristics. Applications of ML in the detection and progress of nanomaterials include quality predicting, designing inversely, synthesizing, characterizing, and collecting information. AI is transforming polymer science and nanotechnology by providing solutions to the intrinsic complexity of these domains [40]. Both disciplines necessitate comprehension of complex interrelations among structure, composition, and function across various scales, rendering the incorporation of AI especially revolutionary. AI improves the efficiency of discovering, optimizing, and applying polymers and nanomaterials by utilizing extensive databases and sophisticated algorithms [41]. Major contributions of AI in polymer design and nanomaterial innovation are tabulated in Table 4.2.

Table 4.2: Key contributions of AI.

Area of contribution	Role of AI	Applications	References
Predictive polymer synthesis	Uses machine learning models to predict polymer properties and behavior before synthesis.	Designing thermally stable polymers for aerospace applications.	[19]
Nanomaterial optimization	AI analyzes datasets to optimize material size, shape, and composition for specific functionalities.	High-efficiency catalysts for fuel cells.	[42, 43]
Smart polymer design	AI proposes self-healing and stimuli-responsive polymers.	Hydrogels for drug delivery that respond to pH changes.	[44]
Green chemistry	AI identifies eco-friendly materials and synthesis methods.	Creating low-energy pathways for polymer synthesis.	[45]
Biocompatibility assessment	AI models predict interactions between polymers and biological systems.	Designing biopolymers for tissue engineering scaffolds.	[46]

4.3.1 Role of AI in polymer design and nanomaterial innovation

AI is essential in the design and invention of polymers and nanomaterials by facilitating data-driven methods to address the extensive combinatorial possibilities present in these domains. By bridging the gap between computational predictions and real-world outcomes, AI will continue to drive innovation at the intersection of polymer science and nanotechnology, making material discovery more efficient, sustainable, and impactful [47]:

1. **Accelerated polymer design:** Design of conventional polymers requires repetitive experiments based on theoretical models. This method is often cumbersome, often requiring a lot of experimentation to achieve desired properties such as strength, flexibility, biodegradation, or thermal resistance. AI disrupts this paradigm by using ML models to predict polymer behavior as well as improve molecular structures. In this context, AI efficiently comprehends the relationships among monomer composition, processing conditions, and associated properties using historical datasets and quickly assesses several potential polymers. There

are certain generative methods, which are used to design new polymer structures for better delivery of medical products [48, 49].

2. **Nanomaterial innovation:** Nanotechnology requires the nanoscale control of material properties and this is extremely challenging in the conventional approaches. AI eliminates these problems by giving analytic forecasts and improving synthesis routes. Current AI models can predict how changing aspects of nanoparticles such as size or surface chemistry would impact properties, including catalytic properties, and thermal and optical behavior [50]. AI also supports the progress of complex structural composite materials such as polymer–nanoparticle materials. Using data from experimental and computer modeling, AI can suggest the best pair of polymers and nanostructures for energy storage, photonics, or environmental clean-up. The ability to integrate different pieces of information given and come up with unprecedented solutions is particularly beneficial in the framework of nanotechnology where the "design space" is vast and multifaceted [51, 52].
3. **Design of sustainable materials:** The combination of AI with polymer science and nanotechnology is enhancing the development of sustainable polymer materials. Specialized AI tools are used to assist in biodegradable polymers design and recyclable composites. These instruments enable the definition of renewable precursors and the improvement of process technology for minimal environmental impact and maximum performance, where necessary [53].

4.3.2 Machine learning for predicting polymer properties and behavior

AI and ML are disrupting polymer science and nanotechnology through technologies that bring added speed and accuracy to research [54]. These tools allow scientists to search through high-dimensional parameters with great accuracy in terms of characteristics and behaviors of all kinds of materials, accelerating the material discovery process. Integration of AI expertise with innovative, computational strategies can unlock further potential for polymers and nanomaterials in various scientific and technological disciplines [55–57]. Furthermore, regarding the complexity of the polymer structure and diversification of their applications, ML has proven to be a useful tool for predicting their properties and behavior:

1. **Data-driven prediction:** ML algorithms are quite capable of identifying patterns in large volumes of data, making them ideal for predicting things like polymer attributes such as the mechanical strength, solubility and glass transition temperature, and thermal conductivity [58]. In this case, various learning methods are employed. These models employ existing experimental or simulation data to more appropriately simulate the characteristics of thus far unexplored polymer systems. Therefore, prediction of viscoelastic properties of polymers with regard

to temperature and pressure has been earlier achieved by using ML methods. Through the incorporation of experimental facts, these models give out insights in the formulation of polymers for fabrication of implants [59, 60].

2. **Behavior in response to external stimuli**: Polymers often exhibit complex character, in response to certain triggers, for instance heat, light, or mechanical stress. The desired behaviors of the customers can be predicted through using historical data and simulations of the ML models. In synthesizing shape-memory polymers, ML predictions are made on the relationship between structure and functionality with regard to the polymer's ability to return to its initial shape after being deformed. Similarly, they may also improve the relationship between cross-linking density and swelling in stimuli-responsive hydrogels, practical for medication delivery or soft robotics [61, 62].
3. **Multiscale modeling**: ML makes data synthesis at various levels, from molecular interactions to macroscopic characteristics, possible. Specimens generated from simulation polymer fragmentations using quantum chemistry may improve mesoscale modeling of polymer blends to indicate macroscale properties via toughness or impact strength. This multiscale paradigm reduces reliance on high-cost simulations, while maintaining accuracy of the forecast [63, 64].
4. **Inverse design**: An especially promising application of ML in polymer science is the category of inverse design, in which specified preferred characteristics are inputs, and from which the system derives the molecular structures most likely to exemplify those characteristics. This approach drastically reduces the time needed for the designing of polymers, resulting in the creation of polymers for novel applications, including flexible electronics, high-performance coatings, or emerging power devices [65].
5. **Performance monitoring in real-time applications**: Real-time polymer behavior is modeled using elements of ML. As an example, wireless sensors in polymeric systems can constantly monitor stress, temperature, or chemical concentration. Analyzing such data, ML methods seek possible failure processes, such as in aerospace materials or biomedical implants, or optimize performance in such applications [66, 67].

4.4 AI and automation in self-driving labs for polymer and nanomaterial research

SDLs are now common in polymer and nanomaterial research, rigorously applying AI and automation to experiments. SDLs reduce the time and effort taken in discovering and optimizing by integrating robots [68], ML, and high flow rate experiments, dealing with challenges related to these materials. The subsequent sections analyze the principles of SDLs, highlighting the significance of their integration with AI and automation in the further development of polymers and nanomaterials through autonomous experiments, synthesis, characterization, and optimization [69, 70]. AI techniques used in polymer and nanomaterial research are tabulated in Table 4.3.

Table 4.3: AI techniques in polymer and nanomaterial research.

AI technique	Application area	Benefits	Challenges	References
Machine learning	Predicting drug efficacy and toxicity	Accelerates drug discovery by modeling biological interactions.	Requires high-quality and diverse clinical datasets.	[71, 72]
Generative adversarial networks	Designing novel therapeutic molecules	Proposes innovative drugs with enhanced therapeutic potential.	Computationally intensive and requires fine-tuning.	[73, 74]
High-throughput data mining	Identifying biomarkers and disease pathways	Enhances precision medicine by analyzing large clinical datasets.	Integration of diverse datasets is complex.	[75, 76]
Neural networks	Diagnosing diseases through imaging data	Improves accuracy in radiology and pathology diagnostics.	High computational requirements for training models.	[77, 78]
Bayesian optimization	Tuning laboratory equipment for precision	Reduces time and resource wastage in experimental setups.	Limited effectiveness for highly complex systems.	[79, 80]
Transfer Learning	Reusing pretrained models for medical datasets	Speeds up the analysis of small or specialized datasets.	May not adapt perfectly to all types of medical data.	[81]

4.4.1 Integration of robotics and AI for autonomous experimentation

Autonomous experiment cycles through robots and AI are intertwined with SDLs to optimize data-driven decision and near-perfect robot execution:

1. **Robotics for great-throughput experimentation**: Robot systems in SDLs perform simple operation tasks with a high degree of precision, such as mixing solutions, evaporating and depositing thin films, or fabricating nanostructures. These systems can support complex processes and advances, which consist of multiple steps, such as synthesis, processing, and characterization. Current robots include complex sensing and feedback systems so that changes can be made in real-time based on trial predictions from AI [82].
2. **AI-driven decision-making**: Robotic working is performed by AI algorithms that consider speculative data while proposing enhancements in the subsequent runs.

RL algorithms are particularly efficient and allow SDLs to find the best strategies all by themselves while experimenting with values within vast parameters seen in polymer and nanomaterial sciences [83]. An SDL specified to produce a polymer with specific mechanical properties may make generic decisions of monomer proportions, polymerization parameters, and curing times by itself. Outcomes are evaluated according to predetermined performance indicators, and the information is used to improve the experimental approach of the system to ideal formulations [84].

3. **Closed-loop experimentation**: Another key feature of SDLs is they being closed loop, in which AI continues varying its models based on experimental outcomes. It forms the basis of a feedback loop, enabling further optimization of the process, saving time in the discovery process. Thus, this skill is critical to sort out complex interrelationships between composition, processing conditions, and structure of polymers and nanomaterials [85].

4.4.2 Autonomous synthesis and characterization of polymers

Robotics and AI are fast changing the ways in which polymers are synthesized and characterized to unlock a range of possible polymers:

1. **Automated polymer synthesis:** Typically, the formation of conventional polymers requires a great deal of effort and time, something that essentially requires expert inputs. Theoretically, to achieve reproducible procedures, accurate reagent dispenses, temperature, and reaction observation, SDLs deploy robotic systems that are familiar with these values. Such systems can utilize combinatorial procedures, exhaustively studying an entire range of monomer types, cross-linking densities, and reaction conditions [86]. SDLs can autonomously synthesize a variety of biodegradable polymers for library construction that could improve their performance for medicine-related applications, such as tissue engineering and drug delivery. AI analysis assesses the properties of the obtained polymers, such as degradability and mechanical characteristics, and determines the best choices [87].
2. **Advanced characterization tools:** Characterization instruments that are called automated spectroscopy systems, incorporated into SDLs, provide detailed structural and characteristic features of polymers. Spectroscopy, rheometry and thermal analysis are employed in MWD, viscoelasticity, and thermal stability measurement frequently. Consequently, the morphology of the samples produces high-dimensional data that is analyzed by AI algorithms to identify the patterns and correlations, to be used in subsequent synthesis stages [88, 89].

4.4.3 AI-driven optimization of nanomaterial fabrication and testing

AI and automation in self-sustaining laboratories, which are in vogue for polymer and nanomaterials research, have revolutionized the experimental setup. By combining robotics, autonomous synthesis, enhanced characterization and AI optimization, SDLs enable high-throughput design for minute and complex parameter spaces to discover novel materials with revolutionary applications [90]. Indeed, due to the unique features of specified size and variability of uses, nanomaterials are among the segments that can significantly benefit from the AI-related optimization features of SDLs:

1. **Autonomous nanomaterial fabrication:** Manufacturing of nanomaterials is fully controlled in SDLs, with the integration of robotics in the production process, and with techniques like vapor deposition, Sol-gel synthesis, or electrospinning. AI controls the process of manufacturing and provides guidelines concerning the conditions for getting the needed particle size, shape, or surface properties. In the synthesis of quantum dots for optoelectronic applications, SDLs can control precursor concentrations, reaction temperature, and annealing time separately. AI measures the optical and electrical properties of the quantum dots in the final product, and subsequently the synthesis parameters, to achieve the best performance.
2. **High-throughput testing and analysis:** Advanced characterization services in SDLs afford high-throughput testing of nanomaterials, porosity, crystallinity, and optical absorbance. Such datasets are processed by the respective AI systems, which identify trends and fluctuations that will be used in subsequent testing [91, 92]. In nanocatalyst investigation, SDLs can characterize the activity and stability of the catalytic system under many scenarios. AI models link these performance measurements to synthesis factors so that potential high-performance catalysts for specific applications, such as CO_2 reduction or hydrogen generation, can be pinpointed [93].
3. **Optimization via multiscale modeling:** Machine learned SDLs integrate the existing research findings with computational simulations for the improvement of various nano-sized material properties. While density functional theory (DFT) calculations reveal atomic behavior and interactions respectively, macroscopic results from molecular dynamics simulations, arise from ML. In polymer–nanoparticle composites, SDLs can use multiple simulation domains to enhance the polymer matrix interactions with encapsulated nanoparticles to increase the material strength or thermal conduction coefficient [94].

4.5 Applications of AI-enhanced self-driving labs in polymer and nanomaterial sciences

The amalgamation of AI with SDLs has heralded a breakthrough epoch in polymer and nanomaterial sciences, providing novel methodologies to expedite discovery and optimize development. These AI-augmented platforms have broadened the horizons of scientific inquiry, facilitating swift progress in intelligent polymers, eco-friendly materials, and high-throughput screening. AI and nanoinformatics have burst out in the recent decade in many fields of nanomaterials application, and are now moving to inverse design and modeling of multi-structure/multi-property relationships. AI has already been used to tackle challenges in the design and discovery of novel nanomaterials to optimize structure and performance by assisting process design and unbiased decision-making to ensure safer and eco-friendlier nanomaterials [95]. We examine these applications in detail, emphasizing the collaborative function of AI and SDLs in influencing the future of materials research.

4.5.1 Accelerating the discovery of smart polymers for advanced applications

Self-responsive polymers that can respond to signals from their environment such as temperature, pH, light, or electric field are critical for new biomedical applications, flexible electronics, and smart coatings [96]. However, their identification and strengthening are quite challenging, given the complex intersection between molecular structures and features of the proteins. AI-augmented SDLs tackle these difficulties by:

1. **Combinatorial design and synthesis**: SDLs, in combination with robotic synthesis systems, can generate large databases of polymers with profound variations in their chemical composition and cross-linking density [97]. AI methods, especially generative models, enable creating novel polymer structures, specific for the intended applications. To develop shape-memory polymers for minimally invasive medical applications such as surgical tools, shape-dependent learning SDL with exact temperature sensitivity has been used [98, 99].
2. **Predictive modeling of properties**: AI helps in the accurate prediction of polymer behaviors in different settings through inputs supported by the analysis of experimental data and the application of ML algorithms. The proposed reinforcement learning methods can be used to increase the level of self-healing in polymers by predicting the impact of cross-linking strategies on healing performance [100].
3. **Real-time feedback mechanisms**: Often, smart polymers require strict control over their reactivity level. SDLs provide feedback loops in real time wherever AI analyzes experiment data; be it thermal, mechanical, or optic; and continuously improves polymer composition. The schema has been critical in the enhancement

of hydrogels for drug delivery systems that discharge therapeutic substances in response to modifications in the physiological conditions [36].

4.5.2 Self-driving labs in the development of sustainable and green materials

The increasing concern with sustainability around the world has increased the need for environmentally friendly materials such as biodegradable polymers and nanomaterials from renewable resources. AI-augmented SDLs are essential in fulfilling this requirement by:

1. **Designing biodegradable polymers**: SDLs can help build pro-ecological polymers as they allow the synthesis and analysis of the obtained formulations based on the renewable monomers such as polylactic acid or polyhydroxyalkanoates independently. AI manipulates factors such as chain length, crystallinity, and copolymer ratios to achieve a certain degree of degradation and mechanical properties for use in packaging or as implant materials [101].
2. **Minimizing environmental impact**: AI-aided SDLs help to enable determination of the synthesis routes for the production of polymers and nanomaterials that are most energy efficient and contain minimal waste. New bio-derived nanocellulose composites for light-weight load-bearing structures have been designed through AI-assisted techniques to lessen the amount of chemicals involved in synthesizing the composite materials without compromising on their performance [102].
3. **Investigation of alternative feedstocks**: The usage of global bio-sourced raw materials requires thorough evaluation of unconventional reactants. SDLs are used to self-appraise the monomers derived out of plants or garbage stream, and then employ AI for anticipating the compatibility with existing polymerization techniques and final qualities. This method has contributed significantly toward the development of sustainable textiles, especially for building and other sectors [103].
4. **Circular material design**: The thematic focus of the AI-augmented SDLs is to develop messages that encourage the circularity of products. In line with this, by using the degradation paths as well as recyclability indicators, AI identifies polymer architectures that enable high performance through several recycle cycles, thus supporting the philosophy of circular economy [85].

4.5.3 AI for high-throughput screening

The indication of new advanced polymers and nanomaterials often requires the evaluation of hundreds of candidate materials, a process that is unrealistic with most experimental methods. Sophisticated SDLs developed from high-throughput screening of capable automation, robotics, and effective predictive modeling tackle this problem. SDLs employ an automated synthesis, testing, and data analysis protocol to enable evaluation of various material candidates at the same time. The use of high-throughput methods in experiments generates large amounts of data that are difficult to process without the use of a software. AI effectively identifies important patterns from this information regarding the relation between the composition, methods of processing, and the outcome metrics [104]. By combining experimental data and AI optimization, SDLs can quickly find better materials for a given application. Also, SDLs possess the ability to adapt experimental strategies in real time. Due to the continual learning models, AI always updates the predictions based on the trial outcomes, only focusing on tests that improve the likelihood of understanding materials with the desired characteristics. This method has shown significant success in the development of polymer blends and hybrid nanomaterials in specific applications, thereby displaying a highly multivariable parameter space.

4.6 Challenges in implementing AI in self-driving labs for polymer science

Many major steps are required as AI is inserted into SDLs for polymer research, allowing research to advance with new methods to discover and optimize materials. However, integration of AI in SDLs faces huge challenges in the technological, operational, and practical domains [68]. The stumbling blocks need to be overcome in order to fully realize the promise of autonomous research in polymer science. These are:

1. **Data integrity and accessibility**: AI models necessitate comprehensive datasets for training and validation; but in polymer science, high-quality, standardized data is often limited:
   - *Restricted experimental data*: There are many polymer systems explored in specialized applications, with many times an exhaustive data set not being available for them. Compounding these difficulties is experimental inconsistency such as differences in synthesis techniques or characterization methodology, since they exacerbate data aggregation.
   - *High-dimensional data*: The properties of polymers tend to be determined by interactions among molecular structure, processing conditions, and environmental fluctuations for high-dimensional data. But the data is prepared for AI and we need to remove the noise and make important features.
   - *Obstacles to data sharing*: Aggregating datasets from across research groups and countries is important, but intellectual property issues and lack of standardized data formats may prevent such aggregation [37].

2. **Intricacy of polymer systems**: AI-based modeling and forecasting suffers from difficulties in modeling and forecasting the complex of polymers:
   - *Nonlinear structure–property relationships*: The correlations of molecular architecture with macroscopic polymer properties exhibit significant nonlinearity and often defy the intuition. The sophistication of these interactions in reality must now be embedded within AI models such that they balance between accurately capturing them and overfitting.
   - *Multiscale modeling*: Events of all sizes, from atomic interactions to macroscopic behavior, result in polymer characteristics. There's a big technical barrier with having to consolidate data from different sizes into one cohesive AI system.
   - *Dynamic behavior*: Time-dependent characteristics found in numerous polymers include viscoelasticity or deterioration. For these dynamic qualities to be acquired as well, you will need time-series data and a sophisticated AI method such as recurrent neural networks, which usually requires specialist knowledge [105].
3. **Experimental automation integration**: Although AI is fundamental to SDLs, its efficacy relies on flawless integration with robotics and experimental hardware.
   - *Reliability of robotic systems*: With high precision and dependability, robotic systems in SDLs are needed to execute tasks like mixing, heating, and analyzing materials. Errors could be generated from hardware faults or anomalies that disrupt the AI-driven feedback loop.
   - *Adaptability to novel techniques*: Innovative methods for polymer synthesis and characterization are continually being brought to the fore. However, to achieve these improvements, considerable effort and resources are needed to modify robotic platforms.
   - *Calibration and validation*: A lot of calibration is also needed on automated systems to make their results consistent in experiments. When working with delicate polymer systems, this is particularly important, because even very small deviations can lead to very large changes in properties [33].
4. **Constraints of algorithms**: The efficacy of AI in SDLs is fundamentally contingent upon the underlying algorithms, which encounter their own restrictions:
   - *Absence of explainability*: A lot of these AI models (especially deep learning algorithms) are said to be 'black boxes' without an easy way to interpret how certain input data will produce a particular prediction. The lack of openness can prevent scientific understanding and undermines confidence in the accuracy of AI-generated findings.
   - *Generalizability challenges*: The data collected for specific polymer systems and used to develop models may be generalized insufficiently to novel or atypical materials, posing the requirement for new training and additional data collection.

   - *Optimization challenges*: In polymer research, optimization techniques can find global minima, but in high-dimensional spaces, which are common in polymer research, optimizations often settle on local minima, missing the global minima. This problem can be alleviated through use of advanced techniques such as Bayesian optimization or evolutionary algorithms, at the expense of increased computational complexity [9].
5. **Financial and resource necessities**: AI-driven research presents ethical and regulatory difficulties, especially in polymer science, where results frequently impact society and the environment directly:
   - *Infrastructure expenditures*: To train and operate AI models efficiently, you require advanced computing infrastructure. Furthermore, robotic platforms and automated characterization instruments are major capital expenditures.
   - *Expertise required*: Specified knowledge of polymer science, data science, and robotics are required for the development and implementation of AI in SDLs. This can be a problem in recruiting and training interdisciplinary teams.
   - *Operational expenses*: At the operating cost level, automated systems, sustained extensive databases, and completed computationally extensive AI algorithms require certain upkeep [12].
6. **Ethical and regulatory considerations**: Specifically in polymer science, where results extremely impact the society and environment, the amount of ethical and regulatory problems linked to AI-driven research is huge. As AI models tend to learn from poor-quality data, they may make poor or detrimental recommendations. AI-powered Software Development Lifecycles (SDLs) pose risks of reduced human oversight, leading to potential mistakes. Roles in autonomous research must be defined to explicit standards. Polymers used in the food packaging or medical devices can only meet strict rules, but then it becomes even more complex to ensure that AI-generated discoveries fit these guidelines [106].

While AI holds great promise for breakthroughs in the polymer research fields of SDLs, several hurdles must be overcome to streamline its use in such a complex and important field. Data availability, the intractability of polymer systems, hardware integration, algorithmic restrictions, financial limits, and ethical considerations constitute the range of obstacles in these studies. To overcome these challenges, scientists, engineers, and policymakers have to make cooperative efforts to generate the comprehensive framework and optimal procedure. If these obstacles are surmounted, AI-driven SDLs could drive unprecedented progress in polymer research and lead to new materials that answer perennial societal needs.

4.7 The future of AI in polymer science and nanotechnology

AI integration with polymer science and nanotechnology will pave the way for material discovery, optimization, and application in new directions. AI has extensive potential in polymer science and nanotechnology for innovation, sustainability, and extraordinary precision in material design. AI is a very promising future for polymer science and nanotechnology, facilitating quick discover, sustained innovation, and progress with precision. AI directly addresses existing problems, opening doors for interdisciplinary collaboration, enabling researchers to fully harness the potential of these materials in order to expedite science, make it more intelligent, and making it far more impactful. Not only will this change future research methodologies, but it will provide practical benefits to industries, ranging from healthcare and energy to environment. Polymers and nanomaterials will transform the role of AI in medical innovation by dealing with issues of material design and personalized treatment to the patient, as well as in healthcare sustainability. The emerging multidisciplinary collaboration will serve as a thrust in healthcare and in the convergence of AI. Polymer science and nanotechnology are poised to transform healthcare [107]. In future, AI-based polymer science and nanotechnology applications in sustainability will play a central role, addressing urgent environmental problems with green chemistry and circular economy principles. Based on the identification of biodegradable polymers and recyclable nanocomposites with the identification of molecular structures and processing parameters, these materials can meet performance standards, while also being environmentally friendly. AI will accelerate this process. The materials created can have reduced ecological footprint and ML algorithms can be used to forecast degradation paths or evaluate recyclability criteria. A cutting-edge development in the science of polymers and nanotechnology are smart materials, polymers that change their property in response to external stimuli. Their advancement will be dominated by AI.

AI models will help in constructing polymers that respond to stimuli such as temperature, pH, or light in a predictable manner, and have applications in drug delivery, sensors, and soft robotics. AI directed SDLs will add to the material synthesis for these materials by screening combinations of monomers, cross-linkers, and processing parameters. AI will drive forward the pace of functional material development in nanotechnology, such as nanoparticle-based drug delivery and quantum dots for optoelectronics. The surface changes and assembly methods of the predictive models will be enhanced to meet the needs of different applications. As far as biomaterial discovery transformation is concerned, innovative tools for detecting artificial polymers and nanomaterials, developed for specific biological and pharmacological purposes, are offered by AI. It can be useful in developing biocompatible polymers, which can be forecasted to be biocompatible by AI algorithms, which examine their molecular structure and their binding to biological systems. This capability facilitates the development of materials for implants, wound dressing, and medication delivery systems with minimal amount of adverse consequences in vivo. These AI models can be used for functionalized nanomaterials for targeted therapy, which enable the design of nanoparticles with functionalized surfaces, improving their specificity for targeting specific tissues or biological machines (receptors) because nanoparticles containing chemotherapeutics can be precisely targeted to tumor locations that reduce systemic toxicity. Pharmaceutical delivery mechanisms with enhanced AI accuracy, which ultimately facilitate the creation of polymers that exhibit predictable responses to stimuli such as temperature, pH, or light, hence enabling applications in drug administration, sensors, and soft robotics, can be built. AI-directed SDLs will enhance the synthesis of these materials by investigating combinations of monomers, cross-linking agents, and processing parameters.

In nanotechnology, AI will propel advancements in functional materials, including nanoparticles for tailored drug delivery and quantum dots for optoelectronics. Predictive models will enhance their surface changes and assembly methods to ensure compliance with various applications. AI provides innovative tools for identifying polymers and nanomaterials designed for particular biological and pharmacological uses. AI algorithms can forecast the biocompatibility of polymers by examining their molecular structure and interactions with biological systems. This capability expedites the advancement of materials for implants, wound dressings, and medication delivery systems, guaranteeing minimal unwanted consequences in vivo. Generative AI models facilitate the construction of nanoparticles with functionalized surfaces, enhancing their capacity to target specific tissues or biological receptors. Nanoparticles containing chemotherapeutics can be precisely targeted to tumor locations, hence reducing systemic toxicity. And AI is helping to push the advancement of polymer-based and nanotechnology-enhanced drug delivery systems for accurate and controlled release of medications. Ways to get smarter polymers that follow physiological cues like pH, temperature, or enzyme activity are crucial for drug delivery systems, hence enabling applications in drug administration, sensors, and soft robotics. AI-directed SDLs will enhance the synthesis of these materials by investigating combinations of monomers, cross-linking agents, and processing parameters.

4.8 Conclusion

The domain of materials science is poised for a transformational revolution due to the amalgamation of AI and SDLs. These developments are revolutionizing how new materials, their properties, and their applications are to be understood, created, and processed. Self-governing laboratories are being powered by AI to enhance complex processes, increase precision, and enhance speed of invention across various sciences such as polymer science and nanotechnology. This chapter has looked into the central position of AI in advancing the autonomous revolution, stressing on material discovery, material optimization, and material execution. This transition is mainly based on the feature of AI for automating experiments, analyze large amount of data, and optimize the results in a dynamic manner. Traditional processes of material generation are often time-consuming, exhausting, and resource-demanding due to random testing. SDLs, however, bring the relatively nontrivial task of ranking to an entirely new level of efficiency. AI supervises robotic systems, necessary in laboratories, so the systems are capable of performing several tasks, for instance, synthesizing polymers and characterizing nanomaterials without human assistance. This helps researchers to focus more on doing high-level analysis and making decisions than spend countless hours doing more of routine or lowly tasks. AI is an impressive domain in terms of the ability to predict and optimize material properties. SDLs can use ML methods to define the mappings between molecular structures and material properties such as mechanical or thermal properties, and biocompatibility. This feature is particularly useful for designing smart materials, such as polymers with memory and functionalized nanomaterials, which change their properties in response to external factors or specific biological environment. By automating such procedures, SDLs can explore an immensely broader number of possibilities, ultimately cutting time and effort by a significant margin in discovering and optimizing new material for applications like drug delivery, tissue engineering, and enhanced diagnostics.

Furthermore, AI-driven optimization guarantees that the created materials are both high-performing and sustainable. In the contemporary world, where environmental issues are critical, the advancement of sustainable materials is imperative. Autonomous laboratories utilizing AI can facilitate the design of biodegradable polymers, hence minimizing waste and decreasing energy usage in material synthesis. Through the integration of real-time data gathering and feedback mechanisms, SDLs enhance their proficiency in optimizing material manufacturing processes, harmonizing scientific innovation with environmental conservation. The independent character of these laboratories also promotes interdisciplinary collaboration. Researchers across several disciplines, including chemistry and bioengineering, may collaborate effectively, utilizing AI-generated insights to develop materials that advance the frontiers of innovation. Biomedical applications currently utilize materials meticulously customized to meet individual patient requirements, including personalized medication delivery systems and tissue scaffolds intended for regenerative treatment. AI facilitates the swift and effective progression of these advancements, surmounting entrance hurdles and expediting the conversion of scientific innovations into practical applications. Nonetheless, despite the significant promise of AI in autonomous laboratories, there are problems that must be confronted. Concerns regarding data quality, ethical implications, and the necessity for interdisciplinary collaboration are pivotal to the successful use of these technologies. Ensuring AI models are transparent, impartial, and inclusive is essential for cultivating trust and enhancing their effectiveness. Moreover, as these laboratories expand, guaranteeing accessibility and affordability is essential for fostering wider scientific progress across various institutions and sectors. Through the integration of automation, data analysis, and predictive intelligence, these laboratories are expediting innovation, enhancing sustainability, and facilitating more personalized and effective applications. As researchers persist in using AI, we can foresee a future in which materials are more intelligent, sustainable, and customized to address the distinct requirements of a swiftly changing scientific environment.

List of abbreviations

AI	Artificial intelligence
ML	Machine learning
DL	Deep learning
SDLs	Self-driving laboratories
HTS	High-throughput synthesis
DFT	Density functional theory

References

[1] Silveira RF, Lima AL, Gross IP, Gelfuso GM, Gratieri T and Cunha-Filho M. The role of artificial intelligence and data science in nanoparticles development: A review. Nanomedicine, 2024;19:1271–83.

[2] Rimban E Intersecting horizons of artificial intelligence and nanotechnology. SSRN Journal [Internet]. 2024 [cited 2024 Dec 24]; Available from: https://www.ssrn.com/abstract=4757599

[3] Izanker SV, Dhole A and Kumar P Navigating the nexus: Exploring the fusion of AI and nanotechnology for cutting-edge advances. 2023 1st DMIHER International Conference on Artificial Intelligence in Education and Industry 40 (IDICAIEI) [Internet]. Wardha, India: IEEE; 2023 [cited 2024 Dec 24]. p. 1–5. Available from: https://ieeexplore.ieee.org/document/10406387/

[4] Gianti E and Percec S. Machine Learning at the interface of polymer science and biology: How far can we go?. Biomacromolecules, 2022;23:576–91.

[5] Tripathy A, Patne AY, Mohapatra S and Mohapatra SS. Convergence of nanotechnology and machine learning: The state of the art, challenges, and perspectives. IJMS, 2024;25:12368.

[6] González-Nilo F, Pérez-Acle T, Guínez-Molinos S, Geraldo DA, Sandoval C, Yévenes A, et al. Nanoinformatics: An emerging area of information technology at the intersection of bioinformatics, computational chemistry and nanobiotechnology. Biol Res, 2011;44:43–51.

[7] Podutwar AA, Chandorkar PU, Chabukswar AR, Polshettiwar SA and Jagdale SC. The Intersection of nanotechnology and biotechnology implications for human health. In: George SC and Tawiah B (Eds.) Nanotechnology in Societal Development [Internet]. 2024; (pp. 271–305). Singapore: Springer Nature Singapore. [cited 2024 Dec 24] Available from https://link.springer.com/10.1007/978-981-97-6184-5_8.

[8] Olawade DB, Ige AO, Olaremu AG, Ijiwade JO and Adeola AO. The synergy of artificial intelligence and nanotechnology towards advancing innovation and sustainability – A mini-review. Nano Trends, 2024;8:100052.

[9] Seifrid M, Pollice R, Aguilar-Granda A, Morgan Chan Z, Hotta K, Ser CT, et al. Autonomous chemical experiments: challenges and perspectives on establishing a self-driving lab. Accounts of Chemical Research, 2022;55:2454–66.

[10] Martin HG, Radivojevic T, Zucker J, Bouchard K, Sustarich J, Peisert S, et al. Perspectives for self-driving labs in synthetic biology. Current Opinion in Biotechnology, 2023;79:102881.

[11] Häse F, Roch LM and Aspuru-Guzik A. Next-generation experimentation with self-driving laboratories. Trends in Chemistry, 2019;1:282–91.

[12] Abolhasani M and Kumacheva E. The rise of self-driving labs in chemical and materials sciences. Nature Synthesis, 2023;2:483–92.

[13] Tom G, Schmid SP, Baird SG, Cao Y, Darvish K, Hao H, et al. Self-driving laboratories for chemistry and materials science. Chemical Reviews, 2024;124:9633–732.

[14] MacLeod BP, Parlane FGL and Berlinguette CP. How to build an effective self-driving laboratory. MRS Bulletin, 2023;48:173–8.

[15] Bennett JA and Abolhasani M. Autonomous chemical science and engineering enabled by self-driving laboratories. Current Opinion in Chemical Engineering, 2022;36:100831.

[16] Lo S, Baird SG, Schrier J, Blaiszik B, Carson N, Foster I, et al. Review of low-cost self-driving laboratories in chemistry and materials science: The "frugal twin" concept. Digital Discovery, 2024;3:842–68.

[17] Sanders LM, Scott RT, Yang JH, Qutub AA, Garcia Martin H, Berrios DC, et al. Biological research and self-driving labs in deep space supported by artificial intelligence. Nature Machine Intelligence, 2023;5:208–19.

[18] Ekins S. The lab of the future: Self-Driving labs for molecule discovery. GEN Biotechnology, 2024;3:83–6.

[19] Tran H, Gurnani R, Kim C, Pilania G, Kwon H-K, Lively RP, et al. Design of functional and sustainable polymers assisted by artificial intelligence. Nature Reviews Materials, 2024;9:866–86.

[20] Li B, Hou B, Yu W, Lu X and Yang C. Applications of artificial intelligence in intelligent manufacturing: A review. Frontiers of Information Technology & Electronic Engineering, 2017;18:86–96.

[21] Xu Y, Liu X, Cao X, Huang C, Liu E, Qian S, et al. Artificial intelligence: A powerful paradigm for scientific research. The Innovation, 2021;2:100179.

[22] Ertel W. Introduction to Artificial Intelligence. 2025; (3rd ed.). Wiesbaden: Springer Fachmedien Wiesbaden. 2025.

[23] Zhang C and Lu Y. Study on artificial intelligence: The state of the art and future prospects. Journal of Industrial Information Integration, 2021;23:100224.

[24] Lu H, Li Y, Chen M, Kim H and Serikawa S. Brain intelligence: Go beyond artificial intelligence. Mobile Networks and Applications, 2018;23:368–75.

[25] Gupta R, Srivastava D, Sahu M, Tiwari S, Ambasta RK and Kumar P. Artificial intelligence to deep learning: Machine intelligence approach for drug discovery. Molecular Diversity, 2021;25:1315–60.

[26] Khalid MY, Arif ZU, Ahmed W, Umer R, Zolfagharian A and Bodaghi M. 4D printing: Technological developments in robotics applications. Sensors and Actuators A: Physical, 2022;343:113670.

[27] Shimizu R, Kobayashi S, Watanabe Y, Ando Y and Hitosugi T. Autonomous materials synthesis by machine learning and robotics. APL Materials, 2020;8:111110.

[28] Rothemund P, Kim Y, Heisser RH, Zhao X, Shepherd RF and Keplinger C. Shaping the future of robotics through materials innovation. Nature Materials, 2021;20:1582–7.

[29] Pollard M. Process development automation: An evolutionary approach. Organic Process Research & Development, 2001;5:273–82.

[30] Stein HS and Gregoire JM. Progress and prospects for accelerating materials science with automated and autonomous workflows. Chemical Science, 2019;10:9640–9.

[31] King RD, Rowland J, Oliver SG, Young M, Aubrey W, Byrne E, et al. The automation of science. Science, 2009;324:85–9.
[32] Selekman JA, Qiu J, Tran K, Stevens J, Rosso V, Simmons E, et al. High-throughput automation in chemical process development. Annual Review of Chemical and Biomolecular Engineering, 2017;8:525–47.
[33] Soldatov MA, Butova VV, Pashkov D, Butakova MA, Medvedev PV, Chernov AV, et al. Self-driving laboratories for development of new functional materials and optimizing known reactions. Nanomaterials, 2021;11:619.
[34] Delgado-Licona F and Abolhasani M. Research acceleration in self-driving labs: Technological roadmap toward accelerated materials and molecular discovery. Advanced Intelligent Systems, 2023;5:2200331.
[35] Volk AA and Abolhasani M. Performance metrics to unleash the power of self-driving labs in chemistry and materials science. Nature Communications, 2024;15:1378.
[36] Hysmith H, Foadian E, Padhy SP, Kalinin SV, Moore RG, Ovchinnikova OS, et al. The future of self-driving laboratories: From human in the loop interactive AI to gamification. Digital Discovery, 2024;3:621–36.
[37] Bayley O, Savino E, Slattery A and Noël T. Autonomous chemistry: Navigating self-driving labs in chemical and material sciences. Matter, 2024;7:2382–98.
[38] Agrawal R, Tilak P, Devand A, Bhatnagar A and Gupta P. Artificial intelligence in nanotechnology. In Mishra D, Yang XS, Unal A and Jat DS (Eds.) Data Science and Big Data Analytics [Internet]. 2024; (pp. 239–48). Singapore: Springer Nature Singapore. [cited 2024 Dec 24] Available from https://link.springer.com/10.1007/978-981-99-9179-2_18.
[39] Chundi P, Gadhamshetty V, Jasthi BK and Lushbough C, (Eds.). Machine Learning in 2D Materials Science. 2023; Boca Raton, FL: CRC Press.
[40] Chugh V, Basu A, Kaushik A, Manshu A, Bhansali S and Basu AK. Employing nano-enabled artificial intelligence (AI)-based smart technologies for prediction, screening, and detection of cancer. Nanoscale, 2024;16:5458–86.
[41] Alshawwa SZ, Kassem AA, Farid RM, Mostafa SK and Labib GS. Nanocarrier drug delivery systems: Characterization, limitations, future perspectives and implementation of artificial intelligence. Pharmaceutics, 2022;14:883.
[42] Zhu X, Li Y and Gu N. Application of artificial intelligence in the exploration and optimization of biomedical nanomaterials. Nano Biomedicine and Engineering, 2023;15:342–53.
[43] Wang Z and Zhu X. Enhancing nanocrystal synthesis: A comparative study of online artificial intelligence optimization and offline high-throughput experimentation in chemical material discovery. ACS Applied Nano Materials, 2024;7:6499–505.
[44] Sagdic K, Eş I, Sitti M and Inci F. Smart materials: Rational design in biosystems via artificial intelligence. Trends in Biotechnology, 2022;40:987–1003.
[45] Lenoir D, Schramm K-W and Lalah JO. Green Chemistry: Some important forerunners and current issues. Sustainable Chemistry and Pharmacy, 2020;18:100313.
[46] Tayubi IA, Jakhete MD, Shetty SB, Verma A, Tiwari M and Kiruba S. Sustainable healthcare AI-enhanced materials discovery and design for eco-friendly and biocompatible medical applications. The Scientific Temper, 2023;14:1066–72.
[47] Rai M, Khan S and Belgamwar A, (Eds.). Nanotechnology Principles in Drug Targeting and Diagnosis. 2023; San Diego: Elsevier.
[48] Zhao Y, Mulder RJ, Eyckens DJ, Houshyar S and Le TC. Advancing antimicrobial polymer development: A novel database and accelerated design *via* machine learning. Polymer Chemistry, 2024;15:4063–76.
[49] Wang R, Fu T, Yang Y-J, Song X, Wang X-L and Wang Y-Z. Scientific discovery framework accelerating advanced polymeric materials design. Research, 2024;7:0406.

[50] Gottardo S, Mech A, Drbohlavová J, Małyska A, Bøwadt S, Riego Sintes J, et al. Towards safe and sustainable innovation in nanotechnology: State-of-play for smart nanomaterials. NanoImpact, 2021;21:100297.
[51] Kraegeloh A, Suarez-Merino B, Sluijters T and Micheletti C. Implementation of safe-by-design for nanomaterial development and safe innovation: Why we need a comprehensive approach. Nanomaterials, 2018;8:239.
[52] Yao W and Ruan Y. Innovative artistic design of nanomaterial products based on intelligent algorithm. In: Ahmed A (Ed.) Journal of Nanomaterials. 2022; (vol. 2022, pp. 3019696).
[53] Hardian R, Liang Z, Zhang X and Szekely G. Artificial intelligence: The silver bullet for sustainable materials development. Green Chem, 2020;22:7521–8.
[54] Li Y and Yue T. Machine Learning for Polymer Informatics. 2024; Washington, DC: ACS Publications.
[55] Jha A, Chandrasekaran A, Kim C and Ramprasad R. Impact of dataset uncertainties on machine learning model predictions: The example of polymer glass transition temperatures. Modelling Simul Mater Sci Eng, 2019;27:024002.
[56] Malashin IP, Tynchenko VS, Nelyub VA, Borodulin AS and Gantimurov AP. Estimation and prediction of the polymers' physical characteristics using the machine learning models. Polymers, 2023;16:115.
[57] Champa-Bujaico E, García-Díaz P and Díez-Pascual AM. Machine learning for property prediction and optimization of polymeric nanocomposites: A state-of-the-art. IJMS, 2022;23:10712.
[58] Upadhya R, Kosuri S, Tamasi M, Meyer TA, Atta S, Webb MA, et al. Automation and data-driven design of polymer therapeutics. Advanced Drug Delivery Reviews, 2021;171:1–28.
[59] Patel RA and Webb MA. Data-driven design of polymer-based biomaterials: High-throughput simulation, experimentation, and machine learning. ACS Applied Bio Materials, 2024;7:510–27.
[60] Patra TK. Data-driven methods for accelerating polymer design. ACS Polymer Au, 2022;2:8–26.
[61] Zhou T, Zhang J, Yuan Z and Xu A. External stimuli mediate collective rhythms: Artificial control strategies. Khanin R, (Ed.). PLoS ONE, 2007;2:e231.
[62] Llargues Asensio JM, Peralta J, Arrabales R, Bedia MG, Cortez P and Peña AL. Artificial Intelligence approaches for the generation and assessment of believable human-like behaviour in virtual characters. Expert Systems with Applications, 2014;41:7281–90.
[63] Eyring V, Gentine P, Camps-Valls G, Lawrence DM and Reichstein M. AI-empowered next-generation multiscale climate modelling for mitigation and adaptation. Nature Geoscience, 2024;17:963–71.
[64] Han C, Zhang P, Bluestein D, Cong G and Deng Y. Artificial intelligence for accelerating time integrations in multiscale modeling. Journal of Computational Physics, 2021;427:110053.
[65] Wang J, Wang Y and Chen Y. Inverse design of materials by machine learning. Materials, 2022;15:1811.
[66] Matos MAS, Pinho ST and Tagarielli VL. Application of machine learning to predict the multiaxial strain-sensing response of CNT-polymer composites. Carbon, 2019;146:265–75.
[67] Cassola S, Duhovic M, Schmidt T and May D. Machine learning for polymer composites process simulation – A review. Composites Part B: Engineering, 2022;246:110208.
[68] Zhu X. AI and Robotic Technology in Materials and Chemistry Research. 2025; (1st ed.). Newark: John Wiley & Sons, Incorporated.
[69] Bhupathyraaj M, Vijaya Rani KR and Essa MM, (Eds.). Artificial Intelligence in Pharmaceutical Sciences. 2023; Boca Raton: CRC PRESS.
[70] Grinin L, Grinin A and Korotayev A. Nanotechnologies, Robotics, Artificial Intelligence and Other MANBRIC Technologies in the Long-Term Development. In Cybernetic Revolution and Global Aging. 2024; (pp. 403–57). Cham: Springer International Publishing. [Internet] cited 2024 Dec 24 Available from https://link.springer.com/10.1007/978-3-031-56764-3_10.
[71] Vo AH, Van Vleet TR, Gupta RR, Liguori MJ and Rao MS. An overview of machine learning and big data for drug toxicity evaluation. Chemical Research in Toxicology, 2020;33:20–37.

[72] Badwan BA, Liaropoulos G, Kyrodimos E, Skaltsas D, Tsirigos A and Gorgoulis VG. Machine learning approaches to predict drug efficacy and toxicity in oncology. Cell Reports Methods, 2023;3:100413.
[73] Zeng X, Wang F, Luo Y, Kang S, Tang J, Lightstone FC, et al. Deep generative molecular design reshapes drug discovery. Cell Reports Medicine, 2022;3:100794.
[74] Lin E, Lin C-H and Lane H-Y. Relevant applications of generative adversarial networks in drug design and discovery: Molecular de novo design, dimensionality reduction, and de novo peptide and protein design. Molecules, 2020;25:3250.
[75] Wang J, Zhang Y, Marian C and Ressom HW. Identification of aberrant pathways and network activities from high-throughput data. Briefings in Bioinformatics, 2012;13:406–19.
[76] Liu Z-P. Identifying network-based biomarkers of complex diseases from high-throughput data. Biomarkers in Medicine, 2016;10:633–50.
[77] Sathiyamoorthi V, Ilavarasi AK, Murugeswari K, Thouheed Ahmed S, Aruna Devi B and Kalipindi M. A deep convolutional neural network based computer aided diagnosis system for the prediction of Alzheimer's disease in MRI images. Measurement, 2021;171:108838.
[78] A A, P M, Hamdi M, Bourouis S, Rastislav K and Mohmed F. Evaluation of neuro images for the diagnosis of Alzheimer's disease using deep learning neural network. Frontiers in Public Health, 2022;10:834032.
[79] Sano S, Kadowaki T, Tsuda K and Kimura S. Application of Bayesian optimization for pharmaceutical product development. J Pharm Innov, 2020;15:333–43.
[80] Kolar D, Lisjak D, Pająk M and Gudlin M. Intelligent fault diagnosis of rotary machinery by convolutional neural network with automatic hyper-parameters tuning using bayesian optimization. Sensors, 2021;21:2411.
[81] Yu X, Wang J, Hong -Q-Q, Teku R, Wang S-H and Zhang Y-D. Transfer learning for medical images analyses: A survey. Neurocomputing, 2022;489:230–54.
[82] Guo Y, Dagnino G and Yang G-Z. Medical Robotics: History, Challenges, and Future Directions. 2023; Singapore: Springer Nature Singapore. [Internet] [cited 2024 Dec 24]. Available from https://link.springer.com/10.1007/978-981-99-7317-0.
[83] Ahmadi A and RabieNezhad Ganji N AI-driven medical innovations: Transforming healthcare through data intelligence. IJBLS [Internet]. 2023 [cited 2024 Dec 24];2. Available from: https://doi.org/10.22034/ijbls.2023.185475
[84] Bagheri M, Bagheritaba M, Alizadeh S, Parizi MS, Matoufinia P and Luo Y AI-driven decision-making in healthcare information systems: A comprehensive review [Internet]. Public Health and Healthcare; 2024 [cited 2024 Dec 24]. Available from: https://www.preprints.org/manuscript/202406.0790/v1
[85] Abolhasani M and Brown KA. Guest editors. role of AI in experimental materials science. MRS Bulletin, 2023;48:134–41.
[86] Tamasi MJ, Patel RA, Borca CH, Kosuri S, Mugnier H, Upadhya R, et al. Machine learning on a robotic platform for the design of polymer–protein hybrids. Advanced Materials, 2022;34:2201809.
[87] Knox ST and Warren NJ. Enabling technologies in polymer synthesis: Accessing a new design space for advanced polymer materials. Reaction Chemistry & Engineering, 2020;5:405–23.
[88] Sabet M. From self-assembly to sustainability: Advanced polymerization techniques for energy, healthcare, and robotics. Polymer-Plastics Technology and Materials, 2024;88:1–31.
[89] Khan ME, Aslam J and Verma C editors Nanocomposites – Advanced Materials for Energy and Environmental Aspects. 2023; Cambridge, MA Kidlington: Woodhead Publishing, an imprint of Elsevier.
[90] Nandipati M, Fatoki O and Desai S. Bridging nanomanufacturing and artificial intelligence – A comprehensive review. Materials, 2024;17:1621.
[91] Li Y, Xia L, Fan Y, Wang Q and Hu M. Recent advances in autonomous synthesis of materials. ChemPhysMater, 2022;1:77–85.

[92] Butakova MA, Chernov AV, Kartashov OO and Soldatov AV. Data-centric architecture for self-driving laboratories with autonomous discovery of new nanomaterials. Nanomaterials, 2021;12:12.
[93] Gongora AE, Saygin V, Snapp KL and Brown KA. Autonomous Experimentation in Nanotechnology. Intelligent Nanotechnology [Internet]. 2023; (pp. 331–60). Elsevier. [cited 2024 Dec 24] Available from https://linkinghub.elsevier.com/retrieve/pii/B9780323857963000123.
[94] Kovachki N, Liu B, Sun X, Zhou H, Bhattacharya K, Ortiz M, et al. Multiscale modeling of materials: Computing, data science, uncertainty and goal-oriented optimization. Mechanics of Materials, 2022;165:104156.
[95] Ananikov VP. Top 20 influential AI-based technologies in chemistry. Artificial Intelligence Chemistry, 2024;2:100075.
[96] Nosrati H and Nosrati M. Artificial intelligence in regenerative medicine: Applications and implications. Biomimetics, 2023;8:442.
[97] Takeuchi I, Lauterbach J and Fasolka MJ. Combinatorial materials synthesis. Materials Today, 2005;8:18–26.
[98] Gregoire JM, Zhou L and Haber JA. Combinatorial synthesis for AI-driven materials discovery. Nature Synthesis, 2023;2:493–504.
[99] Grisoni F, Huisman BJH, Button AL, Moret M, Atz K, Merk D, et al. Combining generative artificial intelligence and on-chip synthesis for de novo drug design. Science Advances, 2021;7:eabg3338.
[100] Toma M and Wei OC. Predictive modeling in medicine. Encyclopedia, 2023;3:590–601.
[101] Sadeghi S, Canty RB, Mukhin N, Xu J, Delgado-Licona F and Abolhasani M. Engineering a sustainable future: Harnessing automation, robotics, and artificial intelligence with self-driving laboratories. ACS Sustainable Chemistry & Engineering, 2024;12:12695–707.
[102] Popescu C, Dissanayake H, Mansi E and Stancu A. Eco breakthroughs: Sustainable materials transforming the future of our planet. Sustainability, 2024;16:10790.
[103] Jenks C, Lee HN, Lewis J, Kagan C, Nealey P, Braun P, et al. Basic Research Needs for Transformative Manufacturing [Internet]. 2020 Mar p. None, 1618267. Report No.: None, 1618267. Available from: https://www.osti.gov/servlets/purl/1618267/
[104] Huser J, Folkers G, Kubinyi H and Mannhold R. High-Throughput Screening in Drug Discovery. 2007; Hoboken: Wiley-VCH.
[105] Basak S and Bandyopadhyay A. Toward Making Polymer Chemistry Autonomous. ACS Applied Engineering Materials, 2024;2:1190–208.
[106] Beaucage PA, Sutherland DR and Martin TB. Automation and machine learning for accelerated polymer characterization and development: Past, potential, and a path forward. Macromolecules, 2024;57:8661–8670. acs.macromol 4c01410.
[107] Taft CA and Canchaya JGS. Overview: Fuel cells, hydrogen energy, science, materials, nanotechnology, artificial intelligence and state of the art. In: Taft CA and De Lazaro SR (Eds.) Progress in Hydrogen Energy, Fuel Cells, Nano-Biotechnology and Advanced, Bioactive Compounds. 2024; (pp. 49–81). Cham: Springer Nature Switzerland. [Internet] cited 2024 Dec 25 Available from https://link.springer.com/10.1007/978-3-031-75984-0_2.

Syed Salman Ali* Hrithik Dey, Atifa Khan, Bhoomika Chaudhary, Yarun, and Mohammad Asif

Chapter 5
AI-powered polymer sequence design: applications and innovations

Abstract: The ability to bridge the theoretical model gap and design sequences for use in the field of materials science has been enabled by the progress of artificial intelligence (AI). The revolution in polymer design, away from a mainly trial-and-error-based approach, has been greatly accelerated by AI-driven approaches such as machine learning, deep learning, and genetic algorithms. Using these techniques, polymer properties are predicted with amazing precision; the search for new sequences and the development of performance metrics for various applications are optimized. However, the integration of AI with computational chemistry, molecular dynamics, and high-throughput experimentation has also greatly sped up the design and synthesis of advanced polymers. Biomedical smart polymers, including polymer-based drug delivery systems (DDS) and tissue engineering scaffolds, as well as advanced polymers for electronics and aerospace and biodegradable plastics are key applications. All things are changing how we do research – from autonomous labs to robotic synthesis to AI-powered validation workflows. However, despite the immense potential, we face challenges: model interpretability, ethical considerations in material design, and data quality. In this paper, we outline some basic principles of AI in polymer science, AI-developed design approaches, applications to real life, and trends, including the role of quantum computing and collaborative advances in the field. These challenges are met with AI-powered polymer research that can generate sustainable and scalable material solutions in multiple industries.

***Corresponding author: Syed Salman Ali,** Lloyd Institute of Management and Technology, Plot No. 11, Knowledge Park II, Greater Noida 201306, Uttar Pradesh, India, e-mail: salmanali.ali32@gmail.com, orcid.org/0000-0001-9163-3353

Hrithik Dey, Lloyd Institute of Management and Technology, Plot No.-11, Knowledge Park II, Greater Noida 201306, Uttar Pradesh, India, e-mail: hrithik.dey@gmail.com, orcid.org/0009-0000-6196-1911

Atifa Khan, Department of Pharmacy, Lloyd School of Pharmacy, Knowledge Park II, Greater Noida 201306, Uttar Pradesh, India, e-mail: atifa9826@gmail.com, orcid.org/0009-0004-8044-8629

Bhoomika Chaudhary, Charak School of Pharmacy, Chaudhary Charan Singh University, Meerut, Uttar Pradesh, India, e-mail: pharm.bhoomika@gmail.com, orcid.org/0009-0001-2106-9424

Yarun, Lloyd Institute of Management and Technology, Plot No. 11, Knowledge Park II, Greater Noida 201306, Uttar Pradesh, India, e-mail: yarunchauhan5@gmail.com

Mohammad Asif, Faculty of Pharmacy, Lachoo Memorial College of Science and Technology, Sector A, Shastri Nagar, Jodhpur 342001, Rajasthan, India, e-mail: writemail2asif@gmail.com

https://doi.org/10.1515/9783111681870-005

Keywords: Polymer sequence design, artificial intelligence, machine learning, quantum computing, deep learning algorithms

5.1 Introduction

Today, new computational methods are being integrated into materials science, revolutionizing the field and accelerating the discovery, design, and optimization of materials. Sophisticated data-driven techniques are replacing traditional empirical approaches to rapidly analyze complex datasets and uncover subtle patterns [1]. Such a paradigm shift provides a faster, more accurate, and more cost-effective approach to identifying and refining materials required to advance scientific knowledge and engineering innovation. With machine learning, a part of AI, researchers will improve their calculations and predictions of material properties, finding ways to optimize processes that lead to the development of groundbreaking materials such as sustainable construction materials and biomimetic composites [2]. By leveraging highly accurate predictions of material properties via a strategic combination of quantum mechanical simulations, such as density functional theory (DFT), with machine learning (ML), the discovery of new materials across different disciplines is accelerated. In terms of the profound impact of AI on materials science, polymers epitomize tremendous adaptability and a vast chemical repertoire, offering substantial challenges due to their virtually boundless chemical landscape [3]. When tackling the unpredictable behavior of complex polymer systems, conventional trial-and-error methods often fail, potentially neglecting promising solutions. Surrogate models based on ML applied to large polymer property datasets can soon enable point predictions of polymer characteristics for novel polymers without the time and resource constraints of traditional methodology. The polymers' intricate structure and dynamic response (e.g., vitrification and semicrystallization) are process-dependent and lack a general theoretical framework. Innovative solutions are needed because a deeper understanding of the interplay between molecular chemistry and materials processing is essential to advance technologies in electronics, medicine, and energy devices [4]. Over two decades, breakthroughs in experimental, theoretical, and computational polymer research have produced tremendous amounts of data that require interdisciplinary collaborations. The roles of data science, ML, and materials informatics include revolutionizing polymer design, for example, encoding polymer chemistry into machine-readable models and predictive modeling. Compared to the study of biomolecules, research on synthetic polymers lacks comprehensive resources. However, contemporary molecular simulations, high-throughput virtual screening, and computational modeling are bridging this gap. Having these approaches allows us to design polymers for specific applications with tailored properties, fostering innovation in energy, medicine, and electronics. Integrating these tools enables researchers to speed up polymer discovery and optimization [5]. Computationally derived structure–property data now join polymer and soft material databases, compensating for restricted experimental information. Rapid property prediction and candidate evaluation are powered by high-performance computing and specialized software combined with molecular simulations. Even in the face of the incredibly intricate development of polymer sequences, artificial intelligence (AI) and deep learning algorithms are catapulting design procedures. This accelerates innovation in bio-inspired materials, metamaterials, and high-efficiency semiconductor materials, which have been licensed out to a number of fields, leading to breakthroughs [6]. The discovery of smart materials with sensitive functionalities between stimulus and actuation has opened the way to new materials with sensory, actuation, and mechanical abilities. These piezoelectric and shape-memory alloys are closing the gap between the worlds of artificial and biological, pushing forward robotics, biomedical systems, and responsive technologies. Controlled and targeted medication release has been made possible with polymer-based DDS. These systems exploit biocompatible polymers conjugated with drugs to selectively concentrate in tumors, leading to the highest efficacy at the least cost. Renewable energy technologies, waste management, water conservation, and eco-friendly building methods are being optimized into sustainable engineering practices, leading toward a more sustainable future [7]. Tools such as the environmental sustainability index (ESI) and the environmental performance index (EPI) are used to assess global sustainability goals. The design of polymer sequences is important for developing sustainable materials. By controlling monomer arrangements in polymer chains, researchers can tailor material properties, making way for innovative materials to be applied in many industries. To unfold material behavior and predict performance, molecular-level simulations are needed. Modular reference models enable the simplification of simulation processes and allow for easier study and innovation of sustainable materials by researchers [8]. Microstructure analysis for material characterization is enhanced with digital image processing. High-accuracy techniques such as convolutional filters are used to extract meaningful features from image datasets and assist with the development and testing of material. Green knowledge management permits eco-innovations by addressing the trade-offs between economic growth and environmental conservation. Resource efficiency, reduced environmental impacts, renewable energy, and advanced materials initiatives – sustainable engineering innovations – provide momentum for progress. Energy storage and conversion are polymerized, advancing biomedical applications. These materials emulate extracellular matrix properties, releasing stored therapeutics under the stimulus and improving drug delivery [9].

5.2 Fundamentals of polymer sequence design

5.2.1 Basics of polymer chemistry and sequence structure

Initially, polymers faced skepticism in the early twentieth century, but they gained recognition after Staudinger introduced the "macromolecular concept" in 1920. The idea that polymers are made of chains of monomers, held together by strong chemical bonds, is of monumental significance. By pioneering this concept, Staudinger won the Nobel Prize in Chemistry in 1953. By the 1930s, advancements in organic chemistry led to the development of new polymers, including polyesters and polyamides. To investigate these synthetic materials, researchers used probabilistic methods and X-ray crystallography [10]. Monomers are linked into macromolecules, forming polymers; polyethylene produces repeating units. Weak van der Waals forces hold bulk polymers together, making them less stiff yet stronger than metals. Relatively low-temperature heating of thermoplastic polymers softens them, allowing the chains to move past one another and harden upon cooling [11]. The characteristics of a polymer are assessed based on its molecular mass, which is evaluated by methods like size exclusion chromatography. Polymers exhibit diverse architectural forms: branched, linear, cross-linked, and networked. Extended, unbranching chains with feeble interchain interactions constitute linear polymers that are thermally moldable [12]. The density and thermal properties of polymeric materials are affected by polymer branching, that is, attaching shorter chains to the main chain. Polymers with cross-linked bonds between chains are robust and more stable. Network polymers, on the other hand, have many 3D connections that make them inflexible and unpliable. The traits of copolymer forms are diverse. They include random copolymers, with a disordered arrangement of monomers, and block copolymers, with repeated sequences of a single monomer type, giving them unique physical properties. Furthermore, graft copolymers have a core backbone with branches of varying monomers, which increases flexibility and durability [13]. For specific uses, alternating copolymers are tailored, wherein the monomers alternate. They are used in many industries for the delivery of pharmaceuticals, tissue regeneration, packaging, and automotive manufacturing. The formation of polymers is a fundamental process based on addition and condensation polymerization. An example of this is polyethylene produced by addition polymerization, where consecutive additions of monomers to a filled polymer are followed by the buildup of repeat units identical to those in the original monomer. In contrast, condensation polymerization produces polymers such as polyesters and nylons through reactions between monomers composed of dual functional groups, with small molecules eliminated in the process. In both cases, various customized polymers can be made with varying characteristics and performance for many applications [14].

5.2.2 Importance of sequence-specific properties

Monomer unit arrangement within a polymer chain is crucial to determining its properties, durability, and efficiency. This in turn makes it unsuitable for use in different areas, such as industrial manufacturing, biomedical research, or materials engineering [15]. The arrangement of monomers has a direct impact on the properties of controlled drug release, biodegradability, and sensor functionality. For example, block copolymers made up of both hydrophobic and hydrophilic segments spontaneously assemble into nanostructures suitable for tissue regeneration and controlled drug delivery. The stability of the film is also affected by the sequence of monomers, which influences resistance to degradation. Rigid structures of sequences enhance thermal stability, whereas functional groups such as aromatic rings augment chemical stability. Additionally, alternating rigid and flexible segments allow a trade-off between strength and elasticity, which is important in automotive and construction materials [16]. Polymers also have intimate sequence dependence in their performance characteristics, such as mechanical strength, elasticity, and optical properties. Like polyurethanes, polymers with alternating flexible and rigid segments exhibit very high elasticity. Furthermore, specific monomer sequences are useful for efficient charge transport in electronic devices. Therefore, it is important to control the sequence of monomers to design polymers with the desired properties, especially for high-performance applications [17].

5.2.3 Traditional methods of polymer design: limitations and challenges

Diverse materials have been systematically developed by conventional polymer design techniques, that is, addition and condensation polymerization; however, they face serious limitations in delivering sophisticated functionalities such as stimuli responsiveness [18]. In the 1980s, shape-memory polymers (SMPs) addressed these limitations by introducing adaptive, responsive characteristics for shape transformations triggered by temperature, light, pH, or magnetic fields. SMPs have also successfully expanded their range of applications, especially for biomedical purposes, where porous SMPs provide optimal geometrical forms for tissue scaffolds or minimally invasive implants [19]. Precise adduct control, construction of complex molecular frameworks, and the introduction of diverse functionalities present challenging problems that established methods struggle to solve. These limitations hinder the growth of advanced materials, such as biodegradable and adaptive polymers that respond to their environment [20]. The use of toxic solvents, unsustainable feedstocks, and energy-consuming processes creates environmental problems for traditional polymerization methods. There is a pressing need to revolutionize polymerization strategies to meet the exacting standards of modern technology and address the escalating demand for customizable, high-performance materials, moving beyond conventional approaches [21].

5.3 Role of artificial intelligence in polymer science

5.3.1 AI algorithms in material design

ML and deep learning are transforming materials science and mechanical engineering through computational methods. These techniques give researchers the ability to accurately predict material properties, design new materials, or discover new mechanisms that diverge from previous thinking. However, as materials become increasingly complex, there remains a need to navigate vast design spaces to optimize mechanical behavior, a task not confined by traditional approaches [22]. ML models utilizing a large dataset of materials that correlates structure, properties, and functions at multiple scales can be employed to overcome this hurdle and enable rapid design space exploration. The success of ML-based material design approaches depends on several key factors: large, well-curated datasets, assembling or generating domain-specific materials science knowledge to preprocess data, and judiciously selecting appropriate ML models. Recent advances in ML techniques have opened up new avenues not only for addressing entrenched problems in mechanics but also for developing groundbreaking materials design strategies [23]. Various AI algorithms used in polymer designing are mentioned in Table 5.1.

Table 5.1: AI algorithms in polymer design.

Algorithm	Purpose	Example
Machine learning (ML)	Predict material properties	Predicting strength, conductivity, and biodegradability of polymers
Deep learning	Identifying patterns in complex datasets	Convolutional neural networks for molecular property predictions
Generative models (GANs and VAEs)	Designing novel polymer sequences	Generating polymer structures with optimized properties
Reinforcement learning	Optimizing synthesis conditions	Adjusting parameters in real time for efficient polymer production

5.3.2 Integrating AI with computational chemistry and molecular dynamics

The integration of ML, quantum computing, and molecular dynamics simulations is redefining computational biology, markedly improving simulation accuracy and speed. These advanced technologies expedite the processing of vast molecular dynamics datasets, enable real-time adjustments to simulation parameters, and illuminate intricate biological processes with unprecedented clarity. Landmark advancements include predictive force fields, adaptive algorithms, and quantum-augmented methodologies, which collectively illuminate the intricacies of molecular behavior [24]. Nevertheless, challenges related to data quality, model interpretability, and computational complexity still exist. To fully harness the potential of these innovations in biomolecular simulations, multidisciplinary solutions are necessary. Next-generation molecular docking simulations are revolutionizing computational chemistry and drug discovery by significantly improving the precision of drug-target interaction predictions [25]. Traditional docking methods struggle with accuracy and speed, but innovative computational methods, such as those inspired by convolutional and recurrent neural networks (RNNs), enhance predictions of binding affinity and interaction modeling. These techniques provide a deeper understanding of molecular interactions, accelerate the exploration of chemical libraries, and enable the identification and optimization of promising compounds and scoring functions [26]. Recent advances in computational techniques enable the discovery of novel drug targets by analyzing vast amounts of omics data. Molecular simulations, essential in various fields such as physics, materials science, and drug discovery, typically rely on resource-intensive methods that utilize manual coding and multiple programming languages. This underscores the interplay between computational methods and molecular simulations, demonstrating how advanced computational platforms provide innovative solutions by enhancing algorithms, programming paradigms, and hardware utilization. Innovative methodologies, including differentiable programming, are revolutionizing molecular simulations by optimizing derivatives, facilitating high-throughput simulations, and improving accuracy, efficiency, and scalability, ultimately opening up new research avenues [27].

5.3.3 Machine learning algorithms

Advanced data analysis techniques, such as ML, utilize algorithms that learn from data to make predictions. These algorithms examine input data, identify trends, and predict outcomes, facilitating tasks like classification and regression. By leveraging large datasets, ML algorithms continually improve predictive accuracy and optimize decision-making efficiency. This enables the rapid processing of vast amounts of data, yielding outputs with greater speed and precision than traditional methods [28]. Recent innovations in data-driven technologies have transformed fields like deep learning, natural language processing, and speech recognition, allowing computers to simulate human-like intelligence with enhanced accuracy. Specialized computational techniques are designed for specific tasks, necessitating careful consideration and selection to achieve optimal outcomes [29]. For instance, certain analytical techniques are well-suited for medical diagnoses, while others excel in financial forecasting. Advances in computing capabilities and the growing availability of extensive datasets have facilitated the broad application of sophisticated data analysis methods across various industries. These methods are especially adept at modeling intricate patterns and relationships, such as image recognition and material property analysis [30]. Supervised learning methods will be taught using a labeled dataset with inputs and their corresponding outputs. Linear regression, decision trees, and neural networks are widely used families of techniques that are applied in finance, healthcare, and marketing for forecasting and making decisions. Guided learning contributes to understanding material properties and composition optimization in materials science. An important characteristic of this approach is that it uses labeled training data to defragment the algorithm [31]. Unsupervised learning consists of studying or learning from data in an unlabeled state, that is, studying or learning patterns, connections, and classifications. Clustering and other dimensionality reduction techniques allow the data to be grouped or the relationships detected for the hidden. It is relevant for outlier detection, data classification, and material characteristic identification. In particular, it is unlike guided learning as it does not need labeled data but finds patterns on its own [32]. All reinforcement learning focuses on training systems in terms of selecting a sequence of actions to optimize a performance measure. These systems are developed to identify the most beneficial choices through experimentation and feedback. Applications requiring real-time decision-making, such as process optimization or synthesis of material by in situ monitoring, will use this approach. Learning and analysis in high-dimensional complex networks of interconnected computational elements permit systems to self-organize and adapt over time [33]. In deep learning, data is processed by networks with layers containing many complex operations, which are trained to identify subtle patterns and relationships in the data. Various network architectures give rise to specific insights, such as convolutional networks for image analysis and recurrent networks for sequential data. Raw data is processed using these techniques, and the results are accurate outputs used to power virtual assistants, anomaly detection, and autonomous systems [34]. An overview illustrating the integration of AI in polymer design is shown in Figure 5.1.

5.4 AI-driven approaches to polymer sequence design

Once, designing polymer sequences was a challenging task requiring specialist prowess and computational facilities. Advancements in data analysis have changed this; however, that means researchers can more quickly identify new polymer structures with desired properties, which simplifies design and exploration. With this shift, it becomes possible to rapidly identify new materials, opening up new possibilities in materials science and engineering [35].

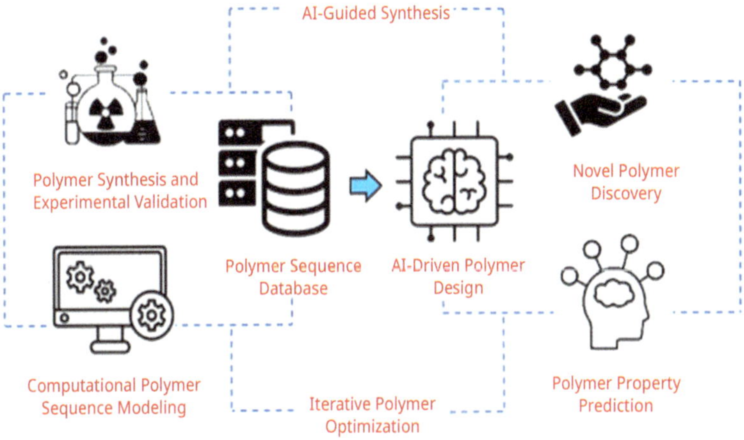

Figure 5.1: AI-driven polymer design workflow.

5.4.1 Predictive models for polymer properties: strength, conductivity, biodegradability

Medical research has been dramatically impacted by having computer simulations run on small molecule diffusion theories to improve the design and application of these theories. Predictive modeling and molecular simulations have a major impact on medical device development and delivery systems, as well as tissue engineering. Membrane methods that optimize biodegradable polymers offer key insights into material behavior and facilitate process streamlining in the development of these materials [36]. The objective is to forecast polymer characteristics such as tensile strength, longevity, and environmental degradability – essential traits for their practical applications. Information from extensive datasets is evaluated using sophisticated analytical techniques to estimate mechanical, electrical, and decomposition properties. Atomic-scale interactions and behaviors are elucidated through molecular dynamics and quantum mechanical modeling. Models for polymer strength, conductivity, and degradation rates take into account factors such as molecular weight and environmental conditions [37]. Despite challenges such as data scarcity and model generalization, multiscale modeling improves prediction accuracy and speeds up custom polymer development, thus streamlining the material design with this integrated approach [38].

5.4.2 Generative models for novel polymer sequence design

Generating novel sequences of polymer monomers using generative models is a tantalizing frontier in research. A dataset of existing sequences and properties is compiled and formatted for statistical analysis to develop novel polymers. The forecasting of sequences with tailored characteristics involves a generative model. The methodology is assessed through generated sequences, predictive models, and experimental validation. The field of material development is revolutionized by generative design techniques, such as GANs and VAEs, which are leading to new material compositions and microstructures created for a set of given performance criteria [39]. These materials research methods take advantage of the latest technology to develop innovative structures, showing great potential to advance materials research. Nevertheless, bottlenecks remain in optimally leveraging the potential of this technology for materials design, including fine-tuning formulas, empirical validation, and computational constraints. Overcoming these barriers and widening applications will likely shape the future of technology-driven materials design [40]. Deep generative models can be very robust, with high-dimensional data converted into low-dimensional (latent) space learned from self-supervised methods. With sequential data like time series, speech, or video, chronological relationships are important, and the original VAE model isolates input data vectors, making it restrictive. More specifically, we address this by adapting VAEs to introduce RNNs or state-space models to capture sequential data, the latent space, and its chronological dependencies [41]. A unified framework combining different architectures, including RNNs for emissions, extended memory networks for handling variable-length data and stochastic processes for latent modeling are offered by dynamical variational autoencoders (DVAEs). DVAEs consist of an encoder that maps input sequences to a probabilistic latent world, preserving chronological relationships, and a decoder that generates output sequences from the latent world conditioned on any preceding time instances [42].

5.5 Applications of AI in polymer sequence design

AI integration into polymer sequence design creates unprecedented opportunities in materials science research and development and enables more eco-friendly and cost-efficient processes in various sectors, including biomedicine, electronics, sustainability, and advanced manufacturing. Industrial applications of AI in polymer sequence engineering are shown in Figure 5.2.

5.5.1 Smart polymers for biomedical applications

Next-generation technologies are transforming DDS by optimizing polymer architectures for targeted, controlled release. Researchers can enhance the biodegradability, biocompatibility, and controlled release of drugs from nanomaterials, improving therapeutic outcomes and minimizing side effects. pH-sensitive polymers, which enable drug

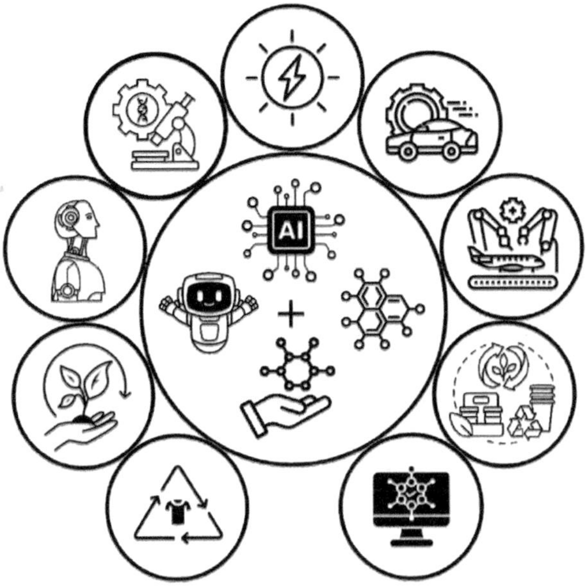

Figure 5.2: Applications of AI in polymer sequence design in various industries.

release triggered by tumor-specific conditions, exemplify how cutting-edge design can create more efficient DDS. As these systems evolve, scientists are addressing challenges related to the biocompatibility, toxicity, and scalability of nanomaterials for clinical use, thereby advancing precision medicine in oncology [43]. Sophisticated computational tools play a crucial role in optimizing tissue scaffolds for applications like bone regeneration, where polymers must satisfy specific mechanical, structural, and biological requirements. By leveraging predictive simulations that forecast polymer interactions with cells, researchers can craft scaffolds that promote cell adhesion, proliferation, and differentiation, yielding more effective tissue engineering solutions. Informed design additions that improve osteogenesis, such as advances in 3D-printed scaffolds coupled with bioactive peptides from extracellular matrices, can help hybrid systems [44].

5.5.2 Advanced polymers for electronics

Using sophisticated simulation tools, conducting polymers can be designed for use in wearable devices, flexible electronics, and energy harvesting in electronics. Guided by computational models that predict the effect of molecular architecture on the conductivity and flexibility of polymers, realized polymers are evaluated for their performance in these applications. Ideal properties for thermoelectric devices based on conducting polymers include nontoxic and flexible attributes that can be produced on a large scale with scalable production routes such as solution processing and roll-to-roll printing [45]. Advanced computational modeling further improves the design of insulating polymers for microelectronics by predicting, for example, dielectric constants, thermal stability, and mechanical strength. The development of high-performance insulating materials is thus enabled, which is necessary for achieving efficient and reliable electrical devices as component sizes decrease and voltages increase. These materials face challenges with space charge accumulation and dielectric breakdown, and advanced simulations are used to address them [46].

5.5.3 Sustainability-driven polymer design

Cutting-edge technologies are at the forefront of developing biodegradable plastics, addressing the global plastic waste crisis. By designing polymers like PLA (polylactic acid) and PHA (polyhydroxyalkanoates) with specific biodegradation rates, experts ensure that plastics decompose efficiently into nontoxic by-products. These polymers are engineered to meet environmental goals while maintaining high-performance standards. Innovative modeling techniques predict degradation rates and optimize polymer structures for safe and effective breakdown under natural conditions, reducing plastic pollution [47]. Recycling challenges in the plastic industry are mitigated through advanced materials science, which enables the creation of polymers that are easier to recycle. By focusing on uniform chemical structures and enhanced material properties, scientifically driven designs facilitate more efficient recycling. These polymers retain their integrity during reprocessing, helping to close the loop in a circular economy by enabling high-quality recycled plastics for reuse in manufacturing [48].

5.5.4 Polymers in aerospace and automotive industries

Advanced polymer research is critical in the aerospace industry, enabling the creation of lightweight and durable materials that withstand extreme conditions found in spacecraft and aircraft. Polyimides and other high-performance polymers are meticulously engineered to boost their thermal and mechanical properties, including resistance to ablation. By optimizing polymer compositions and environmental factors, experts can enhance the efficiency of aerospace components exposed to high temperatures, thereby expanding their applications [49]. The automotive industry leverages innovative materials science to develop sophisticated polymers, including carbon-fiber-reinforced composites, for electric vehicles (EVs). These advanced materials are precision-engineered for optimal mechanical strength, thermal resistance, and lightweight properties, yielding improved energy efficiency and fuel economy. They also create multifunctional materials with special properties (such as self-healing or energy-absorbing) to enhance the performance and safety of automotive components [50]. Various kinds of application of AI in polymer science are listed in Table 5.2.

Table 5.2: Applications of AI in polymer science.

Field	Application	Examples	References
Biomedical	Drug delivery systems, tissue engineering	pH-sensitive polymers for targeted drug release; scaffolds for bone regeneration	[51]
Electronics	Conducting and insulating polymers	Polymers for wearable devices, flexible electronics, and thermoelectric devices	[52]
Sustainability	Biodegradable plastics, recycling-friendly polymers	AI-designed PLA and PHA with tailored biodegradation properties	[53]
Aerospace	Lightweight, high-strength composites	Polymers with enhanced thermal resistance for spacecraft and aircraft applications	[54]
Automotive	Self-healing and energy-absorbing materials	Carbon-fiber-reinforced composites for electric vehicles; crash-resistant polymers	[55]
Energy	Advanced materials for renewable energy systems	Polymers for battery electrolytes, organic photovoltaics, and fuel cell membranes	[56]
Agriculture	Smart polymers for controlled release of agrochemicals	Biodegradable polymer coatings for fertilizers and pesticides	[57]
Food Packaging	Sustainable and functional packaging materials	Biopolymer films with antimicrobial and oxygen-barrier properties	[58]
Textiles	Advanced fibers for smart clothing	Conductive polymers for wearable sensors and temperature-responsive fabrics	[59]
Robotics	Flexible and responsive materials for actuators	Shape-memory polymers for soft robotics and artificial muscles	[60]

5.6 Innovations in AI-powered polymer research

With integration of AI into polymer research, the possibilities of synthesis, discovery, and validation have never been more powerful [61]. The impact of approaching biology computationally and experimentally includes autonomous labs, robotics driven by AI, and high-throughput screening [62].

5.6.1 Autonomous polymer synthesis and characterization labs

Traditional polymer research can be a laborious undertaking involving a hands-on approach and often repetitive experiments to optimize synthesis conditions and achieve desired properties [63]. By contrast, autonomous labs free manual workflows from the bottlenecks introduced by them, implementing AI-driven systems capable of autonomously designing experiments, executing them on robotic platforms, and analyzing results using cutting-edge instrumentation [64]. To give one example, AI predicts the optimal polymerization parameters like temperature, catalysts, and reaction time by analyzing huge datasets from multiple previous experiments, de facto computational simulations, and chemical libraries [65]. By this approach, high precision and reproducibility are assured even for complex reactions, such as atom transfer radical polymerization or reversible addition-fragmentation chain-transfer polymerization [66]. On the characterization side, Fourier-transform infrared spectroscopy (FTIR), nuclear magnetic resonance (NMR) spectroscopy, and scanning electron microscopy (SEM) are seamlessly integrated with AI algorithms for real-time data interpretation when conventional analysis cannot be handled [67]. The University of Glasgow's Chemputer platform is an example of such innovation, which can fully automate chemical synthesis workflows for AI-automated reaction optimization [68]. Remote and autonomous chemical synthesis using cloud-based AI models is demonstrated by IBM's RoboRXN platform. They are more efficient and do so more quickly, largely eliminating material waste and operational costs while increasing systems' environmental and economic sustainability [69]. On the other hand, scalability is also addressed by the lab, which is critical for industrial applications. These labs can identify promising polymer candidates for industrial testing or commercial production with minimal human intervention and can scale them up [70]. High reproducibility afforded by autonomous labs can benefit the synthesis of biocompatible hydrogels for DDS. With the exclusion of multi-objective optimization, these labs have the facilities to consider competing factors such as cost, performance, and environmental impact while trying to optimize them [71].

5.6.2 Combining AI with robotics for polymer discovery

The field of polymer discovery, cooking up advanced polymers from scratch, is an innovative frontier of polymer science, made more precise and automated by the integration of AI with robotics. AI-enabled robotic systems can repeat and perform intricate and mundane tasks such as reagent preparation, reagent mixing, and reaction monitoring with outstanding precision. This precision allows for reduced human error and expedites the finding of novel polymers with tailored properties [72]. For example, robotic platforms can run high-throughput experiments, creating and testing hundreds of polymer formulations in a single day, which is a task that would be virtually impossible for human researchers [73]. Advanced sensors have been fitted to these platforms to monitor temperature, viscosity, and reaction kinetics, which feed real-time data back for analysis by AI systems [74]. Using ML models, the robotic systems will learn to predict how given potential polymer candidates will behave and then use that information to direct which designs to pursue within the robotic systems [75]. For instance, deep learning algorithms running on a set of polymer properties can extract correlations between molecular structure and macroscopic properties, namely tensile strength and electrical conductivity. The insights from these robots dictate how monomers are selected, and how polymerization techniques and reaction conditions are selected to maximize the target attributes [76, 77]. Beyond that, reinforcement learning takes this process a step further and allows robots to adaptively optimize their experimental strategy in an attempt to complete an experiment based on feedback from past experiments [78]. AI algorithms can choose to further develop biodegradable polymers; robotic systems can then execute experiments to identify the degradation profile and mechanical properties of greatest benefit [79]. Additionally, the study of the larger space of possible polymer chemistries expands beyond the scope previously sanctioned by safety considerations due to the safe handling capabilities of robotic systems [80]. While AI and robotics will undoubtedly change and evolve, their introduction in the field of polymer discovery should yield transformative breakthroughs that shorten development time, increase safety, and introduce new vistas in material science [81].

5.6.3 High-throughput screening and validation in AI-powered labs

Maximizing the use of AI in polymer research would not be realized without the critical advancements in high-throughput screening (HTS) and validation, which allow for the very rapid identification and evaluation of thousands of polymer candidates based on their properties and performance [61]. Currently, the screening of polymers is painstaking, laborious, and time-intensive, involving extensive manual labor and trial-and-error experimentation. Instead, HTS uses automated robotic platforms linked with AI-driven analytics that conduct experiments at a scale and speed that was never conceivable [82]. The result is systems capable of synthesizing chemistry, analyzing the results, testing the formulations, and processing product synthetics all in parallel, providing comprehensive data on properties such as thermal stability, elasticity, biodegradability, and electrical conductivity [83]. HTS relies on the role played by AI in screening a larger number of polymer candidates through predictive models. For instance, ML algorithms trained to predict the behavior of new polymers based on datasets of polymer structures and properties can forecast the performance of a new polymer before it has been synthesized [84]. In this prescreening step, the experimental space is narrowed down, allowing HTS platforms to focus on the most promising candidates [85]. VAEs and GANs have inspired the development of additional generative models that further accelerate this process by creating novel polymer sequences best suited for particular applications [86]. With a robotic system, one carries out the screening process once polymers are synthesized using advanced analytical tools such as differential scanning calorimetry (DSC), thermogravimetric analysis (TGA), and dynamic mechanical analysis (DMA) [87]. The generated data from these instruments is quite large, and AI algorithms examine them to find trends, outliers, and optimal formulations. For example, this allows us to cluster polymers based on similar properties and use deep learning models to discover subtle correlations between structural features and performance metrics [67]. The Materials Project at Berkeley Lab is an example of a real HTS application that accelerates the discovery of new energy-related materials using AI and robotics, such as advanced polymers for battery applications [88]. Another essential part of the HTS workflow is validation, which checks the predictions of AI models against real-world performance [89]. Iterative validation makes AI-powered labs effective in experimental data that inform ML models, and predictions are refined on the fly. The closed-loop approach improves not only the accuracy of the models but also their generalization over unseen datasets [90]. HTS has been used, for example, to develop biodegradable plastics with optimal mechanical properties and to screen conducting polymers for flexible electronics. It is the combination of HTS and AI that is most powerful in tackling hurdles in sustainability and cost efficiency [72]. HTS can help scientists design polymers that exhibit excellent performance while having low environmental impact materials that break down into nontoxic by-products or are derived from renewable resources [91]. The growing adoption of AI-powered HTS means advances in polymer research are on the horizon that will lead to the discovery of materials capable of meeting the rapidly changing and interlinked needs of this world [92].

5.7 Validation of AI-designed polymers

5.7.1 Experimental validation of AI predictions

AI algorithms can propose promising polymer candidates from complex datasets, but we need to test them experimentally by confirming whether they will prove feasible and performable. This validation addresses polymers using molecular structures generated by AI and characterizing the properties using state-of-the-art analytical tools [93]. The chemical structure of the synthesized polymers is verified using NMR spectroscopy and mass spectrometry [94]. Thermal stability and phase transitions are assessed by DSC and TGA to ensure that the polymers meet the necessary thermal requirements for specific applications [95]. Validation of the physical robustness of AI-designed polymers, such as tensile and compressive strength, also requires mechanical testing. Therefore, they are used for applications in which one needs to be sure that the polymer after its physical robustness is calculated, can withstand any physical stress [96]. In addition, XRD and atomic force microscopy (AFM) techniques provide information about crystallinity and surface morphology, which are determinants for performance in drug delivery and coating applications [97]. Because AI models usually improve the validation process, they are incorporated into experimental workflows to identify discrepancies between the models and experimental results [98]. The robotic platform further automates validation through the execution of automated high-throughput synthesis and testing of multiple polymer candidates to enable the rapid evaluation of many polymer candidates [99].

5.7.2 Simulation and real-world performance

One of the major impediments in AI-designed polymer research is ensuring that the produced polymer performs as predicted in real-world applications. Often, the computational models also rely on idealized conditions, which may not accurately represent what happens in the environment or incorporate environmental factors, processing constraints, or competing materials well [100]. To bridge this gap, we employ a combination of advanced simulation methods, iterative testing, and adaptive ML models [101]. Molecular dynamics (MD) simulations can predict how polymers respond to stress, heat, or chemical exposure. While these simulations provide useful insights into the system, they are not sufficient to achieve complete accuracy in real-world applications [102]. Accelerated aging tests and field trials are used to evaluate the durability and functionality of polymers under real-life conditions [103]. Outdoor application polymers are tested for UV resistance to simulate excessive sunlight exposure, and biomedical application polymers are tested for biocompatibility and degradation in physiological environments [104]. This iterative process includes the development of advanced polymers for renewable energy systems (organic photovoltaics or battery electrolytes) and the conditions of humidity and temperature variations that occur in real-world applications, which have a profound impact on material performance [105].

5.7.3 Industrial-scale implementation of AI-designed polymers

Scaling up AI-designed polymers from lab synthesis into industrial production is a challenging but necessary step toward deployment in commercial applications. These polymers also depend on the available chemistries and costs of synthesis and are consistent with the existing manufacturing infrastructure [106]. AI helps tackle such challenges by optimizing processes for large-scale production. One such example is that generative design algorithms can propose modifications to polymer structures to be more easily manufactured yet remain efficient. Industrial-scale synthesis necessitates scaling from batch processes to continuous production [107]. This transition is facilitated by AI algorithms, which identify the best reaction conditions, such as temperature, pressure, and catalyst concentrations, to maximize yield and minimize waste. Increasingly, digital twin's virtual representations of industrial processes are being used to model and improve the performance of production lines before they are built [108]. Using these digital tools, manufacturers can anticipate bottlenecks, cut down energy use, and improve overall process efficiency. The application of AI-designed polymers for industrial purposes includes the development of lightweight, high-strength composites for the automotive and aerospace industries [109]. Moreover, using AI tools, PLA has been optimized for scalability and environmental sustainability as a biopolymer derived from renewable sources. Industrial implementation depends on regulatory compliance and quality control [110]. Production is monitored in real time by AI-driven systems that not only meet the specifications but guarantee them. The advanced analytics find and fix deviations from quality standards, lowering the risk of defects and raising customer satisfaction. These implementations demonstrate the revolutionary nature of AI in breaking through the barriers of polymer innovation scaling up [111].

5.7.4 Success metrics: cost, scalability, and sustainability

Ultimately, the success of AI-designed polymers is measured by their ability to balance cost, scalability, and sustainability while achieving superior performance. However, cost-effectiveness is a major metric, with the most advanced polymers likely to fail to enter the market unless they are prohibitively expensive [61]. AI finds efficient synthesis routes, eliminates material waste, and minimizes energy usage during production, which, at the same time, reduces production costs [112]. For example, predictive models can reveal low-cost catalysts or alternative raw materials that meet (or are required to meet) the performance standards. Polymers also must scale, meaning large quantities of the product must be made with minimal quality variation. This is where AI comes in, with the capability of real-time monitoring and control of production processes, leading to consistency in batches [113]. Algorithms in ML can predict and adjust for quality fluctuations in raw materials and ensure that the delivered product meets very strict specifications. With the rise of environmental concerns, sustainability becomes an important success metric. Polymers designed by AI are increasingly body-biodegradable and recyclable [61].

5.8 Challenges and solutions in AI-driven polymer design

5.8.1 Data limitations: quality, quantity, and bias in polymer datasets

Data availability and quality are probably one of the greatest challenges in AI-driven polymer design. The point is that if we want to build polymer AI, we need and want data to train and learn, yet our polymer science does not have a lot of standardized data. Comprehensive datasets from the perspective of experimental data are often scattered across research publications, patents, and proprietary databases [114]. Additionally, the expected performance may be obscured by noise and a reduction in data reliability due to inconsistencies in experimental methods and reporting standards. By creating a knowledge base or knowledge graph, the dataset itself can also be used to learn the categories correctly and transcend the inherent challenges in working with domains where the categories used to represent postings are unreliable or biased [61]. However, current datasets on polymers tend to be highly specific in terms of well-studied properties or applications, resulting in diversity in the training data. This biases an AI model's ability to generalize and find new, truly novel polymers with new properties. For instance, datasets may thoughtlessly overemphasize high-performance polymers used in the aerospace industry and ignore biodegradable polymers for consumer goods, thereby leading AI predictions to favor the upper ends of property space [63]. PubChem and a few other databases are some forward steps, although the field still needs more focus on polymer-specific datasets. Large volumes of standardized data are also generated by automated data logging systems and high-throughput experimentation systems for AI training [115]. Another avenue to enrich datasets and fill gaps in experimental data is to use computational simulations, such as MD or DFT, to generate synthetic data. Data sharing and standardization can be further accelerated by collaborative efforts from academia, industry, and governments [116].

5.8.2 Model interpretability: understanding AI decisions in polymer design

A key challenge in polymer design is the "black box" nature of many AI models, which makes it difficult for scientists to trust and validate AI-driven predictions. Such models are extremely powerful at modeling complex relationships in data; however, the decisions are opaque and hard to understand [117]. In high-stakes applications such as biomedical polymers, this lack of interpretability is problematic because we need to be able to understand why a model predicts a certain property to secure safety and regulatory approval [118]. This challenge has motivated the development of explainable AI (XAI) techniques to make the model decisions more transparent. Feature importance methods, like SHapley Additive exPlanations (SHAP) or local interpretable model-agnostic explanations (LIME), can evaluate which bits of a polymer's structure or composition most directly contributed to a prediction. Attention mechanisms in neural networks can similarly identify important substructures or sequences in a polymer that affect its property [119]. Moreover, visualization tools and a user-friendly interface will allow AI prediction tasks to become accessible to polymer scientists, where domain experts and AI specialists can cooperate [120].

5.8.3 Bridging the gap between theory and practice

Unfortunately, AI-driven polymer design tends to lack alignment between computational predictions and experimental validation. Real-world polymer synthesis and testing still have extra complexities, resulting in AI models working in silico, predicting polymer properties based on theoretical principles and training data [121]. Like AI, polymer properties can vary dramatically due to factors like impurities, synthesis conditions, or environmental variables and are prone to discrepancies between experimental results and AI predictions. We believe that bridging this gap requires the integration of AI with experimental workflows [84]. On the other hand, AI models that propose polymer designs and experimental labs that enable prediction validation feed information back to the model for refinement, representing an emerging promising approach. This iterative process could be simplified and constrained by time and cost through the use of laboratories outfitted with robotics and HTS capabilities, known as automated laboratories [61]. For instance, IBM Research's AI-powered platforms autonomously design and test new material formulations orders of magnitude faster than the human-driven discovery cycle [122]. It is also important for collaboration between polymer and AI researchers to develop models with relevant polymer experimental parameters. Hybrid approaches that use data-driven AI models and physics-based simulations can incorporate real-world variables to make predictions more accurate [123]. Additionally, it can help develop standards and protocols for integrating AI into laboratory workflows to ensure consistency, without which reproducibility will not be assured across research teams and institutions [63]. Key challenges in AI-driven polymer design and their mitigation strategies are listed in Table 5.3.

Table 5.3: Key challenges and solutions.

Challenge	Description	Solution
Data limitations	Lack of standardized, comprehensive datasets	Creation of open-access polymer databases and synthetic data generation
Model interpretability	Opaque AI decision-making processes	Explainable AI (XAI) techniques like SHAP and LIME
Experimental validation	Discrepancies between predictions and real-world performance	Integration of AI with experimental workflows and iterative validation
Ethical and environmental concerns	Bias in datasets, unsustainable design objectives	Incorporating sustainability metrics in AI models; democratizing access to tools

5.9 Advanced computational tools in polymer science

Polymer science now depends on advanced computational tools for unprecedented insights into the design, analysis, and optimization of polymer materials. Understanding polymer behavior at both atomic and molecular levels requires the contribution of MD and quantum chemistry simulations. Researchers can view polymer conformations, interactions, and mechanical properties under various environmental conditions, such as temperature, pressure, and stress, through MD simulations. On the other hand, these simulations are crucial to designing materials that have specifically desired properties, for instance, elasticity in rubber-like polymers or rigidity in structural composites [124]. While quantum chemistry simulations explore the electronic structure of polymers to predict optical, electronic, and chemical reactivity properties, these simulations do not consider any aspect of microstructure. Quantum chemical methods have, for example, been crucial for designing conducting polymers for electronics from an optimization point of view concerning charge transfer and bandgap properties. The integration of these approaches enables a complete characterization of polymer behavior, which leads to the development of high-performance materials for applications such as photovoltaics and drug delivery [125]. Using AI-powered tools, polymer analysis is now happening in real time, characterizing the properties and structure of polymers rapidly and accurately. These tools take advantage of ML algorithms applied to spectroscopic, chromatographic, and microscopic data to automate tasks that would otherwise require prolonged manual effort [51]. Similarly, AI algorithms can read detailed NMR or FTIR spectra and identify sequences of polymers and functional groups in seconds [126]. Further, imaging using AFM or SEM powered by AI can measure surface morphologies and defects with exceptional accuracy. These unique capabilities greatly accelerate the iterative process of polymer design by providing real-time feedback on synthesis and characterization [127]. In addition, AI applications predicting the aging and degradation of polymers are being developed to ensure improved performance and reliability of such polymers in extended use or aerospace components and medical device applications. User-friendly, data-driven solutions to polymer design are emerging as software and platforms in the polymer sequence prediction sector that are transforming the field [61]. ChemML use large databases combined with ML to predict polymer properties from sequence and composition. Advanced algorithms, including graph neural networks and RNNs, are integrated into these tools to learn how polymer structure relates to functionality [128]. To give just one example, generative models such as variational autoencoders are used to propose optimized sequences of polymers for a given property, such as thermal stability or biodegradability. This collaborative research is facilitated by cloud-based platforms that provide sharing of data and computational resources, allowing interdisciplinary teams of polymer innovators to work together [61]. However, such tools are particularly valuable for reducing development cycles and costs for industries looking to deliver new, innovative material solutions. Combining computational power, AI, and user-centered interfaces, these state-of-the-art tools are pushing the frontier of polymer science, enabling breakthroughs across a wide range of scientific fields, from green chemistry to high-tech engineering [63].

5.10 Future directions and emerging trends

The future of AI-driven polymer design is at a crossroads with possibilities for advancements in areas beyond existing limitations and the potential to remove them altogether. A promising direction is the application of AI for multidimensional polymer design, that is, the simultaneous optimization of molecular structure and macroscopic properties [129]. Using ML, coarse-grained simulations, and molecular dynamics, researchers first design polymers with precise atomic-level control and then predict their large-scale performance in various environments. For applications needing exact mechanical, thermal, or optical properties, this approach is particularly useful, for example, as aerospace materials or flexible electronics [130]. Also crucial are the roles of AI in developing sustainable polymer design and supporting the circular economy. Energy use, recyclability, and carbon footprint can be incorporated directly into AI models to ensure the development of biodegradable polymers, materials optimized for recyclability, and eco-friendly synthesis methods. For instance, AI is being used to predict the breakdown pathway of polymers to aid in developing materials that degrade well in nature [61]. A new game changer for the next generation of polymer simulations is quantum computing. Researchers leverage quantum algorithms to solve problems in electronic structure calculations and their molecular interactions with dramatically increased accuracy and speed. The ability to fabricate structures at any arbitrary scale harbors implications for polymer science in general, especially for systems with complex electronic or quantum mechanical properties, such as conducting polymers and photoresponsive materials [131]. Along with this, collaborative AI platforms are rapidly gaining steam and are used in global research efforts in polymer science. The platforms, which are open-access databases, cloud-based AI tools, and collaborative frameworks allow researchers across various fields to share data, algorithms, and insights [84]. Other examples include initiatives like the Materials Genome Initiative, which seeks to speed up the discovery and deployment of advanced materials based on data and AI-powered insights [132]. Democratizing access to cutting-edge tools is what such platforms are doing, but they are also pushing interdisciplinary collaboration, leading to innovation in polymer science. Taken together, these trends will change the face of polymer research to a more efficient, sustainable, and publicly inclusive environment [61].

5.11 Conclusion

Artificial intelligence integration into polymer sequence design has altered the very face of material science, making vast quantities of resources possible at unprecedented levels of accuracy and efficiency. AI-enabled approaches such as machine learning models, generative algorithms, and optimization techniques have been successfully deployed to address long-standing challenges in polymer design, property prediction, sequence discovery, and performance optimization. The transformative potential of AI-driven polymer research is understated by applications derived from biomedical, electronic, automotive, and sustainable industries. Next in this field are innovations such as autonomous synthesis labs and robotics, highlighting the future of this field made possible by seamless, data-driven insights into real-world materials. However, data quality, model interpretability, and ethical concerns are still major hurdles to overcome, and this can only be done through interdisciplinary collaboration and careful planning. If the combination of AI tools and quantum computing with molecular dynamics or global collaborative platforms defines the next frontier of polymer science, then none of this might matter in the end, and very little can be done about it. AI-fueled materials powered by AI are the path to sustainable, scalable, and high-performance materials, where polymers do not just meet today's needs; they also inspire tomorrow's innovations in ways we cannot yet imagine.

List of abbreviations

AI	Artificial intelligence
SHAP	SHapley Additive ExPlanations
LIME	Local interpretable model-agnostic explanations
XAI	Explainable AI
MD	Molecular dynamics
FTIR	Fourier-transform infrared
NMR	Nuclear magnetic resonance
AFM	Atomic force microscopy
SEM	Scanning electron microscopy
PubChem	Public Chemistry Database
IBM	International Business Machines
ChemML	Chemistry machine learning
ML	Machine learning
NLP	Natural language processing
RNN	Recurrent neural network
CNN	Convolutional neural network
VAE	Variational autoencoder
QM	Quantum mechanics

References

[1] Karalis VD. The integration of artificial intelligence into clinical practice. Applied Biosciences, 2024 Jan 1;3(1):14–44.

[2] San Ha N and Lu G. A review of recent research on bio-inspired structures and materials for energy absorption applications. Composites Part B: Engineering, 2020 Jan 15;181:107496.

[3] Schleder GR, Padilha AC, Acosta CM, Costa M and Fazzio A. From DFT to machine learning: Recent approaches to materials science–a review. Journal of Physics: Materials, 2019 May 16;2(3):032001.

[4] Rerup C and Feldman MS. Routines as a source of change in organizational schemata: The role of trial-and-error learning. Academy of Management Journal, 2011 Jun;54(3):577–610.

[5] Huang EW, Lee WJ, Singh SS, Kumar P, Lee CY, Lam TN, Chin HH, Lin BH and Liaw PK. Machine-learning and high-throughput studies for high-entropy materials. Materials Science and Engineering: R: Reports, 2022 Jan 1;147:100645.

[6] Chen G, Shen Z, Iyer A, Ghumman UF, Tang S, Bi J, Chen W and Li Y. Machine-learning-assisted de novo design of organic molecules and polymers: Opportunities and challenges. Polymers, 2020 Jan 8;12(1):163.

[7] Park H, Otte A and Park K. Evolution of drug delivery systems: From 1950 to 2020 and beyond. Journal of Controlled Release, 2022 Feb 1;342:53–65.

[8] Glotzer SC and Paul W. Molecular and mesoscale simulation methods for polymer materials. Annual Review of Materials Research, 2002 Aug;32(1):401–436.

[9] Chapekar MS. Tissue engineering: Challenges and opportunities. Journal of Biomedical Materials Research: An Official Journal of the Society for Biomaterials, the Japanese Society for Biomaterials, and the Australian Society for Biomaterials and the Korean Society for Biomaterials, 2000;53 (6):617–620.

[10] Hummel JP and Flory PJ. Structural geometry and torsional potentials in p-phenylene polyamides and polyesters. Macromolecules, 1980 May;13(3):479–484.

[11] Lepcio P, Ondreas F, Zarybnicka K, Zboncak M, Caha O and Jancar J. Bulk polymer nanocomposites with preparation protocol governed nanostructure: The origin and properties of aggregates and polymer bound clusters. Soft Matter, 2018;14(11):2094–2103.

[12] Graessley WW. Entangled linear, branched and network polymer systems – Molecular theories. In: Synthesis and Degradation Rheology and Extrusion. 2005 Jun 13; (pp. 67–117). Berlin, Heidelberg: Springer Berlin Heidelberg.

[13] Hadjichristidis N, Pitsikalis M, Iatrou H, Driva P, Chatzichristidi M, Sakellariou G and Lohse D. Graft copolymers. Encyclopedia of Polymer Science and Technology, 2010;1:1–38.

[14] Odian G. Principles of Polymerization. 2004 Feb 9; John Wiley & Sons.

[15] Rosales AM, Segalman RA and Zuckermann RN. Polypeptoids: A model system to study the effect of monomer sequence on polymer properties and self-assembly. Soft Matter, 2013;9(35):8400–8414.

[16] Hawkins WL. Polymer degradation. In: Polymer Degradation and Stabilization. 1984; (pp. 3–34). Berlin, Heidelberg: Springer Berlin Heidelberg.

[17] Lutz JF, Ouchi M, Liu DR and Sawamoto M. Sequence-controlled polymers. Science, 2013 Aug 9;341 (6146):1238149.

[18] Arefin AM, Khatri NR, Kulkarni N and Egan PF. Polymer 3D printing review: Materials, process, and design strategies for medical applications. Polymers, 2021 May 6;13(9):1499.

[19] Yakacki CM and Gall K. Shape-memory polymers for biomedical applications. Shape-memory Polymers, 2010;226:147–175.

[20] Shen Y, Fu X, Fu W and Li Z. Biodegradable stimuli-responsive polypeptide materials prepared by ring opening polymerization. Chemical Society Reviews, 2015;44(3):612–622.

[21] Clark JH, Farmer TJ, Hunt AJ and Sherwood J. Opportunities for bio-based solvents created as petrochemical and fuel products transition towards renewable resources. International Journal of Molecular Sciences, 2015 Jul 28;16(8):17101–17159.
[22] Janiesch C, Zschech P and Heinrich K. Machine learning and deep learning. Electronic Markets, 2021 Sep;31(3):685–695.
[23] Guo K, Yang Z, Yu CH and Buehler MJ. Artificial intelligence and machine learning in design of mechanical materials. Materials Horizons, 2021;8(4):1153–1172.
[24] Ollitrault PJ, Miessen A and Tavernelli I. Molecular quantum dynamics: A quantum computing perspective. Accounts of Chemical Research, 2021 Nov 17;54(23):4229–4238.
[25] Putha S. AI-driven molecular docking simulations: Enhancing the precision of drug-target interactions in computational chemistry. African Journal of Artificial Intelligence and Sustainable Development, 2021 Oct 4;1(2):260–300.
[26] Kotkondawar RR, Sutar SR, Kiwelekar AW and Wankhede HS. Performability of deep recurrent neural networks for molecular sequence data. International Journal of Computing and Digital System, 2023 May 6;13(1):1317–1327.
[27] Lavin A, Krakauer D, Zenil H, Gottschlich J, Mattson T, Brehmer J, Anandkumar A, Choudry S, Rocki K, Baydin AG and Prunkl C. Simulation intelligence: Towards a new generation of scientific methods. arXiv Preprint arXiv, 2021 Dec 6;2112:03235.
[28] Wang A, Kapoor S, Barocas S and Narayanan A. Against predictive optimization: On the legitimacy of decision-making algorithms that optimize predictive accuracy. ACM Journal on Responsible Computing, 2024 Mar 20;1(1):1–45.
[29] Lee I and Shin YJ. Machine learning for enterprises: Applications, algorithm selection, and challenges. Business Horizons, 2020 Mar 1;63(2):157–170.
[30] Le T, Epa VC, Burden FR and Winkler DA. Quantitative structure–property relationship modeling of diverse materials properties. Chemical Reviews, 2012 May 9;112(5):2889–2919.
[31] Caruana R and Niculescu-Mizil A. An empirical comparison of supervised learning algorithms. In: Proceedings of the 23rd International Conference on Machine Learning. 2006 Jun 25; (pp. 161–168), Association for Computing Machinery (ACM), New York, NY, USA.
[32] Seeger M Learning with labeled and unlabeled data. 2000.
[33] Morgan D and Jacobs R. Opportunities and challenges for machine learning in materials science. Annual Review of Materials Research, 2020 Jul 1;50(1):71–103.
[34] Choudhary K, DeCost B, Chen C, Jain A, Tavazza F, Cohn R, Park CW, Choudhary A, Agrawal A, Billinge SJ and Holm E. Recent advances and applications of deep learning methods in materials science. Npj Computational Mater, 2022 Apr 5;8(1):59.
[35] Mannodi-Kanakkithodi A, Chandrasekaran A, Kim C, Huan TD, Pilania G, Botu V and Ramprasad R. Scoping the polymer genome: A roadmap for rational polymer dielectrics design and beyond. Materials Today: Proceedings, 2018 Sep 1;21(7):785–796.
[36] Morawietz T and Artrith N. Machine learning-accelerated quantum mechanics-based atomistic simulations for industrial applications. Journal of Computer-Aided Molecular Design, 2021 Apr;35 (4):557–586.
[37] Doan Tran H, Kim C, Chen L, Chandrasekaran A, Batra R, Venkatram S, Kamal D, Lightstone JP, Gurnani R, Shetty P and Ramprasad M. Machine-learning predictions of polymer properties with polymer genome. Journal of Applied Physics, 2020 Nov 7;128(17):171104.
[38] Weinan E and Engquist B. Multiscale modeling and computation. Notices of the AMS, 2003 Oct;50 (9):1062–1070.
[39] Khan S and Awan MJ. A generative design technique for exploring shape variations. Advanced Engineering Informatics, 2018 Oct 1;38:712–724.
[40] Kundu S, Lerner S and Gupta R. Automated refinement checking of concurrent systems. In: 2007 IEEE/ACM International Conference on Computer-Aided Design. 2007 Nov 4; (pp. 318–325). IEEE.

[41] Gu A, Johnson I, Goel K, Saab K, Dao T, Rudra A and Ré C. Combining recurrent, convolutional, and continuous-time models with linear state space layers. Advances in Neural Information Processing Systems, 2021 Dec 6;34:572–585.
[42] Zhang Z, Ye L, Qin H, Liu Y, Wang C, Yu X, Yin X and Li J. Wind speed prediction method using shared weight long short-term memory network and Gaussian process regression. Applied Energy, 2019 Aug 1;247:270–284.
[43] Lim EK, Chung BH and Chung SJ. Recent advances in pH-sensitive polymeric nanoparticles for smart drug delivery in cancer therapy. Current Drug Targets, 2018 Mar 1;19(4):300–317.
[44] Zhao C, Tan A, Pastorin G and Ho HK. Nanomaterial scaffolds for stem cell proliferation and differentiation in tissue engineering. Biotechnology Advances, 2013 Sep 1;31(5):654–668.
[45] Das TK and Prusty S. Review on conducting polymers and their applications. Polymer-plastics Technology and Engineering, 2012 Oct 1;51(14):1487–1500.
[46] Maier G. Polymers for microelectronics. Materials Today: Proceedings, 2001 Sep 1;4(5):22–33.
[47] Naser AZ, Deiab I, Defersha F and Yang S. Expanding poly (lactic acid)(PLA) and polyhydroxyalkanoates (PHAs) applications: A review on modifications and effects. Polymers, 2021 Dec 6;13(23):4271.
[48] Hopewell J, Dvorak R and Kosior E. Plastics recycling: Challenges and opportunities. Philosophical Transactions of the Royal Society B Biological Sciences, 2009 Jul 27;364(1526):2115–2126.
[49] Mitra K. Genetic algorithms in polymeric material production, design, processing and other applications: A review. International Materials Reviews, 2008 Sep 1;53(5):275–297.
[50] Wazeer A, Das A, Abeykoon C, Sinha A and Karmakar A. Composites for electric vehicles and automotive sector: A review. Green Energy and Intelligent Transportation, 2023 Feb 1;2(1):100043.
[51] Omidian H. AI-powered breakthroughs in material science and biomedical polymers. Journal of Bioactive & Compatible Polymers, 2024 Dec;23:08839115241308202.
[52] Chen Y, Zhang X and Lu C. Flexible piezoelectric materials and strain sensors for wearable electronics and artificial intelligence applications. Chemical Science, 2024;15:16436–16466.
[53] Hao LT, Ju S, Hwang DK, Hwang DS, Ok YS, Hwang SY, Kim HJ, Jeon H, Park J, Oh DX and Koo JM. Optimizing bioplastics translation. Nature Reviews Bioengineering, 2024 Apr;2(4):289–304.
[54] Samir S, Gewali JP, Singh T, Debbarma K, Mazumdar S, Lone F and Daniella UE. Introduction to polymer composites in aerospace. CNS&E Journal, 1(4):329–339.
[55] Sahu BB, Moharana S and Behera PK. Elastomeric-based composite materials for engineering applications. In: Polymer Composites: Fundamentals and Applications. 2024 May 4; (pp. 329–355). Singapore: Springer Nature Singapore.
[56] Tripathi BP and Shahi VK. Organic–inorganic nanocomposite polymer electrolyte membranes for fuel cell applications. Progress in Polymer Science, 2011 Jul 1;36(7):945–979.
[57] Pavithran RK, Reddy SG, Kumar BS and Kugabalasooriar S. Enhancing sustainability in agriculture: Natural polymer-based controlled release systems for effective pest management and environmental protection. ES Food & Agroforestry, 2024 Oct 23;18:1276.
[58] Perera KY, Jaiswal AK and Jaiswal S. Biopolymer-based sustainable food packaging materials: Challenges, solutions, and applications. Foods, 2023 Jun 20;12(12):2422.
[59] Hu J, Meng H, Li G and Ibekwe SI. A review of stimuli-responsive polymers for smart textile applications. Smart Materials and Structures, 2012 Apr 18;21(5):053001.
[60] Chen Y, Chen C, Rehman HU, Zheng X, Li H, Liu H and Hedenqvist MS. Shape-memory polymeric artificial muscles: Mechanisms, applications and challenges. Molecules, 2020 Sep 16;25(18):4246.
[61] Tran H, Gurnani R, Kim C, Pilania G, Kwon HK, Lively RP and Ramprasad R. Design of functional and sustainable polymers assisted by artificial intelligence. Nature Reviews Materials, 2024 Aug;19:1–21.
[62] Zhu X. AI and Robotic Technology in Materials and Chemistry Research. 2024 Nov 6; John Wiley & Sons.

[63] Basak S and Bandyopadhyay A. Toward making polymer chemistry autonomous. ACS Applied Engineering Materials, 2024 Apr 18;2(5):1190–1208.
[64] Ferreira da Silva R, Moore II R, Mintz B, Advincula R, Alnajjar A, Baldwin L, Bridges CA, Coffee R, Deelman E, Engelmann C and Etz B. Shaping the Future of Self-Driving Autonomous Laboratories Workshop. 2024 Dec 1; Oak Ridge National Laboratory (ORNL), Oak Ridge, TN (United States).
[65] Ignacz G, Bader L, Beke AK, Ghunaim Y, Shastry T, Vovusha H, Carbone MR, Ghanem B and Szekely G. Machine learning for the advancement of membrane science and technology: A critical review. Journal of Membrane Science and Technology, 2024 Sep;2:123256.
[66] Pallares J, Jaramillo-Soto G, Flores-Cataño C, Vivaldo Lima E, Lona LM and Penlidis A. A comparison of reaction mechanisms for reversible addition-fragmentation chain transfer polymerization using modeling tools. Journal of Macromolecular Science, Part A: Pure and Applied Chemistry, 2006 Sep 1;43(9):1293–1322.
[67] Lu S and Jayaraman A. Machine learning for analyses and automation of structural characterization of polymer materials. Progress in Polymer Science, 2024 May;3:101828.
[68] Nel MA. The Design of Inexpensive and Open-source Laboratory Automation Equipment for Chemical Synthesis. 2021; University of Johannesburg (South Africa).
[69] Pyzer-Knapp EO, Pitera JW, Staar PW, Takeda S, Laino T, Sanders DP, Sexton J, Smith JR and Curioni A. Accelerating materials discovery using artificial intelligence, high performance computing and robotics. Npj Computational Mater, 2022 Apr 26;8(1):84.
[70] Kohn J, Welsh WJ and Knight D. A new approach to the rationale discovery of polymeric biomaterials. Biomaterials, 2007 Oct 1;28(29):4171–4177.
[71] Negut I and Bita B. Exploring the potential of artificial intelligence for hydrogel development – a short review. Gels, 2023 Oct 25;9(11):845.
[72] Xie C, Qiu H, Liu L, You Y, Li H, Li Y, Sun Z, Lin J and An L. Machine learning approaches in polymer science: progress and fundamental for a new paradigm. SmartMat, 2025Feb;6(1):e1320.
[73] Soheilmoghaddam F, Rumble M and Cooper-White J. High-throughput routes to biomaterials discovery. Chemical Reviews, 2021 Jul 2;121(18):10792–10864.
[74] Burke I, Salzer S, Stein S, Olusanya TO, Thiel OF and Kockmann N. AI-based integrated smart process sensor for emulsion control in industrial application. Processes, 2024 Aug 27;12(9):1821.
[75] Pilania G. Machine learning in materials science: From explainable predictions to autonomous design. Computational Materials Science, 2021 Jun 1;193:110360.
[76] Andraju N, Curtzwiler GW, Ji Y, Kozliak E and Ranganathan P. Machine-learning-based predictions of polymer and postconsumer recycled polymer properties: A comprehensive review. ACS Applied Materials and Interfaces, 2022 Sep 14;14(38):42771–42790.
[77] Champa-Bujaico E, García-Díaz P and Díez-Pascual AM. Machine learning for property prediction and optimization of polymeric nanocomposites: A state-of-the-art. International Journal of Molecular Sciences, 2022 Sep 14;23(18):10712.
[78] Kober J, Bagnell JA and Peters J. Reinforcement learning in robotics: A survey. The International Journal of Robotics Research, 2013 Sep;32(11):1238–1274.
[79] Laycock BG, Chan CM and Halley PJ. A review of computational approaches used in the modelling, design, and manufacturing of biodegradable and biobased polymers. Progress in Polymer Science, 2024 Sep;10:101874.
[80] Mestre R, Astobiza AM, Webster-Wood VA, Ryan M and Saif MT. Ethics and responsibility in biohybrid robotics research. Proceedings of the National Academy of Sciences, 2024 Jul 30;121(31): e2310458121.
[81] Harun-Ur-Rashid M and Imran AB. Emerging trends in engineering polymers: A paradigm shift in material engineering. Recent Progress in Materials, 2024 Sep;6(3):1–37.
[82] Taherdoost H. R&D Tools and Technologies. Innovation through Research and Development: Strategies for Success. 2024 Jun 12; (pp. 271–289). Cham: Springer Nature Switzerland.

[83] Yang L, Pijuan-Galito S, Rho HS, Vasilevich AS, Eren AD, Ge L, Habibovic P, Alexander MR, De Boer J, Carlier A and Van Rijn P. High-throughput methods in the discovery and study of biomaterials and materiobiology. Chemical Reviews, 2021 Mar 11;121(8):4561–4677.
[84] Upadhya R, Kosuri S, Tamasi M, Meyer TA, Atta S, Webb MA and Gormley AJ. Automation and data-driven design of polymer therapeutics. Advanced Drug Delivery Reviews, 2021 Apr 1;171:1–28.
[85] Hoffer L, Renaud JP and Horvath D. Fragment-based drug design: Computational and experimental state of the art. Combinatorial Chemistry & High Throughput Screening, 2011 Jul 1;14(6):500–520.
[86] Menon D and Ranganathan R. A generative approach to materials discovery, design, and optimization. ACS Omega, 2022 Jul 24;7(30):25958–25973.
[87] Sukur EF. Thermally conditioned aerospace-grade carbon fiber reinforced polyether ketone composites: Structure, impact response, and thermomechanical performance. Polymer Composites, 2023 Apr;44(4):2530–2544.
[88] Jain A, Ong SP, Hautier G, Chen W, Richards WD, Dacek S, Cholia S, Gunter D, Skinner D, Ceder G and Persson KA. Commentary: The materials project: A materials genome approach to accelerating materials innovation. APL Materials, 2013 Jul 1;1(1):011002.
[89] Pattyam SP. AI in data science for predictive analytics: Techniques for model development, validation, and deployment. Journal of Science & Technology, 2020 Oct 17;1(1):511–552.
[90] Mishra RK, Mishra D and Agarwal R. An artificial intelligence-powered approach to material design. Cutting-Edge Research in Chemical and Material Science, 2024;Volume I:1–30.
[91] Stryczewska HD, Boiko O, Stępień MA, Lasek P, Yamazato M and Higa A. Selected materials and technologies for electrical energy sector. Energies, 2023 Jun 6;16(12):4543.
[92] Ninduwezuor-Ehiobu N, Tula OA, Daraojimba C, Ofonagoro KA, Ogunjobi OA, Gidiagba JO, Egbokhaebho BA and Banso AA. Tracing the evolution of AI and machine learning applications in advancing materials discovery and production processes. Engineering Science & Technology Journal, 2023 Sep 14;4(3):66–83.
[93] Yang Q, Vriza A, Castro Rubio CA, Chan H, Wu Y and Xu J. Artificial intelligence for conjugated polymers. Chemistry of Materials, 2024 Jan 29;36(6):2602–2622.
[94] Weidner SM and Trimpin S. Mass spectrometry of synthetic polymers. Analytical Chemistry, 2010 Jun 15;82(12):4811–4829.
[95] Rahman MR, Hamdan S and Hui JL. Differential scanning calorimetry (DSC) and thermogravimetric analysis (TGA) of wood polymer nanocomposites. In: InMATEC Web of Conferences 2017. (vol. 87, pp. 03013). EDP Sciences.
[96] Landel RF and Nielsen LE. Mechanical Properties of Polymers and Composites. 1993 Dec 14; CRC press.
[97] Nguyen-Tri P, Ghassemi P, Carriere P, Nanda S, Assadi AA and Nguyen DD. Recent applications of advanced atomic force microscopy in polymer science: A review. Polymers, 2020 May 17;12(5):1142.
[98] Myllyaho L, Raatikainen M, Männistö T, Mikkonen T and Nurminen JK. Systematic literature review of validation methods for AI systems. Journal of Systems and Software, 2021 Nov 1;181:111050.
[99] Michael S, Auld D, Klumpp C, Jadhav A, Zheng W, Thorne N, Austin CP, Inglese J and Simeonov A. A robotic platform for quantitative high-throughput screening. Assay and Drug Development Technologies, 2008 Oct 1;6(5):637–657.
[100] Mahrouei M, Mohammadhashemi S, Ashrafi M, Nadimpour M, Pour SZ, Khorramakie H and Yousefinezhad M Frontiers in Engineering: Emerging Technologies and Innovations.
[101] Sharma A, Mukhopadhyay T, Rangappa SM, Siengchin S and Kushvaha V. Advances in computational intelligence of polymer composite materials: Machine learning assisted modeling, analysis and design. Archives of Computational Methods in Engineering, 2022 Aug;29(5):3341–3385.
[102] Caruso MM, Davis DA, Shen Q, Odom SA, Sottos NR, White SR and Moore JS. Mechanically-induced chemical changes in polymeric materials. Chemical Reviews, 2009 Nov 11;109(11):5755–5798.

[103] Qin J, Jiang J, Tao Y, Zhao S, Zeng W, Shi Y, Lu T, Guo L, Wang S, Zhang X and Jie G. Sunlight tracking and concentrating accelerated weathering test applied in weatherability evaluation and service life prediction of polymeric materials: A review. Polymer Testing, 2021 Jan 1;93:106940.
[104] Maraveas C, Kyrtopoulos IV, Arvanitis KG and Bartzanas T. The aging of polymers under electromagnetic radiation. Polymers, 2024 Mar 3;16(5):689.
[105] Shashikumar U, Tsai PC, Wang CT, Lay CH and Ponnusamy VK. Beyond Biomimicry: Innovative Bioinspired Materials Strategies and Perspectives for High-Performance Energy Storage Devices. Process Safety and Environmental Protection, 2024 Sep 4;191:1193–1217.
[106] Ristoski P, Zubarev DY, Gentile AL, Park N, Sanders D, Gruhl D, Kato L and Welch S. Expert-in-the-loop AI for polymer discovery. In Proceedings of the 29th ACM International Conference on Information & Knowledge Management. 2020 Oct 19; (pp. 2701–2708), Association for Computing Machinery, New York, NY 10019-7434, USA.
[107] Qian C, Tan RK and Ye W. Design of architectured composite materials with an efficient, adaptive artificial neural network-based generative design method. Acta Materialia, 2022 Feb 15;225:117548.
[108] Srivastava S, Varshney B, Sharma VP and Ali B. Applications of Artificial Intelligence in polymer manufacturing. Application of Artificial Intelligence in New Materials Discovery, 2023 Jul 5;147:105–122.
[109] Zhai C, Li T, Shi H and Yeo J. Discovery and design of soft polymeric bio-inspired materials with multiscale simulations and artificial intelligence. Journal of Materials Chemical B, 2020;8 (31):6562–6587.
[110] Ortelli S, Costa AL, Torri C, Samorì C, Galletti P, Vineis C, Varesano A, Bonura L and Bianchi G. Innovative and sustainable production of biopolymers. Factories of the Future: The Italian Flagship Initiative, 2019;1:131–148.
[111] Wan J, Li X, Dai HN, Kusiak A, Martinez-Garcia M and Li D. Artificial-intelligence-driven customized manufacturing factory: Key technologies, applications, and challenges. Proceedings of the IEEE, 2020 Nov 23;109(4):377–398.
[112] Abdallah M, Talib MA, Feroz S, Nasir Q, Abdalla H and Mahfood B. Artificial intelligence applications in solid waste management: A systematic research review. Waste Management, 2020 May 15;109:231–246.
[113] Bin Abu Sofian AD, Sun X, Gupta VK, Berenjian A, Xia A, Ma Z and Show PL. Advances, synergy, and perspectives of machine learning and biobased polymers for energy, fuels, and biochemicals for a sustainable future. Energy & Fuels, 2024 Jan 16;38(3):1593–1617.
[114] Gao L, Lin J, Wang L and Du L. Machine learning-assisted design of advanced polymeric materials. Accounts of Materials Research, 2024 Apr 16;5(5):571–584.
[115] Yue T, He J and Li Y. Polyuniverse: Generation of a large-scale polymer library using rule-based polymerization reactions for polymer informatics. Digital Discovery, 2024;3(12):2465–2478.
[116] Lee YS. 'Technology transfer' and the research university: A search for the boundaries of university-industry collaboration. Research Policy, 1996 Sep 1;25(6):843–863.
[117] Hassija V, Chamola V, Mahapatra A, Singal A, Goel D, Huang K, Scardapane S, Spinelli I, Mahmud M and Hussain A. Interpreting black-box models: A review on explainable artificial intelligence. Cognitive Computation, 2024 Jan;16(1):45–74.
[118] Bernard M, Jubeli E, Pungente MD and Yagoubi N. Biocompatibility of polymer-based biomaterials and medical devices–regulations, in vitro screening and risk-management. Biomaterial Science, 2018;6(8):2025–2053.
[119] Mane D, Magar A, Khode O, Koli S, Bhat K and Korade P. Unlocking machine learning model decisions: A comparative analysis of LIME and SHAP for enhanced interpretability. Journal of Electrical Systems, 2024;20(2s):1252–1267.
[120] Uddin MK. A review of utilizing natural language processing and ai for advanced data visualization in real-time analytics. Global Mainstream Journal, 2024 Apr 20;1(4):10–62304.

[121] Han XQ, Wang XD, Xu MY, Feng Z, Yao BW, Guo PJ, Gao ZF and Lu ZY. AI-driven inverse design of materials: Past, present and future. arXiv Preprint arXiv, 2024 Nov 14;2411:09429.
[122] Giro R, Hsu H, Kishimoto A, Hama T, Neumann RF, Luan B, Takeda S, Hamada L and Steiner MB. AI powered, automated discovery of polymer membranes for carbon capture. Npj Computational Materials, 2023 Jul 29;9(1):133.
[123] Wang J, Li Y, Gao RX and Zhang F. Hybrid physics-based and data-driven models for smart manufacturing: Modelling, simulation, and explainability. Journal of Manufacturing Systems, 2022 Apr 1;63:381–391.
[124] Binder K and Paul W. Monte Carlo simulations of polymer dynamics: Recent advances. Journal of Polymer Science, Part B: Polymer Physics, 1997 Jan 15;35(1):1–31.
[125] Ruiperez F. Application of quantum chemical methods in polymer chemistry. International Reviews in Physical Chemistry, 2019 Oct 2;38(3–4):343–403.
[126] Xue X, Sun H, Yang M, Liu X, Hu HY, Deng Y and Wang X. Advances in the application of artificial intelligence-based spectral data interpretation: A perspective. Analytical Chemistry, 2023 Sep 9;95 (37):13733–13745.
[127] Mazerolle S, Breguet JM, Steinecker A, Agnus J, Pérez R and Michler J. Nanomanipulation in a scanning electron microscope. Journal of Materials Processing Technology, 2005 Aug 30;167(2–3):371–382.
[128] Huang X, Ma S, Zhao CY, Wang H and Ju S. Exploring high thermal conductivity polymers via interpretable machine learning with physical descriptors. Npj Computational Materials, 2023 Oct 14;9(1):191.
[129] Badini S, Regondi S and Pugliese R. Unleashing the power of artificial intelligence in materials design. Materials, 2023 Aug 30;16(17):5927.
[130] Kim C, Batra R, Chen L, Tran H and Ramprasad R. Polymer design using genetic algorithm and machine learning. Computational Materials Science, 2021 Jan 1;186:110067.
[131] Micheletti C, Hauke P and Faccioli P. Polymer physics by quantum computing. Physical Review Letters, 2021 Aug 20;127(8):080501.
[132] De Pablo JJ, Jackson NE, Webb MA, Chen LQ, Moore JE, Morgan D, Jacobs R, Pollock T, Schlom DG, Toberer ES and Analytis J. New frontiers for the materials genome initiative. Npj Computational Materials, 2019 Apr 5;5(1):41.

Rahul Sharma, Sonali Sundram*, and Naga Rani Kagithala

Chapter 6
Exploring the relationship between structure and properties in polymer blends using deep learning techniques

Abstract: Polymer blends are versatile substances that combine the properties of different polymers to create advanced materials with tailored functionalities. The complexity of the interactions and the broad design space present a substantial challenge in predicting and regulating the properties of polymer blends based on their molecular structures. Traditional experimental and computational techniques often struggle to thoroughly clarify the intricate connections between structure and properties. Deep learning (DL) has surfaced as an innovative approach to address these challenges, offering powerful resources for analyzing large datasets, uncovering hidden patterns, and precisely forecasting material characteristics. This chapter explores how DL methods influence the understanding of structure–property relationships in polymer blends. In this chapter, the authors provide a summary of the main types of polymer blends – miscible, immiscible, and compatibilized – and the key properties influenced by their molecular structures, such as mechanical, thermal, and optical traits. This chapter examines the application of various DL methods, including convolutional neural networks, recurrent neural networks, and graph neural networks, for predicting polymer blend properties and developing new materials. Special focus is placed on data preparation, feature extraction, and molecular representation methods vital for the effectiveness of DL models in polymer science. Despite the significant potential of DL, obstacles such as data scarcity, model interpretability, and experimental validation persist. Techniques like data augmentation, transfer learning, and high-throughput experimentation are explored as potential solutions to these issues. Furthermore, the chapter highlights the opportunity to combine DL with automation and eco-friendly design principles to revolutionize the development of polymer blends. This chapter seeks to connect structure and characteristics through DL, offering a guide for researchers and practitioners aiming to utilize artificial intelligence-driven methods to advance polymer science and materials engineering. It also explains various common DL techniques and their material applications. Applications include molecular property prediction, molecular production,

***Corresponding author: Sonali Sundram,** Department of Pharmacology, School of Medical and Allied Sciences, Galgotias University, Greater Noida, Uttar Pradesh, India, e-mail: sonaliaim13@gmail.com
Rahul Sharma, Department of Pharmacy, Galgotias University, Greater Noida, India, e-mail: rsrahulsharma9888@gmail.com
Naga Rani Kagithala, Department of Pharmacy, Galgotias University, Greater Noida, India, e-mail: naga.kagithala@galgotiasuniversity.edu.in

https://doi.org/10.1515/9783111681870-006

structure identification, and polymer synthesis. Finally, DL in polymeric materials' future difficulties and prospects are highlighted.

Keywords: Polymer blends, deep learning, structure–property relationships, convolutional neural networks, graph neural networks, recurrent neural networks, molecular design, data augmentation, material properties, sustainable design

6.1 Introduction

6.1.1 Significance of polymer blends in materials science

Materials science has consistently been a significant area of research. Rubber, plastics, batteries, optical fibers, and sensors are among the numerous applications of polymer materials. The intricate chemical structures and diverse structural features of polymers complicate the prediction of their properties and the development of novel polymeric materials. Since the "Materials Genome Initiative (MGI)" was initiated in 2011, a growing number of machine learning (ML) and deep learning (DL) algorithms have been implemented in materials science, owing to the rapid advancement of materials informatics [1]. Polymers are essential in everyday and innovative technologies. In the age of DL, polymer informatics (PI) is essential for polymeric material design and development. However, PI has mostly studied single-component polymers, leaving polymer blends' vast chemical space untouched. They are macroscopically homogenous blends of homopolymers or copolymers [2]. In thermal applications, polymers are employed because of their low cost, high electrical resistivity, mechanical flexibility, corrosion resistance, and lightweight character [3]. The disorganized atomic organization of the majority of amorphous polymers results in a low thermal conductivity (TC) of 0.1–0.5 W/m K [4]. In amorphous polymers, both bound intramolecular interactions (e.g., covalent bonds along the polymer chain) and nonbonded intermolecular interactions (e.g., van der Waals forces, hydrogen bonds, and electrostatic interactions) contribute to the overall TC [5].

DL has made PI vital for polymeric material design and development. Polymer blends are macroscopically homogenous mixtures of homopolymers or copolymers that form novel materials with distinct physical properties. Due to their theoretical and practical importance, these combinations are carefully investigated. They are utilized in various applications, with estimates suggesting that approximately 30% of all polymers are sold as polymer blend products. Polymer blends are significant for their capacity to alter mechanical qualities (e.g., rubber toughening) and to tailor characteristics for particular applications (e.g., biopolymer utilization in food packaging). Chemical structures determine polymer properties; hence, modifications in a polymer's structure result in commensurate changes in its properties, establishing a definitive connection between structure and properties in polymers [6]. PI relies on numerical polymer representations to measure structure–property relationships using DL methods. Thus, polymer mixes must be numerically represented before training the DL model. Materials research relies on polymer mixes to generate materials with improved qualities and many uses.. Specialized polymer blends have possibilities in packaging and medical devices.

Polymer blends are cost-effective and can replace traditional materials in the automotive and aerospace industries due to their adaptability. They also propose that mixing polymers can speed market entry. Mixing polymers has created efficient engineering thermoplastics, which are widely used for their particular features. These studies demonstrate the value of polymer mixes in material development and marketability [7, 8]. Therefore, there are various significant aspects of polymer blends in material science, such as: Cost-effective: Blending allows for the production of materials with desired properties at lower costs by using less expensive polymers alongside high-performance ones [9]. Performance enhancement: Polymer blends can augment particular attributes, including toughness, impact resistance, and thermal stability. Rubber toughening can improve the impact strength of brittle polymers [10]. Recyclability: The blending method enhances plastic recycling by amalgamating recycled materials with virgin polymers to produce new goods, hence advancing sustainability initiatives [10]. Customization in polymers: Manufacturers can modify the properties of polymer blends for particular uses, such as electrical insulation or packaging materials, by choosing suitable polymer combinations. They can utilize ML algorithms to customize polymer mixes for particular applications, such as high-performance coatings or flexible packaging solutions [9]. In advanced manufacturing techniques, the amalgamation of DL with additive manufacturing methods facilitates the generation of intricate shapes and improved material performance in real time [11]. Advancement of sustainable materials: DL facilitates the design of biodegradable and renewable resource-based polymer blends, mitigating environmental issues linked to traditional plastics [11].

6.1.2 Structure–property relationships: a fundamental challenge

Polymer blend structure and properties must be understood to create materials with specified properties. Linking the macroscopic characteristics of polymer blends to their physical and chemical structures is a challenge that can significantly affect their practical uses. Since controlling morphology in polymer blends can improve the durability, transparency, and crack resistance of multiphase polymer materials, polymer material scientists have been focusing on this topic for decades. Ordered structures like as cylinders, lamellae, and spherical domains can be produced by casting solvents and the molecular weights of polymer components [12]. Known chemical approaches help multiphase polymers regulate sub-micrometer-scale morphology. High molecular weight block copolymers, especially multicomponent ones, are difficult to prepare, limiting orderly structures to nanometers [13]. Some monomers are successful with these methods, but their reactivity limits them. Alternatively, many physical approaches regulate micrometer-scale immiscible mix morphology. Most often, immiscible polymer mixtures are stirred at an adjustable speed during mechanical processing. In incompatible polymer combinations, these methods regulate dispersed phase sizes to the micrometer range. The intricacy and nonlinearity of polymer component interactions in blends make structure–property predictions difficult. Blending ratios and processing techniques generate feature variation, making predictive model construction harder. For structure–property correlations, high-resolution microscopy, rheology, and spectroscopy are essential since traditional characterization methods may miss nanoscale interactions and morphologies that affect macroscopic behavior [14]. Recent research has used advanced computational approaches and ML to improve structure–property prediction. These methods can find ideal blending and processing conditions, resulting in customized materials.

6.1.3 The impact of deep learning on advancing polymer research

DL has transformed polymer research, particularly in the area of polymer blends. Researchers can investigate complex connections between polymer mix structure and properties through advanced algorithms and computational capabilities, enhancing material design and optimization. Since AI and materials informatics have advanced, DL algorithms are commonly used in materials science to speed up innovation [15]. Polymers have become more complicated. Traditional ML algorithms struggle with complicated data. DL has enhanced natural language understanding and image recognition [16]. A DL model with additional layers of networks can show more complicated input–output interactions and have a greater hierarchy of attributes. DL works in cheminformatics, bioinformatics, and materials informatics [17]. In materials informatics, polymers or molecular structures need to be transformed into various forms so they can be fed into the appropriate DL models for analysis. Common techniques for modeling molecules include graphs, Morgan fingerprinting (MF), Morgan embeddings, and SMILES (simplified molecular input line entry specification). Red–green–blue (RGB) photographs or experimental data can be used by polymers as model input. Generative adversarial networks (GANs), recurrent neural networks (RNNs), deep neural networks (DNNs), and graph neural networks (GNNs) are some of the well-known DL designs [18]. Material DL predicts properties, builds structures, recognizes features, and evaluates importance. In high-dimensional data, DL can swiftly establish the relationship between ideal material qualities, complicated structure, and environmental conditions. This strong technique should expedite polymeric material discovery.

DNN benefits include the use of molecular or polymer properties as input and output. Direct model inputs from SMILES and molecular maps allow RNNs and GNNs to portray molecular structure–property connections. As library data expands, GNNs forecast molecular properties with greater accuracy and identify superior molecular traits compared to shallow ML. Deep RNN and GAN learning can generate new data by learning structural properties from existing data. As in polymer blends, DNNs, RNNs, and GANs can guide polymer formation. Compared to molecular dynamics (MD) simulation, ML can minimize experiment time, cost, and improve performance and efficiency [19]. DL offers distinct benefits for handling large data and complex problems [20], tackling the lack of molecular property databases using comparable distributions through their distinctive learning mechanisms. Furthermore, efficient DL models for predicting copolymer properties can be developed through linear layer-by-layer expansion of neural networks and mapping discrete polymer sequences to continuous latent spaces. These advancements highlight the potential of DL in accelerating polymer design and development, although challenges remain in model architecture selection and data availability [21].

6.2 Polymer blends

6.2.1 Types of blends: miscible, immiscible, and compatibilized

A polymer blend is a synthesis of two or more polymers merged to create a new material with unique physical characteristics. Polymer blends have been widely investigated in five main categories: thermoplastic–thermoplastic, thermoplastic–rubber, thermoplastic–thermosetting, rubber–thermosetting, and polymer–filler. A simple and cost-effective approach for manufacturing flexible polymeric materials for industrial use, polymer mixing has garnered attention [22]. Polymer blends are categorized based on their miscibility and compatibility, which significantly influence their physical properties and applications. The primary kinds of polymer blends are miscible, immiscible, and compatibilized blends, along with others detailed in Table 6.1. One common way to categorize polymer blends is by whether they are homogeneous (molecularly miscible) or heterogeneous (immiscible). Although certain mixtures, like poly(propylene) (PP)–poly(styrene) (PS) and PP–poly(ethylene) (PE), do not mix, others such as PS–poly(phenylene oxide) (PPO) and poly(styrene–acrylonitrile) (SAN)–poly (methyl methacrylate) (PMMA) are compatible. Typical properties of miscible (single-phase) mixtures include optical transparency and homogeneity at the polymer segment size. When the temperature, pressure, or composition of a mixture changes, it can cause the phases of a single-phase blend to separate [23]. Thermodynamic parameters, through the Gibbs free energy of mixing, ΔG_m, can be used to predict the miscibility of polymer blends. Blends must satisfy the following inequality to be considered miscible:

$$\Delta G_m = \Delta H_m - T\Delta S_m < 0 \tag{1.1}$$

For miscibility (binary blend), the following two conditions must be satisfied: the first condition: ΔGm < 0; and the second condition:

$$\left(\frac{\partial\Delta\, G_{\mathrm{m}}}{\partial\phi_{\mathrm{i}}^{2}}\right)T,\ p > 0 \tag{1.2}$$

ΔG_{m} represents the Gibbs energy associated with mixing, ΔS_{m} represents the entropy factor that measures disorder or unpredictability; it is always positive, thus promoting mixing or miscibility, particularly in solutions with low molecular weight [7]. On the contrary, the enthalpy of mixing (ΔH_{m}) is an essential component for miscibility in polymer solutions since these solutions contain monomers with substantial molecular weight. During mixing, heat can be either absorbed (endothermic) or released (exothermic), and this is represented by ΔH_{m}. In an exothermic mixing process, the system is pushed towards miscibility. Exothermic mixing can only happen when there are very precise interactions between the ingredients [24]. Dipole–dipole, ionic, and hydrogen bonding interactions are the main interactions seen in polymer blends. Various methods such as nuclear magnetic resonance (NMR) spectroscopy, neutron reflectivity, ellipsometry, small-angle neutron scattering, and Fourier transform infrared (FTIR) spectroscopy, are employed to elucidate particular interactions in polymers. The majority of polymer blends is immiscible and necessitates compatibilization to optimize interfacial tension, stabilize morphology, and improve adhesion between phases. Compatibilization can be accomplished by introducing a third component or by reactive blending two polymers with favorable inherent features that are immiscible and incompatible, which can be transformed into blends using compatibilizers. As mentioned in the previous sections, compatibilization requires thermodynamic miscibility factors. Traditionally, single compatibilizing agents have been employed; however, hybrid or multi-agent mixes also show promise. Nano-hybrids are currently in vogue [25]. The investigation will expand with homogeneous compatibilizing agents. To reduce scale, blend system interface modifiers can be integrated into the interface region. Breakup induces nanostructured morphology in mixtures. Compatibilizers are generally selected according to the polymer blend and are applied selectively [26]. Certain blend systems are inappropriate for compatibilizing agents. One-stop compatibilizing agents remain a fallacy in polymer research. Developing a compatibilizing agent for diverse polymer combinations is difficult. In binary or ternary blend systems, the localization of the compatibilizing agent can be unpredictable. The categories of polymer blends – miscible, immiscible, and compatibilized – are illustrated in Figure 6.1. It is essential to create compatibilizers that are recyclable and reusable. Researchers are urged to explore innovative materials, technologies, amalgamation procedures, and characterization.

Table 6.1: Different types of polymer blends with compatibility.

Type of polymer blend	Description	Examples	Features	Miscibility/compatibility	References
Miscible blends	Homogeneous (single–phase) mixtures that are molecularly miscible.	– Poly(styrene) (PS)–poly (phenylene oxide) (PPO) – Poly(styrene–acrylonitrile) (SAN)–poly(methyl methacrylate) (PMMA)	– Optically clear – Uniform at the polymer segmental level – Phase separation induced by temperature, pressure, or composition fluctuations.	Governed by Gibbs free energy of mixing: $\Delta G_m = \Delta H_m - T\Delta S_m < 0$	[22–24, 27]
Immiscible blends	Heterogeneous blends where components are not molecularly miscible.	– Poly(propylene) (PP)–poly(styrene) (PS) – Poly(propylene)–poly(ethylene) (PE)	– Typically exhibit phase separation – Poor interfacial adhesion.	Lack of favorable thermodynamic interaction; typically requires compatibilization to enhance performance.	[22–24, 27, 28]
Compatibilized blends	Immiscible blends made compatible by adding compatibilizers to enhance interfacial properties.	– Nano-hybrids – Hybrid or multi-agent mixtures	– Improved interfacial tension – Stabilized morphology – Enhanced adhesion between phases.	Achieved using a third component (compatibilizer), reactive blending, or nano-hybrids.	[24–26, 29]
Thermoplastic–thermoplastic	Blend of two thermoplastics.	– Polypropylene (PP) blends – Polystyrene (PS) blends	– Enhanced toughness, stiffness, or clarity depending on the combination.	Miscibility varies; compatibilization may be required for immiscible blends.	[22]
Thermoplastic–rubber	Blend of a thermoplastic and a rubber polymer.	– PP–EPDM (ethylene propylene diene monomer) blends	– Improved flexibility and impact resistance.	Often immiscible; compatibilization needed.	[22, 26]

(continued)

Table 6.1 (continued)

Type of polymer blend	Description	Examples	Features	Miscibility/compatibility	References
Thermoplastic–thermosetting	Combination of a thermoplastic and a thermosetting polymer.	– PET (polyethylene terephthalate)–epoxy resin blends	– Balances rigidity with toughness and thermal stability.	Requires careful blending due to different processing conditions and incompatibility.	[22]
Rubber–thermosetting	Blend of rubber and thermosetting polymer.	– Rubber–epoxy resin blends	– Improved impact resistance and elasticity.	Often requires compatibilizers for stable morphology.	[22, 24]
Polymer–filler	Blend of polymers with inorganic or organic fillers.	– Polypropylene–carbon black blends	– Enhanced strength, thermal stability, or conductivity depending on filler.	Compatibility depends on the filler–polymer interaction and surface treatment.	[22, 26, 30]
Miscibility factors	Miscibility determined by thermodynamic parameters, particularly Gibbs free energy (ΔGm).	–	– $\Delta G_m < 0$ for miscibility – ΔS_m promotes mixing, while ΔH_m determines enthalpy of mixing.	Specific interactions like hydrogen bonding, dipole–dipole interactions, and ionic interactions aid miscibility.	[23, 24, 31]
Compatibilization challenges	Issues in creating universal compatibilizers and localization problems in binary/ternary blend systems.	–	– Development of recyclable and reusable compatibilizers is needed. – Nano-hybrids and multi-agent systems show promise but face technical challenges.	Some systems remain unsuitable for current compatibilizers.	[25, 26, 29]

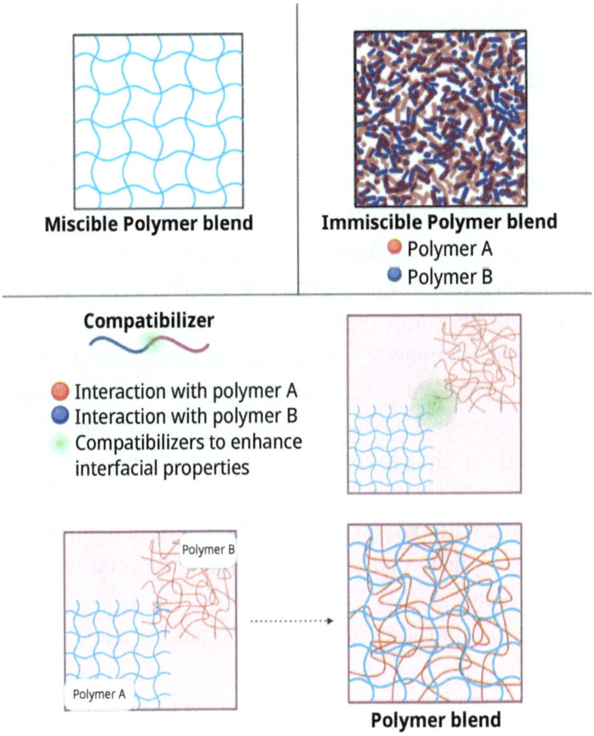

Figure 6.1: Types of polymer blends: miscible, immiscible, and compatibilized blends.

6.2.2 Compatibility in polymer blends

Polymer phase compatibility determines heterogeneous polymer blend properties [32]. Miscibility occurs at very low concentrations due to interfacial tension in polymer systems. Polymer mixes are miscible due to strong phase interactions. When interfacial stresses grow, particle size increases due to phase separation and coalescence of phase-separated particles, which in turn reduces mechanical properties [33, 34]. Compatibilizers are molecules that align at the interfaces of two polymer phases. They have hydrophilic and hydrophobic regions that reduce interfacial tension and increase compatibility between polymer mixes. Particles that are compatible with one another exhibit improved mechanical properties, greater phase stability, and smaller particle sizes that scatter less [7]. Mixtures can be classified as miscible, compatibilized, or uncompatibilized using thermogravimetric, dynamic mechanical thermal, or universal testing equipment.

6.2.3 Key properties influenced by structure

Polymer blends display varied properties determined by their structure and the interactions between components. Comprehending the fundamental concepts and characteristics of polymer blend systems is crucial due to their potential to achieve superior properties, including high elasticity, exceptional toughness, lightweight composition, transparency, cost-effectiveness, and ease of processing, as well as their capacity to meet customer demands. The extent of miscibility and structure is influenced by interfacial tension, which is closely related to mechanical characteristics and polymer–polymer interactions. In poly (lactic acid) blends with different thermoplastics, particle size and distribution markedly influence tensile strength and brittleness, with enhanced component interactions typically resulting in superior mechanical properties [35, 36]. The characteristics and attributes of polymer blends can be enhanced by employing interfacial agents, which impact adhesion between polymers and can change crystallization, melting behavior, and electrical properties.

The following properties associated with polymer blending are outlined below:

1. Supply materials possessing the requisite characteristics at the minimal feasible cost, accessible to the general populace.
2. It can enhance the performance of tailored resin, which is crucial for large-scale applications.
3. Based on its applications, one can readily construct polymer blends with distinctive qualities, such as enhancing specific characteristics by augmenting impact strength or solvent resistance.
4. Polymer blends are appropriate for recycling and biodegradation.
5. Blending technology has significant prospects for the rescue and recycling of polymer waste, which is unfeasible with pure components.
6. Polymer blending significantly enhances processability and product uniformity, and minimizes scrap of the primary desired polymer.
7. Polymer mixing is an effective technique for rapid formulation modifications, such as in yarning applications.
8. In light of environmental concerns, the polymer blends also extend the lifespan of the parent polymer.

6.3 Deep learning in polymer research

6.3.1 Data preparation and representation for DL models

DL potential to speed up the creation and optimization of polymer materials has been a boon to polymer research. The introduction of ML into research processes relies heavily on data representation. However, several data models are very rigid, making it difficult to incorporate experimental and data types commonly seen in polymer science. Polymer property prediction has been a recent focus of DL and data preparation research. Some of the strategies proposed to overcome the difficulties in creating effective DL models include mapping discrete polymer sequences to continuous latent spaces and expanding the network topology linearly layer-by-layer. A weighted sum approach that improves on the representation of single-component polymers is the best representation method for polymer blends [37] and copolymers [38], while the chemical space of polymer blends remains largely unexamined. With the use of experimental literature data and an ML algorithm, Liang et al. developed a database that records the compatibility of polymer mixes. Using compositional data and repeating units, this study is unique among polymer mix informatics.

This result is encouraging; however, further examination of physical properties in polymer blend informatics is necessary. The absence of labeled polymer blend data is an issue. Active learning (AL) and MD simulations could be advantageous. The types of input data or targeted polymer attributes can influence polymer representations. The group contribution approach is briefly reviewed before delving into modern and efficient techniques for polymer representation. The method of group contributions established by Van Krevelan and his team predicts polymer characteristics by calculating a weighted total of fragment contributions, known as a quality structure–property relationship (QSPR) fingerprint [39]. Whether or not ML is used, this method has been used to predict intrinsic viscosity, diffusion in gases and liquids, electrical conductivity, TC, refractive index, glass transition temperature, and dielectric constant [40, 41]. To effectively represent the chemical information of molecules in a format readable by machines, line notations employ text strings. Some examples are the IUPAC International Chemical Identifier, SYBYL Line Notation (SLN), Wiswesser Line Notation (WLAN), and SMILES. Because it can be used by both individuals and robots, SMILES has emerged as a favored technique for representing molecules [42]. SMILES representations of molecules can act as input for generative neural networks in fingerprinting and molecular creation, or as a linguistic structure for text-driven ML methods [43]. The field of data-driven investigations in chemistry and bioinformatics has been transformed by the introduction of SMILES and comparable line notations for molecules. Altered SMILES representations for polymers utilize particular symbols to indicate the endpoints or junctions of repetitive units. Polyethylene is represented as [*] CC [*], where CC indicates the repeat unit and * denotes the joining points between repeat units [15, 44]. This representation may include information about polymer branching, networks, and terminal groups. These studies illustrate the importance of data preparation techniques such as feature extraction.

6.3.2 Summary of deep learning methods: convolutional neural networks, recurrent neural networks, and graph neural networks

Data analysis and prediction are done in DL using multilayered neural networks. DL excels at unstructured data processing. ML methods are being surpassed by DL. It lets computer models learn data properties at different levels [45]. Neural network-based DL techniques gained popularity with the development of high-performance computers. When processing unstructured data, DL gains strength and flexibility by analyzing many features. DL systems process data in layers, extracting and communicating information. Initial layers extract key features, whereas later layers integrate them to create a complete picture [46]. DL can be done with convolutional, recurrent, and graph neural networks.

The preferred method for computer vision problems is CNNs. CNNs are great at illustrating spatial patterns, which helps in extracting various vegetation properties from remote sensing images. It categorizes the image's elements by giving them relative importance and removing biases [47]. It outperforms competing classification approaches in terms of minimal preprocessing. CNN finds the spatial and temporal links in a picture by applying appropriate filters. Some examples of CNN architectures are LeNet, ZFNet, ResNet, VGGNet, GoogleNet, and AlexNet [48, 49]. Semantic segmentation, object detection, and captioning are the three main applications for CNNs. CNNs are made to find spatial elements that best represent the goal class or quantity, such as edges, corners, textures, and abstract forms [50].

RNNs are a class of DL models designed to handle sequential data. RNNs, namely long short-term memory (LSTM) and Gated RNNs, are preferred for sequence processing and natural language tasks [51]. ML has significantly advanced thanks to RNNs, which enable the effective analysis of sequential input [52]. RNNs have unique neural network designs. RNNs handle learning challenges where past information impacts future predictions. Sequential occurrences occur in language modeling when prior words anticipate the next. Because its internal state retains previous inputs, RNNs can manage sequences of inputs, unlike feedforward neural networks [53]. Their sensitivity to context and data sequence is useful for time series forecasting, speech recognition, and natural language processing (NLP). RNNs and other neural network designs have been used to create hybrid models that combine their strengths [47]. By combining CNNs and RNNs, which handle spatial information and capture temporal changes, video analysis works well. In this way, the model can look at both spatial and temporal data [54], which makes it better at finding patterns and making predictions. Additionally, adding attention features to RNNs has made them better at simulating long-distance relationships.

GNNs allow nodes to communicate to show graph links. In recent years, GNNs including GCNs, GATs, and GRNs have excelled in many DL issues [55]. GNNs are graph-domain DL methods. GNN is now commonly utilized for graph analysis due to its effectiveness. GNN variants that aggregate graph structure data using CNNs and graph embedding are given. Thus, they can simulate constituents and their relationships as input and output [56].

6.4 Applications of deep learning in polymer blends

6.4.1 Property prediction: mechanical, thermal, optical, electrical, tensile, thermodynamic parameters, molecular weight, and cross-linking

DL has advanced polymer research, especially in predicting polymer blend mechanical properties. Physical blending controls and improves polymer mix mechanical characteristics. Many experiments are needed to predict mechanical properties from the structure. A DL network (DLN) predicted polymer blend yield strength utilizing structural parameters like blending content, molecular weight, melting point, and polymer density [57]. It has been demonstrated that DNNs can effectively use structural information to forecast the mechanical properties of polymer blends, albeit accuracy could yet be increased [58]. Blend composition, weight formula, density, yield strength, tensile strength at break, melting point, and impact resistance are all provided by the polymer database. The maximum stress a material can withstand before plasticizing is known as its yield strength. By analyzing high-dimensional datasets from molecular simulations and stress–strain curves, CNNs and RNNs can predict mechanical qualities including tensile strength, elasticity, toughness, and hardness [59]. These methods help researchers optimize material performance by linking molecular-level characteristics with macroscopic mechanical behaviors. Table 6.2 lists predictive features.

Polymers are used in thermal applications due to their affordability, electrical resistivity, light weight, mechanical flexibility, and corrosion resistance. While retaining polymer materials' cost-effectiveness and lightweight, increasing the polymer matrix's inherent TC improves the composite's thermal conductivity. Thermal characteristics like T_g, melting temperature (T_m), and thermal conductivity affect polymer blend stability under various situations. DL allows materials science to grow faster by collecting large amounts of data. DL predicted single-chain polymer thermal conductivity by correlating polymer repeating unit molecular architectures with TC [2]. Polymers were characterized as basic, easily accessible numerical representations in predictive models, facilitating on-demand property forecasting.

The TC of the majority of polymers ranges from 0.1 to 0.5 W/(mK), which is inadequate for many high-heat-conduction applications [60, 61]. Low-dimensional polymer chains have increased TC according to MD simulations and empirical findings. Zhang et al. [62] discovered that rigid backbone polymers are thermally stable, while heavier atoms and weaker backbone bonds decrease TC. Yang et al. [63] estimated the graphene–hexagonal boron nitride interfacial thermal resistance using supervised ML and ANN models. It was discovered that ML techniques increase prediction accuracy. DL techniques linked fingerprints to polymer TC. Research shows that ML can speed polymer blend discovery and design with improved thermal and mechanical properties.

The interaction of light with matter is characterized by optical characteristics. They reveal the exposed material's structure, relationships, and uses, making them vital. Researchers start by measuring incoming light reflection, transmission, and absorption [64]. Optics and photonics necessitate refractive index, transparency, and birefringence. Forecasting these features frequently necessitates comprehension of the molecular configuration of polymer blends. Variational autoencoders (VAEs) and generative adversarial networks (GANs) are examples of deep generative models (GMs) that enhance mixture formulations to satisfy specified optical attributes and replicate distributions of optical features. New organic molecules and materials are found using deep GMs, including autoregressive models, VAEs, and generative adversarial networks. Assessing the effectiveness and accuracy of these deep GMs is crucial. Zhavoronkov and associates have introduced Molecular Sets (MOSES) to evaluate drug discovery ML methodologies [65–67]. Molecular generation research can be standardized with this platform, and new ML models can be shared and compared. Additional efforts are required to develop and sustain benchmark platforms for organic molecules and polymers [68]. Visual attribute refractive index: assesses the bending of light as it passes through a substance, grasping optical transparency and efficacy in varied applications, which requires the refractive index [69].

Electronics, including sensors, actuators, and solar systems, use polymer blends with specific electrical properties. DL can explain polymer structure, mix morphology, and electrical properties, including conductivity and dielectric constant. It is possible to use GNNs to represent the intricate chemical interactions that influence the electrical characteristics of molecular structures as graphs [70]. Dielectric constant (AC): denotes the material's ability to store electrical energy in an alternating current (AC) environment. The capacitance of electronic components is affected by the dielectric constant [71]. The ability of a material to convert electrical energy into heat energy is evaluated by the dielectric loss factor. In low-energy-dispersal settings, this quality is crucial. When the dielectric loss factor is connected to the dielectric constant, the resulting tangent shows the material's efficiency in managing electrical energy. Electric conductivity: a material's capacity to conduct electricity. Many electronic and electrical applications require this characteristic. In applications that need surface conductivity, surface resistivity – a material's electrical resistance affects its performance [72]. Volume resistivity: measures the electrical resistance throughout a material's volume, providing understanding of its overall opposition to the flow of electric current. The flow chart based on the application is discussed in Figure 6.2.

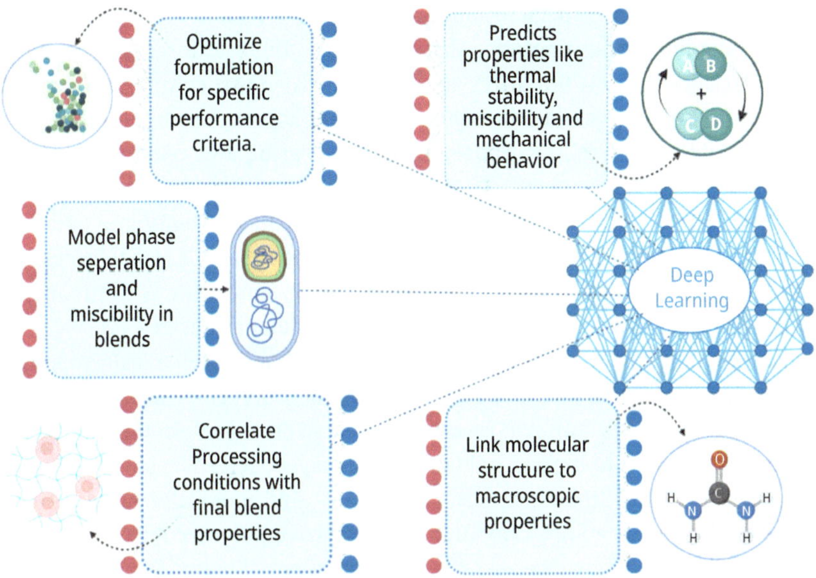

Figure 6.2: Applications of deep learning in polymer blends, with key areas such as property prediction, blend optimization, phase behavior analysis, processing conditions, and structure–property relationships.

Tensile properties are essential for comprehending the mechanical strength and behavior of polymer blends, providing insights into their strength, flexibility, and durability. The qualities include yield strength and elongation at break, which evaluate a material's ability to deform and endure stress before failure. They also include attributes such as tensile modulus, indicating a material's resistance to deformation, and tensile strength at break, representing the maximum stress a material can withstand before failure. The tensile properties inherent to fibers, including tensile elongation and modulus, offer additional understanding of the flexibility and stretch of fiber materials under stress. Furthermore, properties such as yield elongation and yield tensile strength elucidate a material's behavior immediately before the onset of plastic deformation. The dynamic mechanical properties, encompassing storage modulus, loss modulus, and loss tangent, examine material responses to dynamic forces, addressing energy storage, heat dissipation, and overall viscoelastic behavior [70]. The creep properties of a material indicate its performance under prolonged stresses and the stability of its structure over time. The characteristics of dilute solutions and their rheological qualities emphasize the behavior of polymers in solution and during flow, facilitating the processing and use of polymers in contexts that need precision and efficiency [73].

Thermodynamic characteristics, such as solubility parameters, phase separation, and miscibility, govern the stability and compatibility of polymer blends. DL techniques, along with MD simulations, have successfully forecasted phase behavior and miscibility ranges. These projections are crucial for developing polymer blends with enhanced homogeneity and stability under operational settings. The ability to forecast thermodynamic characteristics of new mixtures and conduct high-throughput evaluations through in silico techniques would represent a significant advancement in the quest for optimal, highly stable mixtures in various domains, such as organic photovoltaics, materials science, and pharmaceuticals. To forecast miscibility through computational techniques, two main strategies are employed: the first includes identifying and contrasting the solubility parameters of each phase in the blend (known as Hildebrand and Hansen solubility parameters), whereas the second entails simulating both the mixed phase and the individual pure phases to assess the interaction parameter [74].

Molecular weight distribution greatly impacts physical and mechanical properties of polymer blends. Molecular weight considerably impacts polyethylene's mechanical properties. Polyethylene with 1 million repeats wears best. Strong intermolecular forces and tensile strength reduce branching in high-density polyethylene. Mechanical properties are influenced by crosslinking agents, crystallinity, and molecular weight. High-performance applications require crosslinking to improve polymer mix mechanical and thermal stability. DL models based on experimental crosslinking data can predict appropriate crosslink densities [75, 76]. Lower crystallization temperature polymers have many amorphous areas that reduce mechanical strength. The crystallinity and mechanical characteristics of polyethylene are improved by adding nanofillers as nucleation agents. Degree of polymerization, molecular weight, and crosslinking agents affect polymer or blend mechanical characteristics. Polymers have different effects. As stated, PCL's mechanical characteristics improve with molecular weight above 40,000 g/mol. Poly (ethylene glycol) diacrylate's mechanical qualities improve with molecular weight, but cell growth suffers [81]. The polymer's crystallinity also influences its mechanical properties [77].

Table 6.2: Applications of deep learning in predicting polymer blend properties.

Property	Description	DL techniques used	References
Mechanical	Predicts yield strength, tensile modulus, and toughness; facilitates the optimization of mix composition and molecular interactions.	DNN, CNNs, RNNs	[57–59]
Thermal	Predicts thermal conductivity, glass transition temperature (T_g), and thermal stability; establishes relationships between structure and properties.	Supervised ML, ANN, molecular simulations	[60–63]
Optical	Assesses refractive index, transparency, and birefringence; enhances molecular configuration for optical applications.	VAEs, GANs, deep generative models	[64–67]

Table 6.2 (continued)

Property	Description	DL techniques used	References
Electrical	Predicts conductivity, dielectric constant, loss factor, and resistivity; improves the design of electrical and energy devices.	Graph neural networks (GNNs), DL algorithms	[68–71]
Tensile	Evaluates elongation, modulus, tensile strength, and dynamic mechanical parameters for the optimization of structural performance.	DL models analyzing stress–strain curves	[69–72]
Thermodynamics	Establishes miscibility, solubility, and phase behavior for stable and homogeneous mixtures across diverse domains.	Molecular dynamics, high-throughput screening	[73]
Molecular weight and cross-linking	Affects mechanical and physical properties; models forecast correlations among weight, crystallinity, and crosslinking.	DL trained on experimental data, cross-link modeling	[74–76]

6.4.2 Blend design: generative models and optimization

DL and ML are used in recognizing images and managing intricate nonlinear relationships. When ML is used for inverse design, the model "produces" a structure with the necessary properties when it "acquires" the "representation" of material configurations from data. This is achieved through inverse design employing deep GMs. Deep GMs parameterize the input data as accurately as possible to estimate the probability distribution of the data [78].

When training data is scarce, ML requires generating new data from an existing dataset. A GM captures the true data distribution from the training set and generates new data points with specified variances. The VAE and GAN families of GM algorithms require special attention. Through its encoder, VAE reduces high-dimensional input to a latent space. Regularizing the encoding distribution during training ensures that the latent space has adequate features, allowing the decoder to output similar new data. Encoders create "new features" from "old features," whereas decoders recover data by doing the opposite. A generator and a discriminator compete in the GAN model. Generators create synthetic data, while discriminators verify it. When fully optimized, GANs generate data that is so similar to the original that it is indistinguishable by the discriminator. In numerous polymer and materials informatics applications, VAEs and GANs work well [79].

Optimization is a critical application of ML. The procedure involves the repetitive training of multiple models, which is computationally expensive and may become unmanageable for complex simulations. Optimization approaches can be significantly improved in effectiveness and efficiency by using ML algorithms. Until a satisfactory outcome is obtained, optimization iteratively assesses various models for viable solutions. An optimization issue has three parts: variables, which the algorithm can change; constraints, which limit these parameters; and the objective function, which is the method's purpose. Deterministic optimization techniques ignore uncertainties in variable space and utilize explicit instructions to find solutions, while stochastic algorithms use probability distributions to account for uncertainties [80, 81]. The uncertainties inherent in the problem are explicitly defined and addressed by a set-based deterministic representation of these uncertainties using a novel approach, called robust optimization.

Optimization is done via stochastic algorithms with random objective functions and limitations. Assessing numerous plausible hypothesis functions and establishing their cost functions (squared error function) by identifying design features and restrictions yields an optimal design. The classification of optimization methods is illustrated below in Figure 6.3. Optimization approaches can be markedly enhanced in efficacy and efficiency by the incorporation of ML algorithms [82].

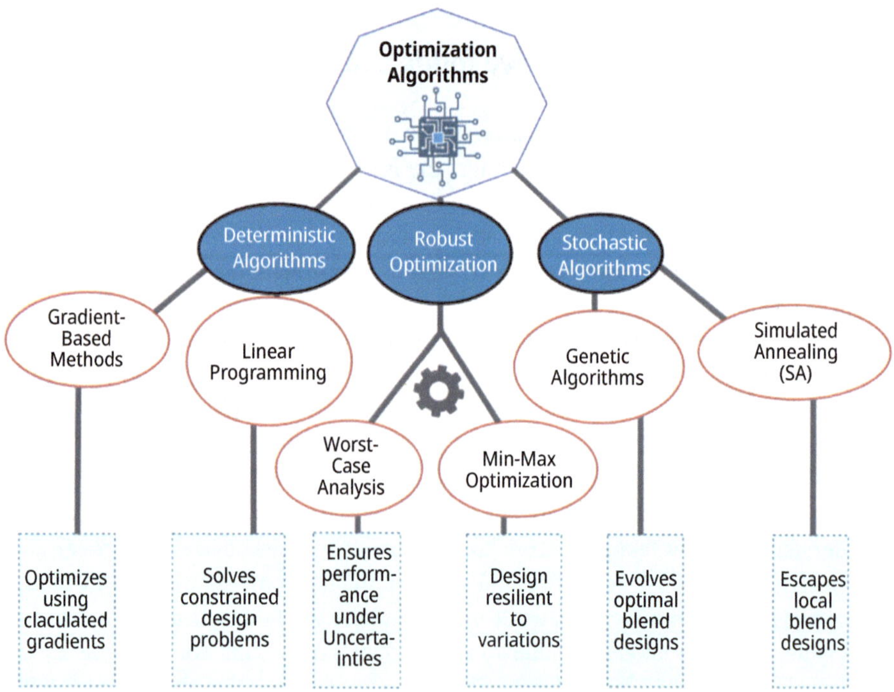

Figure 6.3: Classification of optimization algorithms for polymer blend design with deep learning.

6.4.3 Automation in polymer blend preparation

The widespread implementation of automation and ML for scientific discoveries has ushered in a new epoch of technological progress. The automation of polymer blend preparation improves efficiency, consistency, and precision in the production of polymer materials. In material science, ML algorithms necessitate access to dependable and high-quality training data to discern underlying patterns and accurately predict the desired structures and/or properties. Exclusively depending on academic literature for objective subjects, the algorithms have a bias favoring successful experiments, whereas ML algorithms necessitate a diverse array of both successful and unsuccessful data to develop a more robust problem-solving model. In this context, automation and high-throughput (HTP) experimentation can yield an extensive dataset by swiftly assessing numerous potential chemistries [83]. Consequently, the domain of polymer chemistry has increasingly employed machinery to automate polymerization processes in recent years [84]. The extensive chemical space of potential polymer structures and compositions can leverage the efficiency and precision of automation technologies, while generating high-quality experimental data for ML algorithms. The automated, self-optimizing flow system utilized by Junker's group to synthesize polymers with specific molecular weights and monomer conversions exemplifies the trajectory of future smart polymer synthesis platforms.Due to their extreme control over molecular weight and polymer structure, reversible- deactivation radical polymerizations (RDRP) like ATRP and RAFT are attractive for automation. Multiple monomers and designs can be used in drug delivery systems, surfactants, gas separation, data storage, and DNA replication. Researchers used RDRP to automate the synthesis of high-order multiblock copolymers, which are macromolecular chains with different monomers embedded as homopolymeric segments [85]. This polymer class offers versatility in blocks with diverse chemistries, resulting in unique properties not found in homopolymers or random copolymers. Bates et al. note that block copolymers have evolved over the past 60 years in synthesis and applications. Self-folding sequence-controlled multiblock copolymers [86] may serve as high-capacity data storage platforms or provide biological benefits in biomedical applications. It is impossible to manually investigate all possible structures with different monomers, block numbers, and block lengths [87]. Automated and HTP experiments can streamline the discovery of structure–property relationships in multiblock copolymers, resulting in increased output capacity. A small polymer library is synthesized using automation to study the impact of block numbers, lengths, and sequences on the properties of polymers evaluated by dynamic light scattering (DLS) [88]. Overall, automation in polymer blend preparation has enhanced precision, efficiency, and reproducibility in polymer synthesis, enabling the rapid development of novel materials with tailored properties.

6.4.4 Automation in properties determination

ML is used in polymeric nanocomposites to predict material properties, improve processes, investigate microstructures, and estimate uncertainty from complex production processes. Automation improves and maximizes research and experimentation efficiency. Equipment, software, and computer technologies improve workflow management, data collection, simulations, and decision-making. This strategy is useful in engineering, biology, and social sciences. Novel algorithms for cloud storage and open networking systems are prioritized. Benefits drive experiment automation. Researchers are free to focus on more complicated work that demands in-depth analysis of processes, operator attention, habits, and abilities thanks to the automation of repetitive jobs [89, 90]. This lessens manual effort and improves research accuracy and efficiency, accelerating results. In a dynamic control task, automation level (LOA) and adaptive automation (AA) affect human performance, situational responses, and workload. Fingerprints must include trap states (chemical structures, additives, polymer–nanofiller interfaces, etc.) and morphologies (lineage). Transfer learning, CNN, and inverse design can be utilized for analyzing and predicting structure–property relationships [91]. For polymer nanocomposites, hybrid ML can predict characteristics. Integrating various ML algorithms to improve predicted performance through mutual adjustment and adaptability to novel data is its foundation [92]. Hybrid ML, namely ensemble-based approaches, predicts composite mechanical reactions. Recently developed ML includes adaptive learning. While traditional ML employs training and prediction, adaptive learning incorporates reinforcement learning. Input–output inequalities are connected and learned from. Adaptive ML enhances operational feedback predictions [87].

6.4.5 Natural language processing and big data analytics for the extraction of data from literatures

Improving the discovery of materials necessitates significant automated data retrieval from materials science publications through the use of AI and NLP. The specific characteristics and formats of scientific articles render this approach challenging for lengthy text sections. Various frameworks use commercial (GPT–3.5) and open source (LlaMa 2) large language models (LLMs) and the named entity recognition (NER)–based MaterialsBERT model to automatically extract polymer-property data from entire scientific papers. Materials research uses NLP to independently acquire insights, qualities, and synthesis information from many textual sources and develop hypotheses and designs for novel materials [93, 94]. After receiving the corpus, complicated NLP operations such as text segmentation into tokens, named entity recognition (NER) of significant entities (such as materials, characterization methodologies, or attributes), dependency parsing, and information and organ extraction are conducted.

ML and AI enable DL models like RNNs and large short-term memory structures to detect named items [95, 96].

Transformer-based BERT architecture, which captures contextual and semantic links in scientific writings, has outperformed neural network models in recent years. Researchers created and published MaterialsBERT, a PubMedBERT [97] derived NER model. This model outperforms ChemBERT and MatBERT [98] on public datasets, notably materials science data extraction. The MaterialsBERT-based pipeline amassed approximately 300,000 polymer-property records from 130,000 abstracts, representing the largest project of its kind at that time, and rendered the data accessible [99]. The data extraction method shows that MaterialsBERT can handle several polymer-property abstractions. GPT, LlaMa, Pathways Language Model, and other LLMs are popular in NLP. These flexible models perform well in text categorization, NER, and extractive question answering with little datasets. LLMs succeed because they gain a lot of "knowledge" during semi-supervised pretraining (e.g., masked language modeling to predict the next token in context). Optimized OpenAI Codex and Code LlaMa generate code snippets from natural language. Polymer research uses NLP to retrieve book property data [100]. Inverse design may accurately anticipate synthesis recipes, characterization data, measurement settings, etc., helping researchers develop polymers and identify target material conditions. LLMs may forecast properties, design materials, and offer synthesis recipes in addition to property data extraction [101].

AI and ML are used in PI to develop, design, and discover polymers efficiently. To forecast polymer properties immediately, surrogate models are created utilizing existing polymer data to identify candidates that fulfil target property criteria. Using statistical approaches, synthesizability and prospective (retro)synthesis procedures for a target polymer are being studied. Data-driven approaches are being studied to address the unique chemical and physical variety of polymers at both local and large scales [102]. Scientific literature contains a lot of experimental data, which is manually extracted occasionally. NLP is a revolutionary technology for continuous and dynamic polymer data extraction, but it will take time to obtain polymer data from literature. DFT-based HTP computations are useful for polymer data [103, 104] and classical MD simulations [105]. Recently developed autonomous computational agents with ML modules and HTP computations can generate significant data.

6.5 Challenges and opportunities

6.5.1 Overcoming data limitations with augmentation and transfer learning

Small sample sizes make DL model improvement difficult. Data augmentation and transfer learning overcome DL data shortages. Geometric adjustments, neural style transfer, and GANs improve small datasets. GANs generate fake data with a distribution similar to real data. Transfer learning solves the short training data problem in the second strategy [106, 107]. DL requires lots of data. Short datasets make it difficult to improve these models' learning ability from short samples. The performance of models is hindered by restricted labeled datasets, privacy issues, inadequate generalization, and high annotation expenses. This work thoroughly investigates data augmentation and transfer learning, providing solutions for data scarcity. Data augmentation is one of the most effective methods for enhancing tiny datasets. Incorporated are geometric modifications, kernel filter adjustments, neural style transfer integration, random erasing, GANs, feature space enhancements, and adversarial and meta-learning training methodologies. Transfer learning employs pretrained models to transfer knowledge or retrain on similar datasets. A detailed study shows how integrating these two approaches might improve classification jobs, increasing small datasets. The rise in data availability solves pressing problems created by restricted datasets and makes full use of Big Data in a new age of DL applications [108].

6.5.2 Interpreting and validating DL models in polymer science

Polymer science is evolving because DL and ML allow researchers to forecast polymer features, adjust production settings, and generate new materials. AI algorithms for polymer property prediction emphasize interpretation and transfer learning. Shapley value analysis was used to evaluate each chemical group's prediction contribution to improve model interpretability. AI models represented polymer structures with tokenized SMILES strings. After analyzing these representations, CNNs predicted T_g. Shapley value analysis was used to evaluate each chemical group's prediction contribution to improve model interpretability. Transfer learning is predictive, but direct modeling with polymer-specific data is better, especially for comprehending T_g's complicated interactions. Enhanced interpretability allows the extraction of key scientific advances and improves model reliability, linking theoretical knowledge with real-world facts [109]. These strategies aim to enhance prediction accuracy, simplify models, and speed up polymer discovery. Many individuals discuss about the importance of training data, encoding representations quickly, and using models to understand polymer properties and drive material design.

DL models must be evaluated to ensure dependability and relevance in real-world applications. Cross-validation is a fundamental method that divides a dataset into segments, trains the model on some, and tests it on others to determine the model's generalizability. Researchers can compare DL models to standard prediction approaches to examine accuracy and efficiency improvements using performance metrics like mean absolute error (MAE) or R-squared values [110, 111]. To determine model stability, reliability testing is crucial. By changing aspects like data input or beginning parameters, the model's performance is tested. This ensures accurate predictions in many scenarios [112].

6.5.3 Future prospects: automation and sustainable design

The advancement of polymeric materials has been a significant focus for many years. Experiments and molecular simulations are impractical for material development due to their time and labor demands. As AI and materials informatics advance, materials science uses ML techniques to accelerate development. Polymers have more complex structures. Complex data challenges standard ML. Deep, convolutional, generative adversarial, recurrent, and graph neural networks can learn complex data, making them useful for polymeric material creation. ML has spread across fields as AI has advanced. RF, SVM, and XGB fail difficult materials science tasks [113]. DL does this well. In high-dimensional data, DL can swiftly establish the relationship between ideal material qualities, complicated structure, and environmental conditions [20].

The development of polymer materials is anticipated to advance swiftly with this advanced technology. RNNs and GANs, because of their distinct learning capabilities, can identify structural features from existing data and create new data with comparable distributions, thus addressing the molecular property database issue. Accelerated DL could act as a research objective. RNNs and GANs have demonstrated effectiveness in the creation of molecular structures. A comprehensive assessment of the created compounds is essential, covering their synthetic accessibility score (SAS) and diversity. Reinforcement Learning (RL) has become significant in drone operation, game development, natural language understanding, and many other areas. The algorithm enhances rewards through exploration and choices, balancing the trade-offs between exploring and exploiting to achieve its goal. At present, RL is employed in the generation of molecules. The incorporation of the approach with RNNs enhances the reliability of the generated compounds. RL and RNNs build SMILES-based compounds, which may be more stable or improve prediction outcomes. It can study polymer chemical spatial structure like PolyRL [114] model [115]. Additionally, RL can optimize polymer shape deformation and minimize energy position control of elastomer actuators [114]. Although research on RL-based polymers is now limited, it is anticipated to expand in the future.

6.6 Conclusion

A novel and powerful method to expedite materials design in different industries is ML, especially DL. The expansion of materials databases with chemical structures and properties offers great possibilities for ML-guided organic molecule and polymer creation. DL solved structure–property connection prediction difficulties, revolutionizing polymer research. CNNs, RNNs, and GNNs let academics analyze large datasets, uncover trends, and create customized content. This skill is needed to handle nonlinear interactions and compositional alternatives of polymer blends.

The use of DL has improved property prediction, molecular design, and structural characterization. Data scarcity, model interpretability, and experimental validation remain issues. These issues can be solved with data augmentation, transfer learning, and high-throughput experimentation. Molecular representation techniques, which underpin DL modeling, are also stressed in the chapter. DL enables sustainable and automated polymer blend development while overcoming traditional limitations. Automated and AI-driven synthesis and characterization methods reduce time to market and ensure reproducibility and precision. DL frameworks with sustainable design principles enable the creation of eco-friendly industrial materials.

Reinforcement learning and GMs will likely transform molecular generation and material optimization. DL can revolutionize polymer research and drive industry innovation by bridging theoretical understanding and practical application, enabling the next generation of high-performance, sustainable polymer blends.

Key abbreviations

AI	Artificial intelligence
CNN	Convolutional neural networks
DL	Deep learning
GAN	Generative adversarial network
GNN	Graph neural networks
HTP	High-throughput
MD	Molecular dynamics
MGI	Materials Genome Initiative
RNN	Recurrent neural networks
SMILES	Simplified Molecular Input Line Entry Specification
VAEs	Variational autoencoders

List of abbreviations

AI	Artificial intelligence
AL	Active learning
ATRP	Atom transfer radical polymerization
CNN	Convolutional neural network
DLS	Dynamic light scattering
DL	Deep learning
GAN	Generative adversarial network
GAT	Graph attention network (a GNN variant)
GNN	Graph neural network
HTP	High-throughput (experimentation)
LLM	Large language model
LOA	Level of automation
LSTM	Long short-term memory (a type of RNN)
ML	Machine learning
MD	Molecular dynamics
NER	Named entity recognition
NLP	Natural language processing
PPA	Poly(propylene adipate) (as repeat unit in copolymers)
QSPR	Quantitative structure–property relationship fingerprint
RAFT	Reversible addition–fragmentation chain transfer
RDRP	Reversible-deactivation radical polymerization
RNN	Recurrent neural network
RL	Reinforcement learning
SMILES	Simplified Molecular–Input Line–Entry System
VAE	Variational autoencoder

References

[1] Li D, Liu J and Liu J. NNI-SMOTE-XGBoost: A novel small sample analysis method for properties prediction of polymer materials. Macromolecular Theory and Simulations. 2021; 30(5): 2100010.
[2] Xu J and Luo T. Unlocking enhanced thermal conductivity in polymer blends through active learning. Npj Computational Materials. 2024; 10(1): 74.
[3] Chen H, Ginzburg VV, Yang J, Yang Y, Liu W, Huang Y, et al. Thermal conductivity of polymer-based composites: Fundamentals and applications. Progress in Polymer Science. 2016; 59: 41–85.
[4] Wei X, Wang Z, Tian Z and Luo T. Thermal transport in polymers: A review. Journal of Heat Transfer. 2021; 143(7): 072101.
[5] Kim G-H, Lee D, Shanker A, Shao L, Kwon MS, Gidley D, et al. High thermal conductivity in amorphous polymer blends by engineered interchain interactions. Nature Materials. 2015; 14(3): 295–300.
[6] Basheer BV, George JJ, Siengchin S and Parameswaranpillai J. Polymer grafted carbon nanotubes – Synthesis, properties, and applications: A review. Nano–Structures & Nano–Objects. 2020; 22: 100429.
[7] Parameswaranpillai J, Thomas S and Grohens Y. (Eds.). Polymer Blends: State of the Art, New Challenges, and Opportunities. 2014; Wiley–VCH Verlag GmbH & Co. KGaA.

[8] Weber M. Polymer blends: Materials with versatile properties. Macromolecular Symposia. 2001; 163: 235–250.

[9] Utracki L. Introduction to polymer blends. Polymer Blends Handbook. 2003; 1.

[10] Kruger. Introduction to polymer blends kruger; 2022 [Available from: https://kruger.industries/polymer-blends/.

[11] MATERIALS A Polymer blends and alloys: Tailoring mechanical properties and sustainability 2024 [Available from: https://www.azom.com/article.aspx?ArticleID=23510.

[12] Sehgal B and Kunde GB. Structure property relationship of polymer blends: Effect of grafting on mechanical properties. Asian Journal of Research in Chemistry. 2010; 3: 637–639.

[13] Imagawa A and Tran-Cong Q. Structure–property relationship of polymer blends with co–continuous structures prepared by photo–crosslinking. Macromolecules. 1995; 28: 8388–8394.

[14] Lee KY and Goettler LA. Structure-property relationships in polymer blend nanocomposites. Polymer Engineering and Science. 2004; 44: 1103–1111.

[15] Kim C, Chandrasekaran A, Huan TD, Das D and Ramprasad R. Polymer genome: A data–powered polymer informatics platform for property predictions. The Journal of Physical Chemistry C. 2018; 122(31): 17575–17585.

[16] Li D, Ru Y, Chen Z, Dong C, Dong Y and Liu J. Accelerating the design and development of polymeric materials via deep learning: Current status and future challenges. APL Machine Learning. 2023; 1: 021501.

[17] Zhang Z and Friedrich K. Artificial neural networks applied to polymer composites: A review. Composites Science and Technology. 2003; 63(14): 2029–2044.

[18] Hasebe T. Knowledge–embedded message–passing neural networks: Improving molecular property prediction with human knowledge. ACS Omega. 2021; 6(42): 27955–27967.

[19] Pereira T, Abbasi M, Ribeiro B and Arrais JP. Diversity oriented deep reinforcement learning for targeted molecule generation. Journal of Cheminformatics. 2021; 13(1): 21.

[20] Goodfellow I. Deep Learning. 2016; United States: MIT press.

[21] Himanshu H, Chakraborty K and Patra TK. Developing efficient deep learning model for predicting copolymer properties. Physical Chemistry Chemical Physics: PCCP. 2023; 25: 25166–25176.

[22] Paul D. Control of Phase Structure in Polymer Blends. 1989; (pp. 1–18). Functional polymers: Springer.

[23] Wall FT. Principles of Polymer Chemistry. Flory PJ. 1953; Ithaca, New York: Cornell Univ. Press. 688 pp. Illus. $8.50. Science. 1954.

[24] Liang X, Miyafuji K, Hara K, Mogi K, Nishida R, Liu H, et al. Direct visualization of stretch–induced phase separation in methoxy silyl–terminated polypropylene oxide/epoxy resin–type polymer alloys via AFM nanomechanics: A toughening mechanism. Polymer. 2022; 259: 125343.

[25] He H-L and Liang F-X. Interfacial engineering of polymer blend with Janus particle as compatibilizer. Chinese Journal of Polymer Science. 2023; 41(4): 500–515.

[26] Ajitha A and Thomas S. Introduction: Polymer blends, thermodynamics, miscibility, phase separation, and compatibilization. In Compatibilization of Polymer Blends. 2020; 1: (pp. 1–29). United States: Elsevier.

[27] Ahmadi H, Van Heugten PM, Veber A, Puskar L, Anderson PD and Cardinaels R. Toughening immiscible polymer blends: The role of interface–crystallization–induced compatibilization explored through nanoscale visualization. ACS Applied Materials & Interfaces. 2024; 16(43): 59174–59187.

[28] Raj A, Yousfi M, Prashantha K and Morphologies SC. Compatibilization and properties of immiscible PLA–based blends with engineering polymers: An overview of recent works. Polymers. 2024; 16(13): 1776.

[29] Dorigato A and Fredi G. Effect of nanofillers addition on the compatibilization of polymer blends. Advanced Industrial and Engineering Polymer Research. 2024; 7(4): 405–427.

[30] Jain V, Kumar H and Moona G. A retrospective review of polymer blending for additive manufacturing polymer filament manufacturing: Concept, present and trends. Advances in Materials and Processing Technologies. 2024; 1: 1–18.

[31] Freire JJ and Vlahos C. Enhancing polymer blend compatibility with linear and complex star copolymer architectures: A Monte Carlo simulation study with the bond fluctuation model. Polymers. 2024; 16(12): 1626.
[32] Dorigato A. Recycling of polymer blends. Advanced Industrial and Engineering Polymer Research. 2021; 4(2): 53–69.
[33] Yong WF and Zhang H. Recent advances in polymer blend membranes for gas separation and pervaporation. Progress in Materials Science. 2021; 116: 100713.
[34] Deng Y, Sun H, Hu X, Wang K, Long C, Yu F, et al. Enhanced performance of Polybutadiene/Polyisoprene (PB/PI) blend with asymmetric PB-b-PI diblock copolymers. Macromolecules. 2023; 56(15): 5754–5764.
[35] Paul D and Barlow J. Polymer blends. Journal of Macromolecular Science – Reviews in Macromolecular Chemistry. 1980; 18(1): 109–168.
[36] Shamsuri AA, Abdan K and Md. Jamil SNA. Utilization of ionic liquids as compatibilizing agents for polymer blends–preparations and properties. Polymer-Plastics Technology and Materials. 2023; 62(8): 1008–1018.
[37] Ma R, Zhang H and Luo T. Exploring high thermal conductivity amorphous polymers using reinforcement learning. ACS Applied Materials & Interfaces. 2022; 14(13): 15587–15598.
[38] Kuenneth C, Schertzer W and Ramprasad R. Copolymer informatics with multitask deep neural networks. Macromolecules. 2021; 54(13): 5957–5961.
[39] Bicerano J. Prediction of Polymer Properties. 2002; United States: cRc Press.
[40] Hasnaoui H, Krea M and Roizard D. Neural networks for the prediction of polymer permeability to gases. Journal of Membrane Science. 2017; 541: 541–549.
[41] Huang C, Chen X, Xue Z and Wang T. Effect of structure: A new insight into nanoparticle assemblies from inanimate to animate. Science Advances. 2020; 6(20): eaba1321.
[42] Weininger D. SMILES, a chemical language and information system. 1. Introduction to methodology and encoding rules. Journal of Chemical Information and Computer Sciences. 1988; 28(1): 31–36.
[43] Rogers D and Hahn M. Extended-connectivity fingerprints. Journal of Chemical Information and Modeling. 2010; 50(5): 742–754.
[44] Doan Tran H, Kim C, Chen L, Chandrasekaran A, Batra R, Venkatram S, et al. Machine-learning predictions of polymer properties with Polymer Genome. Journal of Applied Physics. 2020; 128(17): 171104.
[45] Miikkulainen R, Liang J, Meyerson E, Rawal A, Fink D, Francon O, et al. Evolving deep neural networks. Artificial Intelligence in the Age of Neural Networks and Brain Computing. 2024; 1: (pp. 269–287). United States: Elsevier
[46] Wang H and Raj B. On the origin of deep learning. arXiv Preprint arXiv. 2017; 1: 170207800.
[47] Mathew A, Amudha P and Sivakumari S. Deep learning techniques: An overview. Advanced Machine Learning Technologies and Applications: Proceedings of AMLTA 2020. 2021; 1: 599–608.
[48] Le QV. A tutorial on deep learning part 2: Autoencoders, convolutional neural networks and recurrent neural networks. Google Brain. 2015; 20: 1–20.
[49] Yamashita R, Nishio M, Do RKG and Togashi K. Convolutional neural networks: An overview and application in radiology. Insights into Imaging. 2018; 9: 611–629.
[50] Kattenborn T, Leitloff J, Schiefer F and Hinz S. Review on Convolutional Neural Networks (CNN) in vegetation remote sensing. ISPRS Journal of Photogrammetry and Remote Sensing. 2021; 173: 24–49.
[51] Alhajeri MS, Ren YM, Ou F, Abdullah F and Christofides PD. Model predictive control of nonlinear processes using transfer learning-based recurrent neural networks. Chemical Engineering Research and Design. 2024; 205: 1–12.
[52] Shahinzadeh H, Mahmoudi A, Asilian A, Sadrarhami H, Hemmati M and Saberi Y. editors. Deep Learning: A Overview of Theory and Architectures. 2024 20th CSI International Symposium on Artificial Intelligence and Signal Processing (AISP). 2024; New York City, U.S: IEEE.

[53] Mienye ID, Swart TG and Obaido G. Recurrent neural networks: A comprehensive review of architectures, variants, and applications. Information. 2024; 15(9): 517.
[54] Yang J, Qu J, Mi Q and Li Q. A CNN–LSTM model for tailings dam risk prediction. IEEE Access. 2020; 8: 206491–206502.
[55] Zhou J, Cui G, Hu S, Zhang Z, Yang C, Liu Z, et al. Graph neural networks: A review of methods and applications. AI Open. 2020; 1: 57–81.
[56] Veličković P. Everything is connected: Graph neural networks. Current Opinion in Structural Biology. 2023; 79: 102538.
[57] Zeng J-B, Li Y-D, He Y-S, Li S-L and Wang Y-Z. Improving flexibility of poly (L–lactide) by blending with poly (L–lactic acid) based poly (ester–urethane): Morphology, mechanical properties, and crystallization behaviors. Industrial & Engineering Chemistry Research. 2011; 50(10): 6124–6131.
[58] Li R, Wu L and Li B-G. Poly (l–lactide) materials with balanced mechanical properties prepared by blending with PEG–mb–PPA multiblock copolymers. Industrial & Engineering Chemistry Research. 2017; 56(10): 2773–2782.
[59] Fatriansyah JF, Surip SN and Hartoyo F. Mechanical property prediction of poly (lactic acid) blends using deep neural network. 2022.
[60] Wang Z, Carter JA, Lagutchev A, Koh YK, Seong N-H, Cahill DG, et al. Ultrafast flash thermal conductance of molecular chains. Science. 2007; 317(5839): 787–790.
[61] Luo T, Esfarjani K, Shiomi J, Henry A and Chen G. Molecular dynamics simulation of thermal energy transport in polydimethylsiloxane. Journal of Applied Physics. 2011; 109(7): 074321.
[62] Zhang T, Wu X and Luo T. Polymer nanofibers with outstanding thermal conductivity and thermal stability: Fundamental linkage between molecular characteristics and macroscopic thermal properties. The Journal of Physical Chemistry C. 2014; 118(36): 21148–21159.
[63] Yang H, Zhang Z, Zhang J and Zeng XC. Machine learning and artificial neural network prediction of interfacial thermal resistance between graphene and hexagonal boron nitride. Nanoscale. 2018; 10(40): 19092–19099.
[64] Obaid ZS. Physical Properties of PMMA/ZnO Composite Material. Ph. D. dissertations. 2017; University Al–Narhrain Baghdad, Iraq.
[65] Sanchez-Lengeling B and Aspuru-Guzik A. Inverse molecular design using machine learning: Generative models for matter engineering. Science. 2018; 361(6400): 360–365.
[66] Xu Y, Lin K, Wang S, Wang L, Cai C, Song C, et al. Deep learning for molecular generation. Future Medicinal Chemistry. 2019; 11(6): 567–597.
[67] Zhavoronkov A. 2 generative drug. An introduction to generative drug discovery. 2025; 21.
[68] Polykovskiy D, Zhebrak A, Sanchez-Lengeling B, Golovanov S, Tatanov O, Belyaev S, et al. Molecular sets (MOSES): A benchmarking platform for molecular generation models. Frontiers in Pharmacology. 2020; 11: 565644.
[69] Liu J and Ueda M. High refractive index polymers: Fundamental research and practical applications. Journal of Materials Chemistry. 2009; 19(47): 8907–8919.
[70] Malashin IP, Tynchenko VS, Nelyub VA, Borodulin AS and Gantimurov AP. Estimation and prediction of the polymers' physical characteristics using the machine learning models. Polymers. 2023; 16(1): 115.
[71] Zhu L. Exploring strategies for high dielectric constant and low loss polymer dielectrics. The Journal of Physical Chemistry Letters. 2014; 5(21): 3677–3687.
[72] Lekpittaya P, Yanumet N, Grady BP and O'Rear EA. Resistivity of conductive polymer–coated fabric. Journal of Applied Polymer Science. 2004; 92(4): 2629–2636.
[73] Argon A and Cohen R. Toughenability of polymers. Polymer. 2003; 44(19): 6013–6032.
[74] Caddeo C, Ackermann J and Mattoni A. A theoretical perspective on the thermodynamic stability of polymer blends for solar cells: From experiments to predictive modeling. Solar RRL. 2022; 6(9): 2200172.

[75] Augustine R, Dan P, Schlachet I, Rouxel D, Menu P and Sosnik A. Chitosan ascorbate hydrogel improves water uptake capacity and cell adhesion of electrospun poly (epsilon–caprolactone) membranes. International Journal of Pharmaceutics. 2019; 559: 420–426.
[76] Eschbach FO and Huang SJ. Hydrophilic–hydrophobic binary systems of poly (2–hydroxyethyl methacrylate) and polycaprolactone. Part I: Synthesis and characterization. Journal of Bioactive and Compatible Polymers. 1994; 9(1): 29–54.
[77] Mairpady A, Mourad A-HI and Mozumder MS. Accelerated discovery of the polymer blends for cartilage repair through data–mining tools and machine–learning algorithm. Polymers. 2022; 14(9): 1802.
[78] Hiraide K, Hirayama K, Endo K and Muramatsu M. Application of deep learning to inverse design of phase separation structure in polymer alloy. Computational Materials Science. 2021; 190: 110278.
[79] Nguyen D, Tao L and Li Y. Integration of machine learning and coarse–grained molecular simulations for polymer materials: Physical understandings and molecular design. Frontiers in Chemistry. 2022; 9: 820417.
[80] Francisco M, Revollar S, Vega P and Lamanna R. A comparative study of deterministic and stochastic optimization methods for integrated design of processes. IFAC Proceedings Volumes. 2005; 38(1): 335–340.
[81] Sun S. A review of deterministic approximate inference techniques for Bayesian machine learning. Neural Computing and Applications. 2013; 23: 2039–2050.
[82] Kalita K, Mukhopadhyay T, Dey P and Haldar S. Genetic programming–assisted multi–scale optimization for multi–objective dynamic performance of laminated composites: The advantage of more elementary–level analyses. Neural Computing and Applications. 2020; 32(12): 7969–7993.
[83] Burger B, Maffettone PM, Gusev VV, Aitchison CM, Bai Y, Wang X, et al. A mobile robotic chemist. Nature. 2020; 583(7815): 237–241.
[84] Tamasi M, Kosuri S, DiStefano J, Chapman R and Gormley AJ. Automation of controlled/living radical polymerization. Advanced Intelligent Systems. 2020; 2(2): 1900126.
[85] Pan X, Lathwal S, Mack S, Yan J, Das SR and Matyjaszewski K. Automated synthesis of well-defined polymers and biohybrids by atom transfer radical polymerization using a DNA synthesizer. Angewandte Chemie International Edition. 2017; 56(10): 2740–2743.
[86] Bates CM and Bates FS. 50th anniversary perspective: Block polymers? Pure potential. Macromolecules. 2017; 50(1): 3–22.
[87] Champa-Bujaico E, García-Díaz P and Díez-Pascual AM. Machine learning for property prediction and optimization of polymeric nanocomposites: A state–of–the–art. International Journal of Molecular Sciences. 2022; 23(18): 10712.
[88] Jafari VF, Mossayebi Z, Allison-Logan S, Shabani S and Qiao GG. The power of automation in polymer chemistry: Precision synthesis of multiblock copolymers with block sequence control. Chemistry–A European Journal. 2023; 29(53): e202301767.
[89] Kehoe B, Patil S, Abbeel P and Goldberg K. A survey of research on cloud robotics and automation. IEEE Transactions on Automation Science and Engineering. 2015; 12(2): 398–409.
[90] Głowacka J, Derpeński Ł, Frydrych M, Sztorch B, Bartoszewicz B and Przekop RE. Robotization of three–point bending mechanical tests using PLA/TPU Blends as an example in the 0–100% range. Materials. 2023; 16: 6927.
[91] Zhao H, Wang Y, Lin A, Hu B, Yan R, McCusker J, et al. NanoMine schema: An extensible data representation for polymer nanocomposites. APL Materials. 2018; 6(11): 111108.
[92] Sun Y, Li G and Zhang J. Developing hybrid machine learning models for estimating the unconfined compressive strength of jet grouting composite: A comparative study. Applied Sciences. 2020; 10(5): 1612.

[93] Foppiano L, Castro PB, Ortiz Suarez P, Terashima K, Takano Y and Ishii M. Automatic extraction of materials and properties from superconductors scientific literature. Science and Technology of Advanced Materials: Methods. 2023; 3(1): 2153633.
[94] Kononova O, He T, Huo H, Trewartha A, Olivetti EA and Ceder G. Opportunities and challenges of text mining in materials research. Iscience. 2021; 24(3): 102155.
[95] Oka H, Yoshizawa A, Shindo H, Matsumoto Y and Ishii M. Machine extraction of polymer data from tables using XML versions of scientific articles. Science and Technology of Advanced Materials: Methods. 2021; 1(1): 12–23.
[96] Kononova O, Huo H, He T, Rong Z, Botari T, Sun W, et al. Text-mined dataset of inorganic materials synthesis recipes. Scientific Data. 2019; 6(1): 203.
[97] Gu Y, Tinn R, Cheng H, Lucas M, Usuyama N, Liu X, et al. Domain-specific language model pretraining for biomedical natural language processing. ACM Transactions on Computing for Healthcare (HEALTH). 2021; 3(1): 1–23.
[98] Trewartha A, Walker N, Huo H, Lee S, Cruse K, Dagdelen J, et al. Quantifying the advantage of domain-specific pre-training on named entity recognition tasks in materials science. Patterns. 2022; 3(4): 100488.
[99] Mahmood A, Sonakshi G and Shetty P. Polymer scholar: Polymer scholar; 2024 [Available from: https://polymerscholar.org.
[100] Olivetti EA, Cole JM, Kim E, Kononova O, Ceder G, Han TY-J, et al. Data-driven materials research enabled by natural language processing and information extraction. Applied Physics Reviews. 2020; 7(4): 041317.
[101] Gupta S, Mahmood A, Shetty P, Adeboye A and Ramprasad R. Data extraction from polymer literature using large language models. Communications Materials. 2024; 5(1): 269.
[102] Chen L, Pilania G, Batra R, Huan TD, Kim C, Kuenneth C, et al. Polymer informatics: Current status and critical next steps. Materials Science and Engineering: R: Reports. 2021; 144: 100595.
[103] Huan TD, Mannodi-Kanakkithodi A, Kim C, Sharma V, Pilania G and Ramprasad R. A polymer dataset for accelerated property prediction and design. Scientific Data. 2016; 3(1): 1–10.
[104] Chen L, Batra R, Ranganathan R, Sotzing G, Cao Y and Ramprasad R. Electronic structure of polymer dielectrics: The role of chemical and morphological complexity. Chemistry of Materials. 2018; 30(21): 7699–7706.
[105] An Y and Deshmukh SA. Machine learning approach for accurate backmapping of coarse-grained models to all-atom models. Chemical Communications. 2020; 56(65): 9312–9315.
[106] Fayaz S, Ahmad Shah SZ, Ud Din NM, Gul N and Assad A. Advancements in data augmentation and transfer learning: A comprehensive survey to address data scarcity challenges. Recent Advances in Computer Science and Communications. 2024; 17: 22.
[107] Al-Qerem A, Salem AA, Jebreen I, Nabot A and Samhan A. Comparison between transfer learning and data augmentation on medical images classification. 2021 22nd International Arab Conference on Information Technology (ACIT). 2021; 1: 1–7.
[108] Fayaz S, Ahmad Shah SZ, Ud Din NM, Gul N and Assad A. Advancements in data augmentation and transfer learning: A comprehensive survey to address data scarcity challenges. Recent Advances in Computer Science and Communications (Formerly: Recent Patents on Computer Science). 2024; 17(8): 14–35.
[109] Patra TK. Data-driven methods for accelerating polymer design. ACS Polymers Au. 2021; 2(1): 8–26.
[110] Sha W, Li Y, Tang S, Tian J, Zhao Y, Guo Y, et al. Machine learning in polymer informatics. InfoMat. 2021; 3(4): 353–361.
[111] Mohammadzadeh Koumleh S, Hassanpour H, Esmaeili M and Gholami A. Various deep learning techniques for the applications in polymer, polymer composite chemistry, structures and processing. Journal of Chemistry Letters. 2021; 2(4): 157–177.

[112] Choudhary K, DeCost B, Chen C, Jain A, Tavazza F, Cohn R, et al. Recent advances and applications of deep learning methods in materials science. Npj Computational Materials. 2022; 8(1): 59.
[113] Li D, Ru Y, Chen Z, Dong C, Dong Y and Liu J. Accelerating the design and development of polymeric materials via deep learning: Current status and future challenges. APL Machine Learning. 2023; 1(2): 021501.
[114] Ji Q, Chen M, Wang XV, Wang L and Feng L. Optimal shape morphing control of 4D printed shape memory polymer based on reinforcement learning. Robotics and Computer–Integrated Manufacturing. 2022; 73: 102209.
[115] Wiering MA and Van Otterlo M. Reinforcement learning. Adaptation, Learning, and Optimization. 2012; 12(3): 729.

Ruby Srivastava*

Chapter 7
Predicting polymer properties: machine learning insights from Polymer Genome

Abstract: Data-driven and machine learning (ML)-based approaches have recently been employed in materials science to enable rapid predictions of material properties. Polymers, known for their affordability and highly customizable properties, are widely utilized in various technological applications. However, their complex chemical composition and morphology present challenges in their development. The fusion of ML algorithms with vast data resources has opened new avenues for progress in polymer science and engineering. Polymer informatics has emerged as a field aimed at accelerating the performance and optimization of process for novel polymers through ML models trained on reliable data, as these materials exhibit unique physical behaviors across vast length-time scales, particularly in nonequilibrium glassy states, where experimental depictions and simulations are expensive and challenging. With the continuous growth of existing databases, the emergence of new data resources, and ongoing advancements in ML algorithms, the landscape of informatics in polymers is becoming increasingly systematic and broadly applicable. One notable development in this field is *Polymer Genome*, an informatics program, aimed to predict various critical desired polymer properties. This platform, trained on a large polymer dataset, employs surrogate ML models that correlate key polymer features with their properties, leveraging high-throughput density functional theory) alongside experimental data sources from literature and established databases. Despite its robustness and predictive power, these models still face limitations in designing materials that meet specific property or performance objectives.

Keywords: Polymers, machine learning, DFT, informatics, Polymer Genome

7.1 Introduction

In recent years, information-driven and machine learning (ML)-based approaches have been increasingly applied in materials composites to facilitate rapid predictions of physicochemical properties. Polymers, known for their versatility and diverse

Acknowledgment: The author is thankful to her coinvestigator Prof. Gopalan Rajaraman, Indian Institute of Technology, Mumbai, India, for the support.

***Corresponding author:** Ruby Srivastava, Chemistry department, Indian Institute of Technology Bombay, Mumbai, Maharastra, India. e-mail: amitruby1@gmail.com, orcid\org.0000-0002-2367-0176

https://doi.org/10.1515/9783111681870-007

chemical structures, play a crucial role in various applications [1]. ML-based surrogate models, trained on existing polymer-property datasets, provide instant property predictions for new polymers, minimizing the need for lengthy and resource-intensive experiments [2–13]. However, despite the accelerated prediction capabilities, a major challenge remains in compiling a large diversified set of candidate polymers for evaluation. The vast chemical space of polymers presents a complex search problem, making traditional trial-and-error or intuitive methods inefficient and prone to overlooking optimal solutions. Additionally, conventional experimental and computational methods for polymer discovery can be highly resource-intensive.

The increasing availability of computational and experimental data in materials science demanded the development of materials informatics [2–5, 14–16]. This emerging discipline aligns with the "fourth paradigm" of materials discovery, which integrates data-driven approaches alongside traditional experimentation, physics-based models, and computational simulations. Integrated computational materials engineering exemplifies this paradigm by linking multiple length scales to model materials predictively and shorten development times. The fourth paradigm leverages experimental and simulation data to speed up the discovery of new materials, including polymers. Initiatives such as the Materials Genome Initiative (USA), the NOMAD Laboratory in Europe, and Japan's Materials Research by Information Integration Initiative are driving a paradigm shift in functional material discovery by establishing essential infrastructures for materials informatics [17]. The fifth paradigm is recognized as the driving force behind knowledge integration across various fields, enabling intelligence-driven work within the computational community through the widespread presence of ML systems.

Polymers are the vital and complex classes of materials, with wide use from everyday plastic packaging to advanced technological components like high-energy-density capacitors, Li-ion battery electrolytes, polymer light-emitting diodes, and photovoltaic materials [18–25]. Recent advancements in rational polymer design, achieved through experiment-computation synergies, indicate promising opportunities for ML and informatics-based methods in polymer industries. The vast diversity of polymer structures – spanning atomic connectivity, chain packing, and morphology – contributes to their wide-ranging properties, making them ubiquitous in various industries [26]. Materials discovery often relies on either "chemical intuition" or high-throughput screening (HTS) methods. For example, evaluating the gas permeation properties of novel polymer membranes necessitates polymer synthesis, membrane fabrication, and permeation testing, all of which demand significant optimization efforts (see Figure 7.1).

This upfront investment obstructs high-throughput research aimed at discovering new polymer materials for gas separation. ML methods have demonstrated significant success in chemistry and materials discovery, offering an alternative for streamlining this process. To effectively implement ML in polymer science, high-quality datasets must be compiled from diverse sources, including literature, expert-annotated databases, handbooks, standards, and product datasheets. These datasets contain information from experimental measurements, theoretical calculations, physical modeling,

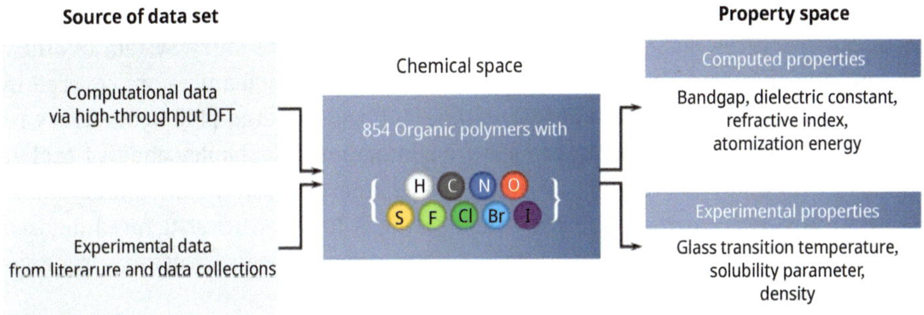

Figure 7.1: An overview of the polymer dataset utilized for developing property prediction models. The figure is adapted from Sha, W. et al. InfoMat.2021; 3:353–361. [80]

simulations, and expert-labeled properties. Considerable efforts have been made to collect and standardize dispersed data from multiple sources. Techniques such as entity recognition-based large language models [27] enable the extraction of polymer names, compositions, and properties [28] (see Figure 7.2).

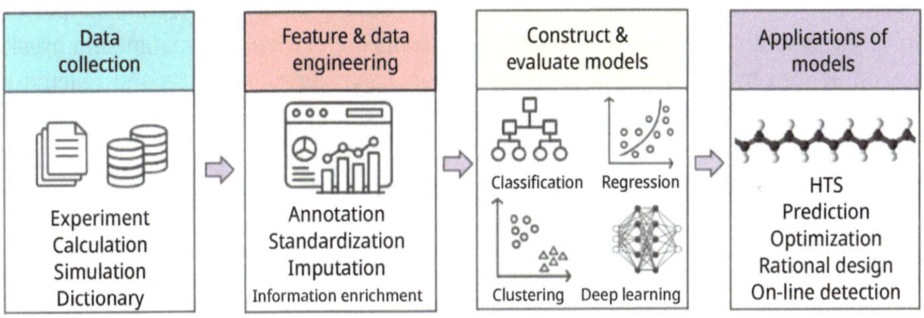

Figure 7.2: Figure 7.2: A schematic illustration of the four stages (data collection, features and data engineering, construction and evaluation of models, and applications of models) of applying an ML approach inpolymer science. This figure is adapted from Xie, C. SmartMat, 2025; 6:e1320. [81]

Large-scale density functional theory (DFT) computations and molecular dynamics simulations further enrich dataset quality [29]. Publicly accessible datasets, such as protein data bank [30], PubChem [31], ChemSpider [32], Reaxys [33] and QM9 [34], especially polymers, have PolyInfo [35], Polymer Genome [4], and polymer property database [36] providing unified representations of molecular compositions and their associated chemical and physical properties. Several benchmark datasets, such as PI1M, which was derived from approximately 12,000 polymers in PolyInfo and expanded to include ~1 million auto-generated polymers, facilitate ML model training [37]. Efforts like Poly Name2Structure (PN2S) enable the conversion of polymer names into structural representations [38], while datasets such as MoleculeNet support benchmarking of new ML approaches [39]. The intrinsic properties are computed, based on linear notation of chemical structures, such as Simplified Molecular Input Line Entry System (SMILES) [40] and International Chemical Identifier [41]. Various other features are covered by using tools like Dragon [42], RDKit [43], PaDEL [44], and Mordred [45]. Redundancy reduction techniques, such as recursive feature elimination and Shapley additive explanations (SHAP) analysis [46] help maintain model robustness and computational efficiency. Ultimately, the goal of feature engineering is to construct structured datasets that facilitate models development based on composition–process–structure–property–performance (CPSPPr) relationships.

Polymer science involves complex interactions, including bonded and nonbonded interactions such as bond lengths, bond angles, dihedrals, van der Waals forces, electrostatic interactions, and hydrogen bonding. These interactions contribute to the overall enthalpy of a system, while translational, rotational, and conformational entropy factors dictate thermodynamic equilibrium or steady-state conditions. Material properties manifest at three levels: ingredient, material, and device. Ingredient properties, such as glass transition temperature (T_g), melting temperature (T_m), boiling temperature (T_b), and degradation temperature (T_d), depend on phase, state, and external conditions like pH, ionic strength, strain rate, and electric fields. Material properties arise from synergistic interactions between polymer components, making them distinct from simple additive ingredient properties [47]. Chemical reactions involve bond formation and breaking, influenced by factors such as sequence, topology, stereochemistry, and reactivity. The dispersion of ingredients within a system determines whether it is homogeneous or heterogeneous, and computational models assist in understanding energy and mass transport, fluid dynamics, relaxation, and curing behaviors.

The complexity of polymer property prediction is attributed to three major challenges: (i) statistical distribution rather than uniform small molecules, (ii) synergistic contributions across monomers, segments, chains, aggregates, and phase domains, and (iii) the accumulation of small perturbations into large-scale responses, which influence the quantitative structure–property relationship (QSPR).

Performance, often confused with material properties, pertains to the functional capabilities of a polymer-based material or device. It encompasses multiple physicochemical, biological, thermal, mechanical, optical, electrical, acoustic, and magnetic properties, alongside relaxation and dissipation dynamics under external stress. Performance metrics on various aspects are determined based on standardized measurement protocols by established organizations such as International Organization for Standardization, American Society for Testing and Materials, DIN, and Great Britain GB [48–51]. Various sources of uncertainty, including compositional impurities, measurement errors, and polymer polydispersity, result in data distributions that need careful consideration in ML modeling. By integrating ML and informatics approaches, researchers can accelerate polymer discovery, optimize materials processing, and enhance the predictive accuracy of polymer property models, ultimately contributing to advancements in materials science and engineering.

7.2 Polymer Genome

Recently, Ramprasad et al. [3] developed Polymer Genome, employing ML techniques to model different properties of polymers based on polymer fingerprints. By simply inputting the polymer's chemical formula as a SMILES string, the program can predict properties such as glass transition temperature, dielectric properties, solvent interactivities, and refractive index, among others [2–13]. Several studies have applied ML to predict the gas permeability and glass transition temperature for the polymers [8, 10]. Insights and principles derived from data-driven and physics-based approaches align with the general CPSPPr framework, which spans the entire polymer lifecycle. There are various other databases (publicly accessible) databases, which are given in Table 7.1 with the name, description, and uniform resource locator (URL).

Among the four primary material classes (metals, inorganic non-metals, organic polymers, and composite materials), polymers are the most rapidly evolving, and shaping modern life. The Polymer Genome project predicts properties such as density, solubility, refractive index, glass transition temperature (T_g), band gap, and dielectric constant, integrating extensive DFT computations and molecular fingerprints for homo- and copolymers, as well as conjugated polymers with potential applications in energy technologies [11, 12]. The project predicts dozens of properties categorized into multiple groups, including electronic, dielectric, optical, mechanical, thermal, physical, thermodynamic, and solubility properties. The powerful capabilities of ML in image recognition for chemical structures and phase domains enhance property prediction and high-resolution segmentation, improving the speed and accuracy of characterization methods [52, 53]. The models can be accessed via the user-friendly online platform, Polymer Genome (www.polymergenome.org), which continues to expand systematically to enhance its applicability across various technological domains.

ML techniques are widely applied in the design and discovery of metal-organic frameworks (MOFs) [54], pharmaceuticals [55], and solid-state electrolytes for Li-ion batteries [56]. They also facilitate the production scale-up of polymer nanocomposite dielectrics for energy storage devices [57] and aid in the pyrolysis of biomass to extract bio-based compounds [58]. ML methods support tasks such as dimensionality reduction, pattern recognition, data parsing, structure prediction, interpretation, engineering design, QSPR modeling, and causative relationship analysis [59, 60]. Significant efforts have been directed toward force field-based calculations, including Chemistry at HARvard Molecular Mechanics (CHARMM) [61], Assisted Model Building with Energy Refinement (AMBER) [62], optimized potentials for liquid simulations [63], MARTINI [64], and reactive force field [65], spanning quantum mechanics to atomistic and coarse-grained modeling. These approaches require self-consistent solutions for topological structures and parameters, balancing computational cost and accuracy. Based on simulation trajectories, multiple properties and kinetic patterns can be effectively extracted using ML algorithms, such as correlating relaxation motions in glassy thermoset polymers with shear modulus [66], analyzing grain size distributions in 3D patterns [67], classifying

Table 7.1: Different types of polymer blends with compatibility (table adapted from Xie, C. SmartMat, 2025; 6:e1320).

Type of polymer blend	Description	Examples	Features	Miscibility/compatibility	References
Miscible blends	Homogeneous (single-phase) mixtures that are molecularly miscible	– Poly(styrene) (PS)–poly(phenylene oxide) (PPO) – Poly(styrene-acrylonitrile) (SAN)–poly(methyl methacrylate) (PMMA)	– Optically clear – Uniform at the polymer segmental level – Phase separation induced by temperature, pressure, or composition fluctuations	Governed by Gibbs free energy of mixing: $\Delta G_m = \Delta H_m - T\Delta S_m < 0$	[22–24, 27]
Immiscible blends	Heterogeneous blends where components are not molecularly miscible	– Poly(propylene) (PP)–poly(styrene) (PS) – Poly(propylene)–poly(ethylene) (PE)	– Typically exhibit phase separation – Poor interfacial adhesion	Lack of favorable thermodynamic interaction; typically requires compatibilization to enhance performance	[22–24, 27, 28]
Compatibilized blends	Immiscible blends made compatible by adding compatibilizers to enhance interfacial properties	– Nano-hybrids – Hybrid or multi-agent mixtures	– Improved interfacial tension – Stabilized morphology Enhanced adhesion between phases	Achieved using a third component (compatibilizer), reactive blending, or nano-hybrids	[24–26, 29]
Thermoplastic-thermoplastic	Blend of two thermoplastics	– Polypropylene (PP) blends – Polystyrene (PS) blends	– Enhanced toughness, stiffness, or clarity depending on the combination	Miscibility varies; compatibilization may be required for immiscible blends	[22]

Thermoplastic-rubber	Blend of a thermoplastic and a rubber polymer	– PP-EPDM (ethylene propylene diene monomer) blends	– Improved flexibility and impact resistance	Often immiscible; compatibilization needed	[22, 26]
Thermoplastic-thermosetting	Combination of a thermoplastic and a thermosetting polymer	– PET (polyethylene terephthalate)–epoxy resin blends	– Balances rigidity with toughness and thermal stability	Requires careful blending due to different processing conditions and incompatibility	[22]
Rubber-thermosetting	Blend of rubber and thermosetting polymer	– Rubber–epoxy resin blends	– Improved impact resistance and elasticity	Often requires compatibilizers for stable morphology	[22, 24]
Polymer-filler	Blend of polymers with inorganic or organic fillers	– Polypropylene–carbon black blends	– Enhanced strength, thermal stability, or conductivity depending on filler	Compatibility depends on the filler–polymer interaction and surface treatment	[22, 26, 30]
Miscibility factors	Miscibility determined by thermodynamic parameters, particularly Gibbs free energy (ΔG_m)	–	– $\Delta G_m < 0$ for miscibility – ΔS_m promotes mixing, while ΔH_m determines enthalpy of mixing	Specific interactions like hydrogen bonding, dipole–dipole interactions, and ionic interactions aid miscibility	[23, 24, 31]
Compatibilization challenges	Issues in creating universal compatibilizers and localization problems in binary/ternary blend systems	–	– Development of recyclable and reusable compatibilizers is needed – Nano-hybrids and multi-agent systems show promise but face technical challenges	Some systems remain unsuitable for current compatibilizers	[25, 26, 29]

global structures using local environments for copolymers, and detecting topological assembly defects in off-lattice simulations of rod-like molecules confined in a square box [68].

Certain properties, like atomization energy, depend mainly on atomic constituents and short-range bonding, while others, such as T_g, are strongly influenced by morphological features such as chain rigidity and branching. Polymer Genome is built using Python and standard web languages, including PHP and HTML. Users can input the repeat polymer units or its SMILES representation, or draw the chemical structure of the repeat unit of a polymer. Once submitted, the platform predicts polymer properties along with uncertainty estimates and presents the results in a structured table. Users receive polymer names with corresponding SMILES and repeat unit information, alongside a customizable selection of properties. Selecting any polymer from the list provides extensive details, including its name, class, 3D structural visualization with availability of its atomic coordinates, and ML-determined properties.

QSPR and HTS provide an energetic perspective to describe structural responses from molecular to macro scales. This approach correlates molecular design and structural characteristics with diverse optical, electronic, acoustic, and magnetic properties. Particle-based insights into transport, binding, adsorption, and diffusion properties complement wave-based or field models that assess polymer interactions through transmission, diffraction, scattering, and refraction. By connecting molecular fingerprints, chemical topology, and aggregated structural characteristics, ML can predict key thermal and mechanical properties such as T_g, specific heat, and thermal conductivity. These properties influence polymer design, preparation, operation, and recycling. Thermal analysis detects chemical and physical changes under temperature variations, measuring variations in size, dimensions, mass, heat, optical, electrical, acoustic, and magnetic properties. Mechanical properties, including tensile, compressive, bending, flexural, impact, fatigue, abrasion, rheology, and damping characteristics, involve bulk and surface interactions [69].

Intelligent materials, such as shape-memory, self-healing, and vitrimer polymers, exhibit stimuli-responsive behaviors governed by similar QSPR models. These materials integrate sensors and actuators at the device level to detect environmental changes, physical stress, temperature, pH, electric current, or light. ML approaches have surpassed traditional continuum mechanics in modeling deformation under applied loads for soft actuators [69]. ML techniques have also advanced controlled-release delivery systems by analyzing polymer interactions with cargo substances, cells, and tissues, considering dosage forms such as tablets, capsules, powders, injections, gels, granules, liposomes, and suspensions. These models evaluate preparation, encapsulation, release profiles, pharmacokinetics, and liberation, absorption, distribution, metabolism, excretion, and toxicity characteristics. Furthermore, ML has contributed to studying microplastic distributions and their environmental, ecological, and health impacts. Research has focused on the degradation and settlement of polycarbonates [70], biochar and resin-based organic pollutant remediation [71], waste slurry settling rates [72], flocculation dewatering polymers [73], and the role of nanoplastics in diatom growth [74].

7.3 Limitations

The primary challenges in applying ML to the discovery of polymers stem from both scientific and nonscientific factors. Polymer properties are significantly influenced by various constituents such as branching, molecular weight distribution, copolymerization, additives, and processing conditions. Additionally, inconsistencies in naming conventions make it difficult to establish a standardized polymer database that serves as the foundation for a polymer informatics framework. Although Polymer Genome currently reports Bayesian uncertainty for property predictions, further deeper examination of the uncertainties and possible errors related to the fundamental experimental measures remains crucial. One significant challenge lies in the scarcity of benchmark datasets needed for rigorous assessments, which can be attributed to several factors:

(i) Ambiguous definitions with high uncertainty extending beyond monomer representation
(ii) Complex multimodal and long-tailed distributions for molecular weight, size, and dynamic relaxations
(iii) Limited understanding of how minor stimuli can trigger substantial changes in polymer properties
(iv) The intricate interplay between enthalpy and entropy, along with the challenge of quantifying interactions and entropic contributions
(v) Synergistic responses occurring across different scales, from bulk materials to interfaces, surfaces, and mass transport within matrix-embedded phase domains

Future improvements in this area may come from advancements in dataset construction, standardized data-sharing protocols, and HTS platforms. Additionally, integrating experimental and simulation-based approaches, leveraging ML models to drive autonomous laboratories, and accelerating scientific data collection using natural language processing and transformer-based models for named entity recognition and relationship extraction could significantly enhance the field.

The continuous advancement of ML presents an exciting opportunity for automating database generation, potentially minimizing human involvement in the process. Notable achievements in this domain include IBM's Watson project and the automated extraction toolkits developed by M. C. Swain and J. M. Cole at Cambridge University, which facilitate the retrieval of thermodynamic properties such as melting temperature and nuclear magnetic resonance spectral measurements for small molecules [75]. To enhance accuracy, these automated systems can be complemented by human-curated reviews of low-certainty data, reducing – though not eliminating – the need for manual effort. Beyond streamlining polymer informatics, automated database creation could also ease the workload of researchers by improving data accessibility and validation for new models. Despite these advancements, polymer databases continue to face challenges that predate the advent of computing. These challenges include accurate material description, nuances in property definitions, the context of underlying measurements, and the completeness of reported information. The limitations of existing polymer data repositories arise partly from these difficulties and resolving them will necessitate both fundamental research and agreement within the polymer science community. Reducing challenges in database creation and fostering reproducibility, and promoting discussions are essential steps toward making polymer informatics a reality.

Although polymer properties are often considered well-defined, this is not always the case. Properties can generally be categorized into three types: fundamental, application-specific, and phenomenological. Fundamental properties require no additional contextual information. In contrast, application-specific properties depend on the measurement method, but a standardized framework for performing and interpreting these measurements exists [76–79]. Phenomenological properties, however, require extensive contextual details – including sample characteristics, measurement methods, and data analysis procedures. For such cases, providing raw data along with analysis workflows can enhance polymer informatics predictions and improve reproducibility. However, the more contextual information included, the more difficult it becomes for algorithms to automatically capture and process it. Additionally, challenges related to accuracy, precision, data reliability, and missing information further complicate database development.

To address this, several incentives can be considered:

1. **Mandates** by funding agencies requiring data sharing as part of research projects.
2. **Journal policies** enforcing data submission alongside published research.
3. **Direct benefits** to researchers, which could take two forms – voluntary contributions for the greater scientific community (similar to Wikipedia) or the provision of resources that make data sharing seamless.

One promising approach is the development of platforms almost identical to the Galaxy Project [80], where researchers can share analysis, activities, develop functions, and simplify data deposits. By both removing the burden of data submission and improving researcher efficiency, such platforms could accelerate the adoption of polymer informatics. Since implementing these strategies may take time, a dual approach – combining entry of voluntary data with data extraction automation – should be prioritized to build comprehensive polymer databases and support further advancements in the field.

7.4 Conclusion

The field of polymer science is at a pivotal moment, with groundbreaking opportunities emerging through advances in computer science. The vast repository of scientific data embedded in literature can now be unlocked through automated database creation, a process that can be significantly enhanced by systematic data contributions. Initiatives like the Materials Genome Initiative have propelled data-driven informatics approaches, enabling faster property prediction and materials design using surrogate models trained on reliable historical data. The integration of data-driven insights with physics-based understanding via ML is rapidly advancing, fostering deeper fundamental knowledge and facilitating the discovery of novel polymer materials [80, 81]. Strategies and algorithms for addressing diverse challenges in polymer science continue to evolve at a remarkable pace. The innovation in polymer materials will flourish alongside advancements in ML, driving more efficient material design and promoting sustainable applications. Polymers currently account for a substantial portion of global material investments, with their market expected to grow at twice the rate of the U.S. gross domestic product. To maximize the benefits of this growth, the polymer science community with informatics tools as Polymer Genome must take the lead in organizing, curating, and leveraging both experimental and computational data, ultimately shaping the future of the industry and contributing to the global economy.

Funding

DST/WIDUSHI-B/CS/2024/35.

List of abbreviations

ML	Machine learning
DFT	Density functional theory
SMILES	Simplified Molecular Input Line Entry System
CPSPPr	Composition–process–structure–property–performance
QSPR	Quantitative structure–property relationship
HTS	High-throughput screening

References

[1] Van Krevelen D. Properties of Polymers: Their Correlation with Chemical Structure. 1997 URL; Elsevier. https://books.google.com/books?id=RQp9vgEACAAJ.

[2] Mannodi-Kanakkithodi A, Chandrasekaran A, Kim C, Huan TD, Pilania G, Botu V and Ramprasad R. Scoping the polymer genome: A roadmap for rational polymer dielectrics design and beyond. Materials Today. 2018; 21(7): 785796. doi: https://doi.org/10.1016/j.mattod.2017.11.021.

[3] Kim C, Chandrasekaran A, Huan TD, Das D and Ramprasad R. Polymer genome: A data-powered polymer informatics platform for property predictions. Journal of Physical Chemistry C. 2018; 122(31): 1757517585. doi: doi.org/10.1021/acs.jpcc.8b02913.

[4] Mannodi-Kanakkithodi A, Treich G, Huan TD, Ma R, Te Eri M, Cao Y, Sotzing G and Ramprasad R. Rational co-design of polymer dielectrics for energy storage. Advances in Materials. 2016; 28: 62776291.

[5] Huan TD, Boggs S, Teyssedre G, Laurent C, Cakmak M, Kumar S and Ramprasad R. Advanced polymeric dielectrics for high energy density ap plications. Progress in Materials Science. 2016; 83: 236.

[6] Chen L, Kim C, Batra R, Lightstone JP, Wu C, Li Z, Desh Mukh A, Wang Y, Tran H, Vashishta P, Sotzing G, Cao Y and Ramprasad R. Frequency-dependent dielectric constant prediction of polymers using ma chine learning. Npj Computational Materials. 2020; 6: 61.

[7] Lightstone JP, Chen L, Kim C, Batra R and Ramprasad R. Refractive in dex prediction models for polymers using machine learning. Journal of Applied Physics. 2020; 127(21): 215105.

[8] Zhu G, Kim C, Chandrasekarn A, Everett JD, Ramprasad R and Lively RP. Polymer genome based prediction of gas permeabilities in polymers. Journal of Polymer Engineering. 2020; 40(6): 451–457. doi: 10.1515/polyeng-2019-0329.

[9] Chen L, Huan TD, Quintero YC and Ramprasad R. Charge injection barriers at metal/polyethylene interfaces. Journal of Materials Science. 2016; 51(1): 506–512. doi: 10.1007/s10853-015-9369-2.

[10] Kim C, Chandrasekaran A, Jha A and Ramprasad R. Active-learning and materials design: The example of high glass transition temperature polymers. MRS Communications. 2019; 9: 860866. doi: 10.1557/mrc.2019.78.

[11] Mannodi-Kanakkithodi A, Huan TD and Ramprasad R. Mining materials design rules from data: The example of polymer dielectrics. Chemistry of Materials. 2017; 29: 21):90019010. doi: 10.1021/acs.chemmater.7b02027.

[12] Venkatram S, Kim C, Chandrasekaran A and Ramprasad R. Critical assessment of the hildebrand and hansen solubility parameters for polymers. Journal of Chemical Information and Modeling. 2019; 59(10): 41884194. doi: 10.1021/acs.jcim.9b00656.

[13] Nasreen S, Treich GM, Baczkowski ML, Mannodi Kanakkithodi AK, Baldwin A, Scheirey SK, Cao Y, Ramprasad R and Sotzing GA. A material genome approach towards exploration of Zn and Cd coordination complex polyester as dielectrics: Design, synthesis and characterization. Polymer. 2018; 159: 95–105. doi: https://doi.org/10.1016/j.polymer.2018.10.017.

[14] Rajan K. Materials informatics. Mat Today: Proceedings. 2005; 8(10): 38–45. doi: https://doi.org/10.1016/S1369-7021(05)71123-8.

[15] Kim E, Huang K, Saunders A, McCallum A, Ceder G and Olivetti EC. Materials synthesis insights from scientific literature via text extraction and machine learning. Mater. 2017; 29: 9436–9444.

[16] Segler MHS, Preuss M and Waller MP. Planning chemical syntheses with deep neural networks and symbolic AI. Nature. 2018 Mar 28; 555(7698): 604–610. doi: 10.1038/nature25978.

[17] De Pablo JJ, Jackson NE, Webb MA, *et al.* New frontiers for the materials genome initiative. Npj Comput Mater. 2019; 5: 41. doi: https://doi.org/10.1038/s41524-019-0173-4.

[18] Huan TD, Batra R, Chapman J, *et al.* A universal strategy for the creation of machine learning-based atomistic force fields. Npj Comput Mater. 2017; 3: 37–54. doi: https://doi.org/10.1038/s41524-017-0042-y.

[19] Wessling M, Mulder MHV, Bos A, Van der Linden M, Bos M and Van der Linden WE. Modelling the permeability of polymers: A neural network approach. Journal of Membrane Science. 1994; 86: 193–198. doi: https://doi.org/10.1016/0376-7388(93)E0168-J.

[20] Park JY and Paul DR. Correlation and prediction of gas permeability in glassy polymer membrane materials via a modified free volume based group contribution method. Journal of Membrane Science. 1997; 125: 23–39. doi: 10.1016/S0376-7388(96)00061-0.

[21] Robeson LM, Smith CD and Langsam M. A group contribution approach to predict permeability and permselectivity of aromatic polymers. Journal of Membrane Science. 1997; 132: 33–54.

[22] Yampolskii Y, Shishatskii S, Alentiev A and Loza K. Group contribution method for transport property predictions of glassy polymers: Focus on polyimides and polynorbornenes. Journal of Membrane Science. 1998; 149: 203–220. doi: 10.1016/S0376-7388(98)00152-5.
[23] Hasnaoui H, Krea M and Roizard D. Neural networks for the prediction of polymer permeability to gases. Journal of Membrane Science and Technology. 2017; 541: 541–549. doi: 10.1016/j.memsci.2017.07.031.
[24] Ryzhikh V, Tsarev D, Alentiev A and Yampolskii Y. A novel method for predictions of the gas permeation parameters of polymers on the basis of their chemical structure. Journal of Membrane Science and Technology. 2015; 487: 189–198.
[25] Hou W, Xiao Y, Han G and Lin JY. The applications of polymers in solar cells: A review. Polymers (Basel). 2019; 11(1): 143. doi: 10.3390/polym11010143.
[26] Roman HE. Polymers in physics, chemistry and biology: Behavior of linear polymers in fractal structures. Polymers. 2024; 16: 3400. doi: https://doi.org/10.3390/polym16233400.
[27] Dagdelen J, Dunn A, Lee S, *et al*. Structured information extraction from scientific text with large language models. Nature Communications. 2024; 15: 1418. doi: https://doi.org/10.1038/s41467-024-45563-x.
[28] Gupta S, Mahmood A, Shetty P, *et al*. Data extraction from polymer literature using large language models. Commun Mater. 2024; 5: 269. doi: https://doi.org/10.1038/s43246-024-00708-9.
[29] Salo-Ahen OMH, Alanko I, Bhadane R, Bonvin AMJJ, Honorato RV, Hossain S, Juffer AH, Kabedev A, Lahtela-Kakkonen M, Larsen AS, et al. Molecular dynamics simulations in drug discovery and pharmaceutical development. Processes. 2021; 9: 71. https://doi.org/10.3390/pr9010071.
[30] https://www.rcsb.org/
[31] Kim S, Chen J, Cheng T, et al. PubChem 2025 update. Nucleic Acids Research. 2025; 53(D1); D1516–D1525. doi: 10.1093/nar/gkae1059.
[32] Pence HE and Williams A. ChemSpider: An online chemical information resource. Journal of Chemical Education. 2010; 87(11): 1123–1124. doi: 10.1021/ed100697w.
[33] https://www.reaxys.com/.
[34] Nandi S, Vegge T and Bhowmik A. MultiXC-QM9: Large dataset of molecular and reaction energies from multi-level quantum chemical methods. Scientific Data. 2023; 10: 783. doi: https://doi.org/10.1038/s41597-023-02690-2.
[35] Polymer Database (PoLyInfo). http://polymer.nims.go.jp/index_en.html (accessed Mar. 2017).
[36] Holding S. Polymers: A property database. Chroma. 2010; 72: 587. doi: https://doi.org/10.1365/s10337-010-1620-z.
[37] Ma R and Luo T. PI1M: A benchmark database for polymer informatics. Journal of Chemical Information & Modeling. 2020; 60(10): 4684–4690. doi: 10.1021/acs.jcim.0c00726.
[38] Lin C, Wang PH, Hsiao Y, Chan YT, Engler AC, Pitera JW, Sanders D, Cheng J and Tseng YJ. Essential step toward mining big polymer data: PolyName2Structure, mapping polymer names to structures. ACS Applied Polymer Materials. 2020; 2(8): 3107–3113. https://doi.org/10.1021/acsapm.0c00273.
[39] Wu Z, Ramsundar B, Feinberg EN, Gomes J, Geniesse C, Pappu AS, Leswing K and Pande V. MoleculeNet: A benchmark for molecular machine learning. Chemical Science. 2017 Oct 31; 9(2): 513–530. doi: 10.1039/c7sc02664a.
[40] Weininger D. SMILES, a chemical language and information system. 1. Introduction to methodology and encoding rules. Journal of Chemical Information and Computer Sciences. 1988; 28(1): 31–36. https://doi.org/10.1021/ci00057a005.
[41] Download InChI version 1 (software version 1.05) for Standard and Non-Standard InChI/InChIKey (January 27, 2017). http://www.inchi-trust.org/downloads/ (accessed Mar. 2017).
[42] Bouton CM and Pevsner J. DRAGON and DRAGON view: Information annotation and visualization tools for large-scale expression data. Current Protocols in Bioinformatics. 2003 Aug; Chapter 7:Unit 7.4. doi: 10.1002/0471250953.bi0704s02.

[43] Bento AP, Hersey A, Félix E, *et al*. An open source chemical structure curation pipeline using RDKit. Journal of Cheminformatics. 2020; 12: 51. doi: https://doi.org/10.1186/s13321-020-00456-1.
[44] Martín-Miguel I, Escudero-Tena A, Muñoz D and Sánchez-Alcaraz BJ. Performance analysis in padel: A systematic review. Journal of Human Kinetics. 2023; 89: 213–230. doi: 10.5114/jhk/168640.
[45] Moriwaki H, Tian YS, Kawashita N, *et al*. Mordred: A molecular descriptor calculator. Journal of Cheminformatics. 2018; 10: 4. doi: https://doi.org/10.1186/s13321-018-0258-y.
[46] Wang H, Liang Q, Hancock JT, *et al*. Feature selection strategies: A comparative analysis of SHAP-value and importance-based methods. Journal of Big Data. 2024; 11: 44. doi: https://doi.org/10.1186/s40537-024-00905-w.
[47] Murad MS, Hamzat AK, Asmatulu E, *et al*. Flame-retardant fiber composites: Synergistic effects of additives on mechanical, thermal, chemical, and structural properties. Adv Compos Hybrid Mater. 2025; 8: 31. doi: https://doi.org/10.1007/s42114-024-01111-1.
[48] https://www.iso.org/.
[49] https://ub.rptu.de/en/search/books-journals-more/din-norms-patents.
[50] Kim HL, Park SM, Cho IJ, et al. Standardized protocol of blood pressure measurement and quality control program for the korea national health and nutrition examination survey. Clinical Hypertension. 2023; 29: 28. doi: https://doi.org/10.1186/s40885-023-00252-7.
[51] Nesselroade JR and Molenaar PCM. On standardized measurement in behavioral science. Journal for Person-Oriented Research. 2022; 8(2): 43–51. doi: 10.17505/jpor.2022.24854.
[52] Mannodi-Kanakkithodi A, Pilania G, Huan TD, Lookman T and Ram Prasad R. Machine learning strategy for the accelerated design of polymer dielectrics. Scientific Reports. 2016; 6: 20952. doi: 10.1038/srep20952.
[53] Huan TD, Mannodi-Kanakkithodi A and Ramprasad R. Accelerated materials property predictions and design using motif-based ngerprints. Physical Review B. 2015; 92: 014106.
[54] Demir H, Daglar H, Gulbalkan HC, Aksu GO and Keskin S. Recent advances in computational modeling of MOFs: From molecular simulations to machine learning. Coordination Chemistry Reviews. 2023; 484: 215112.
[55] Yang X, Wang Y, Byrne R, Schneider G and Yang S. Concepts of artificial intelligence for computer-assisted drug discovery. Chemical Reviews. 2019; 119(18): 10520–10594.
[56] Zheng Y, Yao Y, Ou J, et al. A review of composite solid-state electrolytes for lithium batteries: Fundamentals, key materials and advanced structures. Chemical Society Reviews. 2020; 49(23): 8790–8839.
[57] Hu H, Zhang F, Luo S, Chang W, Yue J and Wang CH. Recent advances in rational design of polymer nanocomposite dielectrics for energy storage. Nano Energy. 2020; 74: 104844.
[58] Li F, Li Y, Novoselov KS, et al. Bioresource upgrade for sustainable energy, environment, and biomedicine. Nano-Micro Letters. 2023; 15(1): 35.
[59] Rickman JM, Lookman T and Kalinin SV. Materials informatics: From the atomic-level to the continuum. Acta Materialia. 2019; 168: 473–510.
[60] Zhang K, Gong X and Jiang Y. Machine learning in soft matter: From simulations to experiments. Advanced Functional Materials. 2024; 34(24): 2315177.
[61] Brooks BR, Brooks CL 3rd, Mackerell ADJ, Nilsson L, Petrella RJ, Roux B, Won Y, Archontis G, Bartels C, Boresch S, Caflisch A, Caves L, Cui Q, Dinner AR, Feig M, Fischer S, Gao J, Hodoscek M, Im W, Kuczera K, Lazaridis T, Ma J, Ovchinnikov V, Paci E, Pastor RW, Post CB, Pu JZ, Schaefer M, Tidor B, Venable RM, Woodcock HL, Wu X, Yang W, York DM and Karplus M. CHARMM: The biomolecular simulation program. Journal of Computational Chemistry. 2009 Jul 30; 30(10): 1545–1614. doi: 10.1002/jcc.21287.
[62] Case DA, Aktulga HM, Belfon K, Ben-Shalom IY, Berryman, Brozell JT, Cerutti SR, Cheatham DS III, Cisneros TE, Cruzeiro GA, Darden VWD, Forouzesh TA, Ghazimirsaeed N, Giamba¸su M, Giese G, Gilson T, Gohlke MK, Goetz H, Harris AW, Huang J, Izadi Z, Izmailov S, Kasavajhala SA, Kaymak KMC, Kolossváry I, Kovalenko A, Kurtzman T, Lee TS, Li P, Li Z, Lin C, Liu J, Luchko T, Luo R, Machado M,

Manathunga M, Merz KM, Miao Y, Mikhailovskii O, Monard G, Nguyen H, O'Hearn KA, Onufriev A, Pan F, Pantano S, Rahnamoun A, Roe DR, Roitberg A, Sagui C, Schott-Verdugo S, Shajan A, Shen J, Simmerling CL, Skrynnikov NR, Smith J, Swails J, Walker RC, Wang J, Wang J, Wu X, Wu Y, Xiong Y, Xue Y, York DM, Zhao C, Zhu Q and Kollman PA. Amber 2024. 2024; San Francisco: University of California.

[63] Jorgensen WL, Maxwell DS and Tirado-Rives J. Development and testing of the OPLS all-atom force field on conformational energetics and properties of organic liquids. Journal of the American Chemical Society. 1996; 118(45): 11225–11236.

[64] Marrink SJ, Risselada J, Yefimov S, Tieleman DP and De Vries AH. The MARTINI force field: Coarse grained model for biomolecular simulations. Journal of Physical Chemistry B. 2007; 111(27): 7812–7824.

[65] Chenoweth K, Van Duin ACT and Goddard WA. ReaxFF reactive force field for molecular dynamics simulations of hydrocarbon oxidation. Journal of Physical Chemistry A. 2008; 112: 1040–1053.

[66] Elder RM, Zaccone A and Sirk TW. Identifying nonaffine softening modes in glassy polymer networks: A pathway to chemical design. ACS Macro Letters. 2019; 8(9): 1160–1165.

[67] Statt A, Kleeblatt DC and Reinhart WF. Unsupervised learning of sequence-specific aggregation behavior for a model copolymer. Soft Matter. 2021; 17(33): 7697–7707.

[68] Walters M, Wei Q and Chen JZY. Machine learning topo logical defects of confined liquid crystals in two dimensions. Physical Review Engineering. 2019; 99(6): 062701.

[69] Sedal A, Wineman A, Gillespie RB and Remy CD. Comparison and experimental validation of predictive models for soft, fiber-reinforced actuators. The International Journal of Robotics Research. 2021; 40(1): 119–135.

[70] Wu D, Zhang D, Liu S, et al. Prediction of polycarbonate degradation in natural atmospheric environment of china based on BP ANN model with screened environmental factors. Chemical Engineering Journal. 2020; 399: 125878.

[71] Zhao Y, Fan D, Li Y and Yang F. Application of machine learning in predicting the adsorption capacity of organic compounds onto biochar and resin. Environmental Research. 2022; 208: 112694.

[72] Qi C, Fourie A, Chen Q, Tang X, Zhang Q and Gao R. Data driven modelling of the flocculation process on mineral processing tailings treatment. Journal of Cleaner Production. 2018; 196: 505–516.

[73] Qi C, Ly HB, Chen Q, Le TT, Le VM and Pham BT. Flocculation-dewatering prediction of fine mineral tailings using a hybrid machine learning approach. Chemosphere. 2020; 244: 125450.

[74] Guo Y, Ma W, Li J, et al. Effects of microplastics on growth, phenanthrene stress, and lipid accumulation in a diatom, phaeodactylum tricornutum. Environmental Pollution. 2020; 257: 113628.

[75] Swain MC and Cole JM. ChemDataExtractor: A toolkit for automated extraction of chemical information from the scientific literature. Journal of Chemical Information and Modeling. 2016; 56: 1894–1904.

[76] Matsen MW. The standard Gaussian model for block copolymer melts. Journal of Physics. Condensed Matter. 2002; 14: R21–R47.

[77] Bates FS, Schulz MF, Khandpur AK, Forster S, Rosedale JH, Almdal K and Mortensen K. Fluctuations, conformational asymmetry and block copolymer phase behaviour. Faraday Discussion. 1994; 98: 7–18.

[78] Fredrickson GH and Helfand E. Fluctuation effects in the theory of microphase separation in block copolymers. The Journal of Chemical Physics. 1987; 87: 697–705.

[79] Miquelard-Garnier G and Roland S. Beware of the flory parameter to characterize polymer-polymer interactions: A critical reexamination of the experimental literature. European Polymer Journal. 2016; 84: 111–124.

[80] Sha W, Li Y, Tang S, et al. (2021) Machine learning in polymer informatics. InfoMat 3(4):353–361. doi: https://doi.org/10.1002/inf2.12167.

[81] Xie C, Qiu H, Liu L, et al. (2025) Machine learning approaches in polymer science: Progress and fundamentals for a new paradigm. SmartMat 6(1):e1320. https://doi.org/10.1002/smm2.1320.

Isha Kalbhor, Shantanu Chakote, Rushikesh Dabhade, Pranay Sharma, Sushant Gaikwad, Raj Agarwal, and Satish Polshettiwar*

Chapter 8
AI-driven development: innovating biomedical polymer materials

Abstract: In the era of artificial intelligence (AI), polymeric biomaterials alter efficiency and precision in material design and properties. AI-driven models are leveraged for predicting, optimizing and developing novel biomaterials. A data-driven model for enhancing material properties may improve the development of polymeric biomaterials. This chapter depicts various AI-driven models, techniques, their advantages, and disadvantages for a robust and adequate AI model for innovating polymeric biomaterials. Advanced computational tools with AI-driven models for novel polymeric biomaterials may help researchers understand and redefine the landscape of biomedical polymers for innovating, patient-centric solutions in modern medicine.

Keywords: Biomaterials, artificial intelligence, deep learning, machine learning, polymers, reinforcement learning, predictive modeling

---

Conflict of interest: The authors do not have any conflict of interest.

***Corresponding author: Satish Polshettiwar**, Department of Pharmaceutical Sciences, School of Health Sciences and Technology, Dr. Vishwanath Karad MIT World Peace University, Pune 411038, Maharashtra, India, e-mail: satish.polshettiwar@mitwpu.edu.in
Isha Kalbhor, Department of Pharmaceutical Sciences, School of Health Sciences and Technology, Dr. Vishwanath Karad MIT World Peace University, Pune 411038, India, e-mail: isha.kalbhor@mitwpu.edu.in
Shantanu Chakote, Department of Pharmaceutical Sciences, School of Health Sciences and Technology, Dr. Vishwanath Karad MIT World Peace University, Pune, 411038, India, e-mail: shantanu.chakote@mitwpu.edu.in
Rushikesh Dabhade, Department of Pharmaceutical Sciences, School of Health Sciences and Technology, Dr. Vishwanath Karad MIT World Peace University, Pune,411038, India, e-mail: rushikesh.dabhade@mitwpu.edu.in
Pranay Sharma, Department of Pharmaceutical Sciences, School of Health Sciences and Technology, Dr. Vishwanath Karad MIT World Peace University, Pune 411038, India, e-mail: pranay.sharma@mitwpu.edu.in
Sushant Gaikwad, Department of Pharmaceutical Sciences, School of Health Sciences and Technology, Dr. Vishwanath Karad MIT World Peace University, Pune, 411038, India, e-mail: sushant.gaikwad@mitwpu.edu.in, orcid.org/0000-0003-0284-8374
Raj Agarwal, Department of Pharmaceutical Sciences, School of Health Sciences and Technology, Dr. Vishwanath Karad MIT World Peace University, Pune, 411038, India, e-mail: raj.agarwal1@mitwpu.edu.in

https://doi.org/10.1515/9783111681870-008

8.1 Introduction

The current state of biomedical polymer research has been greatly influenced by the convergence of materials science and artificial intelligence (AI). The pharmaceutical industry is changing due to this intersection, particularly in drug systems for formulation, delivery, and discovery. Innovation in these key areas of the pharmaceutical industry is being driven using AI techniques in the design of biomedical polymers [1]. The rise of Industry 4.0, which is defined by the dominance of advanced digital technologies such as robots, big data analysis, and the Internet of Things (IoT), makes it easier to integrate AI-based design and production processes from start to finish. AI computer algorithms, for example, can optimize polymer formulas, and IoT sensors continuously monitor manufacturing to assure consistency and quality. Customized medication is made feasible by this partnership between AI and Industry 4.0, which enables a more effective, productive, and environmentally friendly approach for creating biomedical polymers and cutting-edge medical equipment [2, 3]. With the inclusion of AI in polymer designing, there is the potential to tap into vast design possibilities that are otherwise not achievable using traditional means [4]. ML algorithms can easily design novel polymer structures with special properties for targeted applications, for example, drug delivery devices to be implanted. This is of tremendous significance and plays an important role in addressing the growing need for specialized medical devices [5, 6]. Cooperation is an essential force to propel biomedical polymer innovation in solving new demands in the medical field [7, 8]. The continued emergence of such a technology is significant in its power to redefine pharmaceutical practice as well as alter the patient's outcome with a more compounded medical landscape [9].

8.2 AI techniques in polymer discovery

The inclusion of AI in polymer discovery has changed the field of research, offering endless opportunities for innovation, optimization, and efficiency. Use of such techniques will enhance our ability to predict, optimize, and formulate within the field of polymer science.

- **Machine learning (ML)** models have already shown a lot of potential in the field of research, as these algorithms act as the backbone of AI-driven advancements in polymer science. ML models can evaluate different sets of data, even those that may have complicated patterns, which can be applied in predictive modeling, high-throughput screening, and generative polymer design. It will drastically cut down on the time and expense involved in conventional experimentation [10]. The prediction of polymer properties and design optimization was achieved through the use of multiple ML models.

- **Supervised learning (SL)** includes different types of training models based on labeled datasets, enabling it to learn known input–output relationships. Different types of algorithms, such as support vector machines (capture nonlinear relationships between molecular descriptors and polymer properties), random forests (uses a combination of decision trees, effectively handling the complex interactions between features), and gradient boosting (techniques such as XGBoost and LightGBM were used to enhance predictive accuracy), can be employed for property prediction and optimization. These techniques can be key to predicting properties like glass tensile strength and transition temperature [11]. For example, SVMs and RF can understand complex nonlinear relationships between polymer properties and molecular descriptors. Polymer property prediction, support vector machines achieved an R^2 score of 0.85 for glass transition temperature, whereas it displayed a 0.80 R^2 value for tensile strength. It elaborates about the capability of understanding molecular interactions responsible for thermal and mechanical properties, which makes it a valuable tool in polymer discovery [12].
- **Unsupervised learning (UL)** includes principal component analysis (PCA) (aids in reducing the dimensionality of a large polymer dataset, which helps in feature selection and visualization), k-means clustering (groups polymers based on structural and physiochemical similarities, allowing for identification of novel polymer classes), and autoencoders (can be used for feature extraction and representation learning, enabling efficient classification). These algorithms deal with unlabeled data and play a crucial role in uncovering hidden structures or patterns without predefined outcomes, which allows researchers to explore unknown relationships [13].
- **Reinforcement learning (RL)** includes generative adversarial networks (employed to generate new polymer structures with potentially useful properties), variational autoencoders (aids in exploring the latent space of polymer structure, leading to generation of innovative polymer candidates), and Bayesian optimization (optimizes polymer performance by effectively adjusting processing settings with minimal experimental work) [14]. RL incorporates decision-making into the learning process, allowing models to discover the best plan of action via trial and error. RL has demonstrated potential in optimizing polymer formulations and process conditions. For example, Reiser et al. [15] used active learning to design the polymer molecular weight distributions, gaining major improvements in computational efficiency and accuracy. Also, RL can be used to find the best polymer for specific use by continuously improving models based on feedback from previous results.

8.2.1 AI and ML for biomaterial applications

AI-driven advancements in polymer science are making their way in drastically with advanced techniques like computer simulations, high-throughput screening, and predictive models. There are a lot of techniques and algorithms that can be employed for biomaterial application figure 8.1, which were explained earlier. Table 8.1 summarizes different techniques that are used with respect to biomaterial applications.

Table 8.1: Different AI techniques with their biomaterial applications.

AI technique	Application area in biomaterials
Supervised learning (RF and SVM)	Property prediction, classification of materials by biocompatibility, and mechanical performance modeling
Unsupervised learning (PCA and k-means)	Biomaterial clustering based on structural and functional properties, and feature extraction for predictive modeling
Deep learning (CNNs and RNNs)	Microstructure analysis, predicting in vivo behavior, and processing condition optimization.
Reinforcement learning (RL)	Optimizing synthesis processes and adaptive formulation tuning for smart biomaterials.
Generative models (GANs, VAEs)	Designing novel polymers with tailored degradation rates, mechanical properties, and bioactivity

8.2.2 Computational techniques and high-throughput screening

Virtual testing and high-throughput screening: AI-powered virtual testing and high-throughput screening have also accelerated the search for novel polymers. These methods quickly assess and optimize candidate polymers using automated workflows and molecular dynamics simulations. They accomplish this by merging real-world experimental validations with insights from AI. According to Pyzer-Knapp et al. [13], integrating AI with HTS not only expedites the discovery process but also enhances the reliability of predictions by incorporating real-time data analysis and optimization.

8.2.3 Predictive modeling for polymer properties

Predictive modeling serves as the foundation for AI-mediated advancements in polymer science and provides remarkable capabilities for the prediction of material behavior and design optimization. CNNs are capable of analyzing microstructure images and abstracting hierarchical features for effective property predictions. GNNs describe molecular structure through graphs, exposing complex interactions in atoms and bonds to accurately predict polymer behavior [15]. Generative models operate in tandem with predictive modeling by generating new polymer structures with desired properties. GANs, as described by Goodfellow, engage in sequence minimax games to produce viable and diverse alternative designs. VAEs, while also generative models, differ from GANs in that they utilize a probabilistic method for navigating latent space, extend the possibilities for polymer design, and enable completely new material synthesis [16]. While progress has been made, there are still issues to overcome, particularly with data quality and algorithm designs. According to Himanen et al., "most datasets for polymers are either small, do not exhibit consistency, or do not have proper annotation"; these are all genuine issues for training good models. Future studies should aid in the cleaning and enriching of datasets, and most importantly, the collection of good experimental data, at appropriate resolutions for model training and validation. The refinement of methodologies and the development of hybrid models that leverage multiple ML techniques might also offer advantages in performance and help tackle long-standing issues related to overfitting and generalization [17]. The application of AI to polymer discovery represents a new frontier in material science that provides higher levels of predictive capability, faster innovations, and designs that are optimal. Through SL, UL, and RL algorithms, experienced practitioners can engage in partnerships with high-throughput screening for new polymers and utilize predictive models that will liberate the untapped promise of polymers for the benefit of scientists, and for use in initiatives across the sectors that utilize these materials and are so vital to all our endeavors. As we contend with a set of current and emerging issues, AI in polymers will change the landscape of our materials and their design and performance.

8.2.4 Data collection and preprocessing

To build robust ML models, multiple meaningful datasets are required. Figure 8.2 captures the overall polymer discovery process with further involvement of ML. The process of feature engineering was an integral step in converting raw data into descriptive variables that characterize the inherent chemical and structural properties of polymers [18]. Various molecular descriptors, including conventional topological indices and newer molecular fingerprints, were created using the RDK tools. In addition to the above, additional polymer-specific features, including chain length, branching topologies, and crosslinking density, were pulled from structure data to further enrich the feature set. In addressing the potential challenges stemming from working with a high-dimensional feature space and improving model interpretability, methods were used that preserved important information but reduced redundancy and noise in the data.

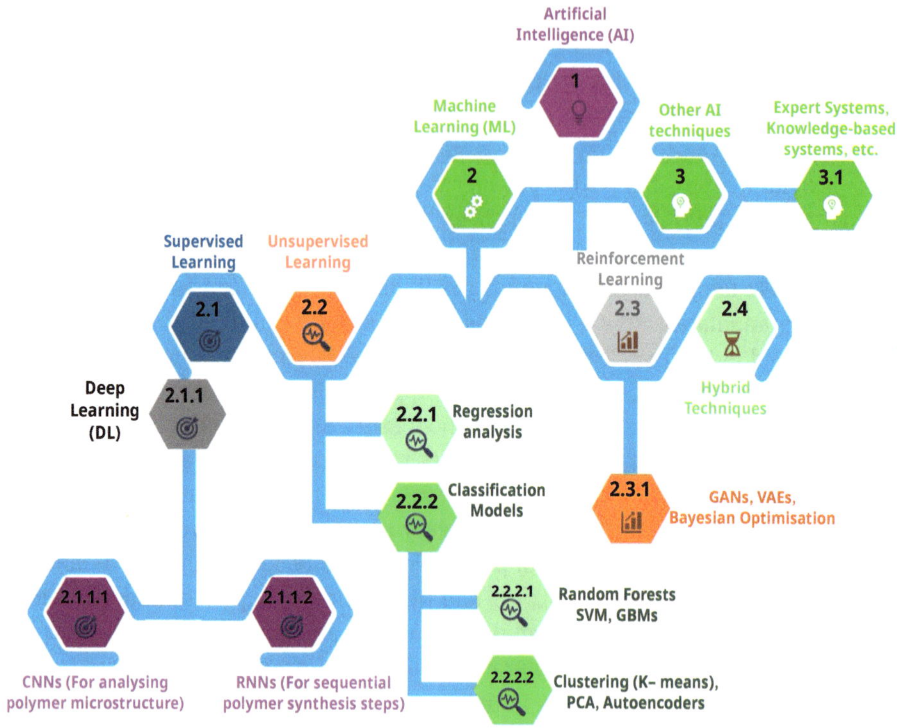

Figure 8.1: Classification of the hierarchy for AI, ML, and DL for polymer science applications.

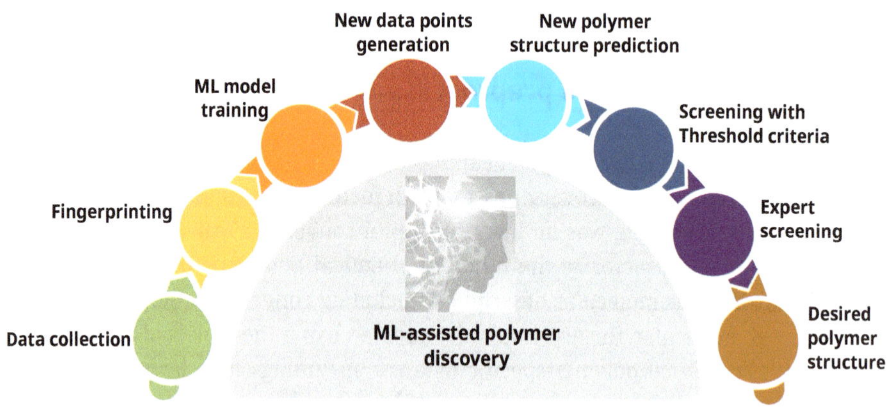

Figure 8.2: Workflow of ML-assisted polymer discovery.

8.3 Applications for AI in biomedical polymers

8.3.1 Drug delivery system

The drug delivery system has experienced unprecedented developments over the years that have allowed it to deliver therapeutic agents in a targeted and effective way. One of the largest benefits of applying AI in healthcare delivery systems is its ability to handle and analyze huge amounts of data. This kind of analysis enables researchers to know how different components of a formulation interact, their concentrations separately, and how they are all responsible for drug delivery efficacy [19].

AI can analyze many sources of data, including chemical structures, physicochemical descriptors, and bioactivity data, to find key associations. With trend identification and pattern discovery, AI could play a vital role in making efficient drug delivery systems and thus making them more safe and stable [20]. Utilizing the strength of AI along with large sets of data, scientists can hasten the development of safe and effective drug delivery systems, improving patient outcomes and advancing the field of pharmaceutical sciences [21].

8.3.2 Intelligent nanocarriers

Colloidal drug carrier systems with particle sizes less than a micrometer are referred to as nanocarriers. AI methods are often suggested as solutions to a range of nanotechnology issues. These are the development of nanosystems, nanocomputing, and the use of AI approaches to the development of simulation methods at the nanoscale. Several AI-based methods are being suggested to predict the effectiveness of drug combinations using pharmacological synergies. Employing various types of nanoparticles to enhance drug localization at tumor sites, potential avenues for AI optimized optimization in combination drug delivery are being explored in this field [22]. One of the potential areas of drug delivery research is nanocarrier design and optimization for targeted drug delivery, and AI algorithms can greatly enhance their performance. Scientists can develop predictive models that consider a variety of variables like the physicochemical properties of nanocarriers, pharmacological characteristics, and biological parameters using AI algorithms [23]. For optimal drug delivery, these models can be used to select the appropriate nanocarrier composition, size, surface modification, and other formulation characteristics. Combinatorial drug design is maximized through an AI-driven feedback system control (FSC) platform, specifically in treating cancer with nanocarriers. Surface properties of nanocarriers play a very vital role in determining their stability, bioavailability, and cellular dispersion and absorption within the body. AI supports the analysis and prediction of the effects of surface alterations, for example, ligand conjugation or the deposition of targeting molecules, on cell–nanocarrier interaction. It decides the best ligands or method using molecular and cellular data. The aim in combining automation and AI is to increase the tailoring of targeted therapeutic nanocarriers for cell types and individual patients [24].

8.3.3 Smart drug delivery

AI has an important function in the design of sophisticated drug delivery systems that respond to physiological cues or health conditions [25]. AI programs facilitate continuous monitoring and control of drug release from these delivery systems. Through the evaluation of tailored patient data, for example, blood drug concentration, physiological metrics, or biomarkers, AI programs can provide expert advice regarding drug dosage, scheduling, and release rates. They can constantly analyze the collected data and adjust the drug release profile accordingly. Additionally, AI assists in the formulation of injectable drug delivery systems using thermosensitive hydrogels, which are sensitive to temperature changes. With ML algorithms, researchers can forecast the behavior of intricate polymer compositions under various conditions of the surrounding environment. With this capability, controlled drug delivery is formulated for prolonged periods. Furthermore, the combination of AI and wearable technology, biosensors, or sensors make it possible to deliver drugs for the purpose of delivering customized and dynamic therapies, resulting in enhanced therapeutic responses [26].

8.3.4 Tissue engineering scaffolds

Tissue engineering scaffolds play a critical role in the practice of regenerative medicine, as they provide the structural support required for cell attachment, proliferation, and differentiation [27]. The incorporation of AI into scaffold design, biomaterial optimization, and predictive modeling has greatly enhanced the efficiency and clinical translation of tissue-engineered constructs. In various applications AI can be leveraged for tissue engineering applications.

8.3.4.1 AI-based material selection and design

AI employs ML algorithms to forecast the mechanical and biological characteristics of polymers in terms of variables such as molecular structure, cross-linking density, and degradation rates. AI models determine the interaction between biomaterials and cells, forecasting biocompatibility, immunogenicity, and degradation patterns. AI platforms like ChemSpider interpret chemical compositions to screen the best candidate for scaffold production.

8.3.4.2 Prediction of scaffold mechanical properties

AI greatly promotes the molecular and chemical optimization of tissue engineering scaffolds by predicting material properties, directing scaffold design, and optimizing chemical functionalization. ML models evaluate large datasets to estimate elasticity, porosity, and load-bearing capacities, while AI algorithms assess the effect of motifs and functional groups on cell adhesion, protein adsorption, and overall scaffold performance [28].

8.3.4.3 In silico prediction

Finite element modeling (FEM), in association with design strategies such as triply periodic minimal surfaces and signed distance field (SDF), enables in silico prediction of mechanical response. FEM simulations help fine-tune the pore and structure dimensions to achieve the required scaffold architecture, thus improving both mechanical and degradation properties [28].

8.3.4.4 Hydrogel and composite scaffold design

AI forecasts the self-assembly of peptide hydrogels, which are popular in tissue engineering for their biocompatibility and biodegradability. AI assists in comprehending and tuning the nontrivial interaction between peptide sequence and hydrogel properties.

8.3.4.5 Optimization of 3D printing parameters

AI could optimize 3D printing parameters for tissue engineering scaffolds, resolving issues involving material selection, process control, and minimizing defects. AI methods can improve the accuracy, efficiency, and reliability of 3D-printed scaffolds, in turn resulting in enhanced tissue regeneration and clinical performance [29].

8.3.4.6 Polymeric bioinks

AI models evaluate the printability of different polymeric bioinks based on critical parameters, such as viscosity, curing behavior, and mechanical stability. Identification of suitable printing conditions for new materials is usually a time-consuming and resource-scarce endeavor, requiring copious experimentation. AI can help mitigate this challenge by analyzing the underlying behaviors existing within a given dataset [30].

8.3.4.7 Process parameter optimization

AI algorithms adjust numerous process parameters such as nozzle diameter, extrusion rate, temperature, and pressure to produce scaffolds with exact dimensions [31]. The potential of AI to process large amounts of complex data, perform complex calculations, automate processes, learn from history, and optimize operations on an ongoing basis makes it a crucial resource [32].

8.3.4.8 Patient-specific scaffolds

AI promotes the creation of patient-specific scaffolds through controlling and adjusting printing parameters, offering feedback, and having sufficient data for reproducibility. Incorporating AI helps predict, adjust, and manage parameters, which reduces the scope for errors and improves customization of scaffolds [31].

8.4 Medical diagnostics and biosensors

Medical diagnostics are not only crucial for the care of individual patients but also for the prevention, screening, and monitoring of many diseases and conditions [33]. They cover a broad range of tools and methods, ranging from laboratory tests (in vitro diagnostics) to imaging tests [34]. AI systems utilize ML algorithms to efficiently and accurately evaluate medical images. This essential feature helps with the early diagnosis of hard-to-diagnose diseases. Early detection of such conditions is important, as it can lead to instant interventions, thus having the potential to save lives and enhance treatment outcomes. Furthermore, AI has opened new avenues in the areas of quantification and image segmentation. Using sophisticated algorithms, AI can identify regions of interest in medical images, such as tumors, vessels, and cells, with remarkable accuracy. Such image segmentation plays a pivotal role in treatment planning since it enables medical professionals to precisely define intervention areas, optimize surgical procedures, and deliver therapies that are directed to specific regions [35].

AI system helps radiologists identify breast cancer more precisely, rapidly, and uniformly. Such AI systems learn thousands of de-identified mammograms and hence recognize the subtle signs that represent indications of cancer, perhaps detecting warning signs that may not be recognized by experts. It is used to examine CT scans to detect lung nodules, which can signal lung cancer. This technology has been promising in enhancing rates of detection and helping radiologists detect both benign and malignant nodules [36].

Previously, applications of AI in nodule detection utilized conventional computer-aided design (CAD) strategies, but subsequent developments have utilized advanced

DL methodologies such as CNNs [37]. Earlier research on CAD for pulmonary nodule diagnosis with chest X-ray has reported high false-positive rates and sensitivity as low as 52%. But the use of AI in combination with contemporary methods for detecting pulmonary nodules in radiographs has improved sensitivity to a level of 70–90%, comparable to that of radiologists, while reducing false-positive rates to 0.3–2.0 per radiograph [38].

AI shows substantial promise in improving the performance and efficiency of radiologists in detecting and analyzing lung nodules. Improvements in AI algorithms and a rise in imaging volumes will propel the creation and use of AI in clinical environments. Depending on its implementation, AI can benefit patients by mitigating uncertainty from indeterminate lung nodules but also potentially risk harm by introducing unnecessary diagnostic work. In addition, AI creates new challenges for radiologists, including higher IT demands and potential liability concerns. Developing strong technological and regulatory infrastructures will enable the incorporation of AI algorithms into clinical practice, ultimately resulting in beneficial impacts on patient health [39].

The incorporation of AI into medical imaging has significantly improved the field of personalized medicine. Through the analysis of medical images along with patient information, AI algorithms can produce insights specific to each patient. This ability allows for the creation of customized treatment protocols that consider individual differences in anatomy, physiology, and disease profiles [40].

A personalized model of healthcare not only optimizes the efficacy of therapies but also minimizes the likelihood of undesirable side effects, leading to improved patient outcomes and an improved quality of life [41]. AI-based algorithms significantly increase the function of biosensors through enhanced interpretation of information [42].

Polymers used in biosensors are bound to specific biomarkers, while AI algorithms treat the sensor data to recognize patterns and predict biological changes, making the sensors more sensitive and accurate. ML techniques and DL, cleanse sensor data from noise and enhance the quality of signals, thus setting the trustworthiness of biosensors. Computer simulations accelerated by AI accelerate research and development of new biomedical polymers by offering quicker prototyping and testing. Researchers employ computer software to simulate the interaction between different polymer materials and cells, tissues, and body fluids with diseases and other health conditions, minimizing big animal tests and accelerating the design of new diagnostic devices based on polymers [41]. For example, nanocomposites that combine polymers with materials like graphene and carbon nanotubes have AI-assisted analysis to maximize their electrical conductivity and surface area for the enhancement of the sensitivity of biosensors.

The ability of AI to forecast and optimize the properties of nanocomposites is realized as enhanced biosensor performance in terms of sensitivity, selectivity, repeatability, and stability. AI is necessary for the development of smart biosensors with the capability of adapting to changing biological environments. Glucose sensors based on polymers for diabetic patients can be improved using AI algorithms that measure glucose levels and predict future variations to allow active condition management. This adaptability encompasses wearable sensors that measure health parameters such as heart rate and oxygen saturation continuously, giving immediate feedback to clinicians and patients. These smart biosensors also comprise operational polymers that respond to physiological parameters, such as pH or glucose concentration, to deliver medication or provide feedback. AI improves these systems by predicting optimal rates of drug release and monitoring the effectiveness of the treatment in real time. AI assists in the design of long-term stable implantable sensors with high mechanical and biological compatibility. The sensors employ flexible polymers to obtain a tissue mimetic structure, which allows real-time monitoring of physiological data from within the body. Short-term implantable devices have been developed to allow high throughput drug sensitivity testing in solid tumors. The implantable microsystems for personalized anticancer therapy project aims to create implantable wireless sensors to track tumor oxygen levels in real time. Table 8.2 provides the advantages of the application of AI and ML in biomaterials.

Table 8.2: AI and ML applications in biomaterials.

Application	Description	Benefits
Material design	AI algorithms forecast material characteristics, thereby refining designs for uses	Accelerated development and enhanced material efficacy
Biosensor development	AI improves sensitivity and specificity of biosensors by evaluating sensor data and anticipating biological variations	Enhanced diagnostic precision and continuous monitoring
Tissue engineering	AI refines the design and production of scaffolds, promoting tissue regeneration	Tailored scaffolds and better tissue integration
Drug delivery systems	AI fine-tunes drug release profiles and bioavailability, thereby increasing treatment effectiveness	Enhanced drug targeting and reduced side effects

8.5 Advantages of AI-driven development in biomaterials and manufacturing

There are various advantages of AI in polymer manufacturing and development, which speed up the process and become an integral part of the industries [43, 44]. Certain advantages of the AI in polymer manufacturing and the ML are given below.

8.5.1 Advantages in biomaterial research

1. **Process optimization for manufacturing**
   AI proved as an efficient and cost-effective tool in manufacturing biomaterials using optimization parameters such as shaping, molding, 3D printing, and heat treatment.
2. **Improved materials design and their synthesis**
   AI analyzes large material properties databases and their correlation to generate new materials designs of specific characteristics. This helps in innovating materials having good performance and biocompatibility. In fields like tissue engineering and drug delivery this advantage is valuable.
3. **Better interpretation and image analysis**
   AI can analyze medical images, biomaterial images, help in diagnosis, monitor, and treat the disease while evaluating material performance and integrity. This is especially used for analyzing complex data.
4. **Quicker material finding and predictive modeling**
   Traditional trial and error in the experiment is considerably reduced by this predictive capability [43]. AI can utilize large datasets of material properties and accelerated material discovery by predicting biocompatibility, degradation rates, and characteristics of biomaterials.
5. **Overcoming obstacles in the study of biomaterials**
   AI is beneficial in various aspects of biomedical polymer manufacturing, but there are challenges such as quality and limited data availability [44]. However, improvement in model interpretation and the development and standardization of datasets are gradually solving these issues.

8.5.2 Advantages in polymer manufacturing

1. **Enhanced manufacturing processes**
   Application of AI in polymer production process includes various stages, which are maintenance, production scheduling, monitoring, and production manufacturing. Hence AI is useful in improving overall efficiency and optimization.
2. **Improved structural design and their properties**
   AI can discover novel materials and enhance accessibility of existing ones. AI improves structural design and polymeric characteristics.
3. **High optimization capacity and 24/7 availability**
   AI algorithms provide high optimization and are available whenever required, which helps in maximum production and reducing downtime
4. **Advanced analytical features and fast decision-making**
   AI provides advanced analytical capabilities and facilities faster decision-making in polymer manufacturing, in the end increasing productivity and efficiency in polymer manufacturing.

5. **Application across diverse industries**
   AI greatly impacts polymer manufacturing in industries such as food, cosmetics, healthcare, dentistry and agriculture. For example, AI helps in managing quality control, supply chain, and cleaning process in food industry. In dentistry AI supports bio implants planning and the development of new dental materials.
6. **Addressing challenges in polymer manufacturing**
   There are various advantages of AI in polymer manufacturing. Despite the advantages polymer production processes face AI- related challenges, which include high initial costs, potential job displacement, and the need for clear production strategies. But the efficient production quality and the long-term benefits overcome these obstacles.

8.5.3 Advantages in custom manufacturing

1. **Flexible production methods**
   Without human involvement, AI itself is flexible in optimizing manufacturing processes and maintains optimal performance instantly responding to changes in production and material properties. It is an essential advantage for low volume and customized manufacturing.
2. **Improved machining precision**
   AI-enhanced precision helps in decreasing errors in computer numerical control (CNC) machining by optimizing tool paths, resulting in higher-quality products and reduced material waste.
3. **Real-time quality control**
   AI-driven vision system conducts real time quality control, identifying manufacturing defects early and reducing waste and rework.
4. **Predictive analysis**
   To predict customer demand AI uses real-time data and history trends to minimize waste and optimize inventory levels. It is beneficial in the rapidly evolving automotive industry.
5. **Predictive maintenance**
   AI algorithm predicts equipment failures, enabling planned maintenance during off peak hours, reducing downtime, lowering maintenance cost, and extending equipment life span.
6. **Addressing challenges in custom manufacturing**
   Adding AI into custom manufacturing presents challenges such as system compatibility, data security, and workforce adaptation. However, phase integration, employee training, and cloud-based solutions help mitigate these issues.

8.6 Generative AI for novel polymer structure

The models use machines to design new polymer forms and predict the results they will have [45]. This type of model is data-powered, and it learns from the data, not just theory. It can induce novel types of molecular structures optimized for high strength and toughness, with progress sustained only through circulation within a supportive ecosystem, or through biologically degradable intelligence committed to maintaining engagement with emerging possibilities. And it makes good paraphernalia for many different activities in industry as well – healthcare systems overview of generative models: Machine knowledge of polymer structures has been applied to predicting outcomes for medical applications such as drug delivery, tissue engineering, and self-healing materials, thereby enhancing the efficacy of new materials research and design. The "new beast" of machine knowledge models called generative AI can produce innovative, synthetic data which is very similar to realistic data. Many fields, like chemistry and knowledge of polymers have been transformed by these models.

8.6.1 Generative adversarial network (GAN)

GANs are also being increasingly utilized in designing biomedical polymers wherein two neural networks (the generator and discriminator) are used for data generation and classification, respectively. This approach is convenient for designing new biomedical polymer architectures. While doing these tasks, the discriminator analyzes the candidate polymers that the generator suggests for possible biomedical uses. The development cycle of new medical devices can be accelerated by using GANs to refine polymer characteristics by verifying that the produced polymers are biocompatible and functional.

GANs are a type of deep generative model first proposed by Goodfellow et al., designed to generate high-quality synthetic data by capturing essential structural features from input data and associating them with desired properties. The foundation of GANs stems from the two main building blocks, the generator and the discriminator, which are fixed at opposition with each other. The generator has to extract locally meaningful features to output realistic data unlike the discriminator that tries to identify if the sample has been taken from the real dataset or generated by a model. GANs are trained such that their generated samples are indistinguishable from true data by the discriminator. The goal of GANs is to generate data that is indistinguishable from real samples to the discriminator. In the field of biomedical polymers, GANs have been used to generate molecular structures, and their adversarial training mechanism can be used to solve complex nonlinear formulation problems in various fields and can also be applied to UL tasks. These capabilities enable GANs to contribute to data generation and material design processes within polymer science.

8.6.2 Common models of GANs

Many improved GAN models were proposed to achieve more stable and better performance. As an example, deep convolutional GANs (DCGANs) use convolutional layers in place of fully connected layers, leading to a near-symmetrical structure of the generator and the discriminator. What this design accomplishes is substantially more stable training and higher quality samples. The Kullback-Leibler divergence is often used to ensure the generated distribution becomes close to the true distribution, thus preventing the mode collapse problem, which is common in standard GANs, leading to more diverse and realistic output.

8.6.3 Graph neural network (GNN)

GNNs have found extensive applications in biomedical polymer research since they can model complex molecular structures and capture the relationships between atoms and bonds [46]. GNNs rely on passing messages back and forth between nodes and edges before updating a node; all information from neighboring nodes and edges is aggregated before it is used to update the node representation, where the nodes represent atoms and edges represent bonds in a molecular graph, GNNs are a strong way to represent the structure of molecules. Although SMILES strings offer compact representations of polymer monomers, critical information like hybridization states, stereochemistry, and 3D connectivity is largely lost. Molecular graphs, on the other hand, capture these structural relationships directly, and thus are better fit for machine learning.

As GNN research progresses, various GNN variants, including graph convolutional networks (GCNs), gated graph neural networks (GGNNs), and message passing neural networks (MPNNs), have been used for molecular representation. These models have traditional molecular feature extraction methods, such as Morgan fingerprints, by providing richer molecular representations. When combined with techniques like extreme gradient boosting (XGB), GNNs further improve predictive performance. Deep learning models typically require large datasets, but in polymer research, particularly for newly discovered polymers, available datasets are often sparse. This data scarcity hinders the model's ability to fully capture the latent structural features of polymers, making accurate property prediction difficult. To overcome this limitation, Guo et al. proposed the Meta-MGNN model, which applies few-shot learning techniques to enhance GNN performance in low-data scenarios.

8.6.4 Deep neural network (DNN)

Basic ML models are increasingly inadequate for analyzing the vast and complex data associated with biomedical polymers, especially given the expanding range of applications and data types. Advanced machine learning techniques are becoming insufficient for analyzing data related to biomedical polymers due to their increasing complexity and volume. In biomedicine polymers, input molecular structural descriptors are fed into the model, while output target features are biocompatibility or mechanical strength. Metaheuristic optimization strategies such as genetic algorithms (GA) and Bayesian optimization (BO) are often applied to adjust the rest of the hyperparameters such as the number of hidden layers, neurons per hidden layer, and learning rates, to enhance model accuracy and stability. The model is continuously retrained with updated hyperparameters, leading to higher stability and improved accuracy of the model.

8.6.5 Recurrent neural networks (RNNs)

Recurrent neural networks (RNNs) stand out for their ability to integrate various types of sequential data, such as polymer breakdown, drug release activity, and chemical sequence analysis for neural network applications in polymers. RNNs can model the interdependence of the structure of a polymer monomer and its properties effectively. The sequence of atomic bonding within a polymer has an important impact on its structure-property relationships [47]. Molecular structures can be represented using sequences that treat each chemical fragment as a character within a sequence. Given their capability to capture long-term dependencies within sequences, RNNs can explore the intrinsic influence of atomic arrangements on the properties of polymers. This enables the construction of highly accurate predictive models for polymer properties by feeding SMILES strings as input and predicting specific polymer features as outputs.

8.6.7 Variational autoencoders (VAEs)

VAEs are a class of deep latent space-generative models that have shown considerable success across numerous biomedical informatics applications, including polymer structure prediction, medical image classification and segmentation, protein design, molecular design, and multi-omics data analysis. Polymers, critical across numerous sectors, including in pharmaceuticals and biomedical devices, face significant limitations in terms of the diversity of commercially available materials. Considerable time and financial investment have been directed toward developing novel biomedical polymers to address the shortcomings of existing medical-grade materials. ML, particularly VAEs, offers an unprecedented opportunity to accelerate the discovery process by reducing the reliance on traditional trial-and-error synthesis approaches.

VAEs can facilitate the rapid design and optimization of polymer properties using minimal experimental data, significantly shortening development timelines. High-throughput screening techniques can be employed for directly assessable properties, such as water uptake or cytotoxicity, enabling rapid dataset generation for model training. These high-throughput approaches, when combined with data-efficient techniques such as Bayesian optimization or active learning, can further accelerate polymer discovery processes. High-throughput methods have already demonstrated promise in polymer excipient design and antibacterial polymer discovery within regenerative medicine. Applying machine learning to biomedical polymer development can benefit from datasets generated through such methods, particularly when properties like mechanical strength or processing characteristics require larger sample sizes. However, high-throughput polymerization techniques have primarily focused on chain-growth polymerizations (including ring-opening polymerization and reversible addition-fragmentation chain transfer polymerization), which are widely used in medical polymers. There remains a notable lack of high-throughput techniques for step-growth polymerizations, which are critical for producing polymers such as poly (lactic acid) (PLA), polyurethanes, and nylons used in sutures, medical tubing, and short-term implants. Developing scalable high-throughput techniques for these processes would significantly enhance the available data for ML-driven biomedical polymer development [48, 49].

8.7 Other approaches for polymer development

AI-driven molecular modelling techniques, such as density functional theory and Monte Carlo simulations, play a crucial role in simulating the behavior of polymer molecules at the atomic scale. These simulations help researchers understand how different monomers and polymerization techniques influence key material properties such as tensile strength, thermal stability, and biodegradability.

8.7.1 Data-driven approaches

An increasing number of correlations between molecular features and characteristic material characteristics have been observed via machine learning models, generated from very large datasets of the structure and properties of polymers. These approaches may be used for predicting and developing properties of polymeric biomaterials.

8.7.2 AI-driven molecular modelling in polymer design

Based on the literature survey, the AI-driven molecular modelling for polymer design can be used for the following purposes:

1. **Polymer property prediction:** Applied researchers may develop predictive models for a variety of polymer properties (tensile strength, melting point, and glass transition temperature) using machine learning approaches. Computational simulations or experimental data may be employed to train these models.
2. **Monomer screening and design:** AI algorithms can screen many potential monomers and decide whether they can be used to produce sustainable polymers. The AI system can also tailor monomer combinations for improving properties of interest, such as biodegradability or recyclability, to achieve the desired characteristics.
3. **Structure–property relationships:** ML models might derive structure-property relations, which otherwise remain unbreakable. Downloaded data about models, predictions, and materials approach can also provide insight into how polymer-specific structures impact the performance of these materials in the end-use applications [49].
4. **Innovative polymer structure:** Self-healing polymers are polymers that can repair themselves when they are damaged. They depend on natural mechanisms that are similar to human skin, and have an innate capacity to repair themselves. Reversible polymer chemistry of polymers in the past has emerged as a promising route, with many examples of self-healing polymer systems being reported. Called "mendomers" or "dynamers," these special polymers are composed of reversible bonds, formatted to respond to environmental conditions, including heat, light, and acidic conditions. This unique behavior has demonstrated effective self-healing polymeric action. Therefore, the potential for a mendomer to continuously heal itself resembles the restoration of the original material. The reversible bonds in mendomers can be engineered to act as weak links that break under stress, thereby preserving the stability of permanent bonds – a crucial aspect for self-healing processes [50]. The market trend in self-healing materials, particularly in the USA, is driven by the presence of multinational companies investing in ongoing research and development to discover cutting-edge technologies. The application of advanced expertise to expand product portfolios assists in capturing untapped markets by increasing the customer base [51]. These materials reduce the frequency and maintenance required for damage repair, which is expected to be a key driver of market growth [52].

8.8 Ethical considerations and regulatory challenges for AI in polymer science

Integrating AI and ML in polymer science provides new horizons of opportunities and innovations [53, 54]. But this convergence also poses some regulatory challenges that need to be prudently addressed. This discussion will emphasize data privacy and security, the ethical implications of AI in developing materials, and building regulatory frameworks.

8.8.1 Data privacy and security

AI for polymer science depends highly on large datasets for properties, synthesis, and applications of the materials. Sensitive data may be contained within these data that may pose a threat to data security and privacy. So, any data to do with how you formulate a chemical, how it's manufactured, and what you use it for, especially in sensitive areas like health care or national security, could be targeted for unauthorized access or misuse.

Data breaches can be so harmful; therefore, they require tight security and regular security audits. Beyond that, anonymization or pseudonymization methods need to be used to safeguard individuals' or organizations' privacy while inserting their data into AI models. It is crucial to be transparent regarding the usage of data and how to deal with the data [57]. Using only relevant data protection rules, such as the General Data Protection Regulation or the California Consumer Privacy Act and complying to them is critical.

8.8.2 Ethical implications of AI in material developmen

There are various ethical challenges in the deployment of AI for material design and development. The possibility of bias in AI algorithms is one of the biggest problems. It is important to point out that these recommendations and predictions can reinforce and even exaggerate the existing biases in case the training data reflects existing biases [55]. The creation of information that is harmful to certain societal groups or the environment can have unforeseen adverse impacts. For example, instead of promoting sustainable production, a biased algorithm may optimize for cost-effectiveness, producing products with adverse environmental impacts [57]. The potential for AI to speed up the development of multiple uses of materials raises another ethical concern. But compounds that are attractive to law enforcement will also have possible uses that should raise concerns, such as their potential to be turned into weapons or used wrongfully. Risks associated with dual-use materials need to be carefully considered to prevent unwanted effects.

8.8.3 Establishing regulatory frameworks

Considering the rapid progress of AI in polymer science, a proper regulatory framework to deal with potential hazards and ethical issues must be built. So, this framework should be adaptable enough to handle the changing nature of AI technology, while providing sufficient security for individuals, society, and the environment [54]. There are several vital decisions that must be made when developing these frameworks:

- Algorithmic account and transparency: Here, to enable examination and accountability, laws must require the transparency of algorithms used in AI systems. This could help in locating and reducing any possible errors or mistakes in the models [55].
- Data management and safety guidelines: For the protection of sensitive data used in AI models, specific data management and security guidelines are needed. So basically, these guidelines must be followed for data collection, processing, storage, and exchange [56].
- Risk assessment and reduction: Thorough risk assessment procedures that identify possible risks and suggest ways for reducing them should be part of regulatory frameworks for AI-driven material development. This would prevent unintended consequences and ensure responsible innovation [56].
- Ethical review boards: It is necessary to develop ethical review boards, especially for the production of AI-driven products that can make sure that moral issues are taken into account during the stages of research and development while ensuring fair monitoring [55].

Policies concerning AI in polymer science require changes to meet new challenges. Effective policymaking, ethical consideration, and industrial engagement need collaboration to ensure that potential harms are mitigated, ethical standards are met, and the benefits of AI are maximized. A more developed response to the challenges and risks of a swiftly evolving environment is supplied by more proactive and flexible planning and policy making [55].

8.9 Future directions

With a variety of possible developments in the works, the future of AI-driven polymer design is bright:

8.9.1 Integration of AI with high-throughput experimentation

Finding new sustainable polymers will be accelerated by combining AI with high-throughput screening methods, which will allow for the efficient testing of several polymer candidates in a short time [58].

8.9.2 Multiscale modeling

Future AI models could use information from several scales, such as macroscopic material properties or molecular dynamics simulations, to provide a more thorough knowledge of polymer behavior.

8.9.2 Quantum computing

The development of quantum computing can transform molecular simulations, making it feasible to model even more complicated polymer systems and speeding the creation of sustainable polymers.

8.9.3 Enhanced machine learning models

The accuracy and reliability of ML models will continue to improve as more extensive and standardized datasets become available. Advances in processing power and algorithms will enable the development of increasingly sophisticated models, facilitating the exploration of complex material behaviors and interactions. For example, combining ML with high-throughput screening techniques can significantly accelerate the discovery of novel materials with desired properties [59].

8.9.4 Lipid–polymer hybrid nanoparticles (LPHNPs)

LPHNPs represent a transformative class of nanodrugs with substantial therapeutic potential in translational biomedical research. These hybrid systems offer distinct advantages over conventional delivery approaches, including enhanced selectivity for targeted delivery, improved biological responses, and better safety profiles. Most LPHNPs exhibit prolonged circulation times, allowing for effective accumulation at the desired target site. The increasing focus on combinatorial cancer therapies further highlights the significance of LPHNPs, which act as multifunctional platforms capable of both simultaneous imaging and drug delivery (theranostics). Additionally, with the rise of gene therapy, LPHNPs are being actively explored as nonviral gene delivery vectors, demonstrating versatility across a wide range of therapeutic applications.

Given the rapid advancements over the past decade, LPHNPs are expected to gradually replace conventional liposomes and polymeric nanoparticles in therapeutic delivery systems. The exponential growth of research focused on hybrid systems, especially studies investigating diverse surface modifications and tunable cargo loading techniques, further underscores their potential. Ongoing research is expected to lead to even more innovative applications, solidifying the versatility of LPHNPs for safe and targeted drug delivery [59, 60].

8.9.5 Polymer science and nanotechnology integration

Although polymer science has a rich history spanning several decades, it continues to captivate scientists worldwide. The integration of nanotechnology with polymer science holds the potential to revolutionize numerous fields by harnessing the unique benefits of both disciplines. Through nanotechnology, polymers can be engineered at the nanoscale, introducing fascinating optical, electrical, and magnetic properties, thereby enhancing their versatility for a wide range of applications.

Polymer science offers extensive research opportunities in diverse areas, including polymer electronics, biopolymers, mechanophores (molecular units that alter polymer properties in response to mechanical forces), smart and reactive polymers, self-cleaning polymers, and self-healing polymers. Polymer nanoparticles hold considerable promise as future building materials, raw materials for 3D printing, packaging materials, polymer electrolytes, polymer lasers, flame-retardant materials, advanced catalysts, and components for next-generation electrochemical devices. The future of polymer science is bright, with transformative innovations anticipated at the interface of nanotechnology and polymer science.

8.10 Conclusion

The integration of AI into biomedical polymer materials has revolutionized the field, offering innovative solutions for the development of advanced materials with enhanced biocompatibility, mechanical properties, and targeted functionalities. AI-driven methodologies have demonstrated their ability to optimize material design, streamline synthesis processes, and accelerate the discovery of novel biomaterials tailored for medical applications. This chapter may be helpful for researchers, biomedical engineers, and regulatory authorities. AI-driven methodologies may further be used for predictive analysis of mechanical and biological properties, targeted drug delivery, screening purposes, personalized medicine, and novel biomaterials design.

List of abbreviations

AI	Artificial intelligence
CAD	Computer-aided design
CNN	Convolutional neural network
CNC	Computer numerical control
DC	Deep convolutional
DNN	Deep neural network
IoT	Internet of Things
GAN	Generative adversarialn
GNN	Graph neural network
FSC	Feedback system control
FEM	Finite element modeling
LPHNP	Lipid–polymer hybrid nanoparticles
ML	Machine learning
MPNN	Message passing neural network
SL	Supervised learning
UL	Unsupervised learning
RL	Reinforcement learning
RF	Random forest
SVM	Support vector machine
PCA	Principal component analysis
RNN	Recurrent neural network
SDF	Signed distance field
VAEs	Variational autoencoders
XGB	Extreme gradient boosting
3D	Three-dimensional

References

[1] Vora LK, Gholap AD, Jetha K, Thakur RR, Solanki HK and Chavda VP. Artificial intelligence in pharmaceutical technology and drug delivery design. Pharmaceutics. 2023 Jul 10;15(7):1916.

[2] Raza MA, Aziz S, Noreen M, Saeed A, Anjum I and Ahmed Raza SM. Artificial intelligence (AI) in pharmacy: An overview of innovations. INNOVATIONS in Pharmacy. 2022;13(2):10–24926.

[3] Ulery BD, Nair LS and Laurencin CT. Biomedical applications of biodegradable polymers. Journal of Polymer Science, Part B: Polymer Physics. 2011 Jun 15;49(12):832–864.

[4] Gokcekuyu Y, Ekinci F, Guzel MS, Acici K, Aydin S and Asuroglu T. Artificial intelligence in biomaterials: A comprehensive review. Applied Sciences. 2024 Jul 28;14(15):6590.

[5] He Z, Chen W, Zhou Y, Weng H and Shen X. The importance of AI Algorithm combined with tunable LCST smart polymers in biomedical applications. Frontiers in Computing and Intelligent Systems. 2023;6(3):92–95.

[6] Xie S. Perspectives on development of biomedical polymer materials in the artificial intelligence age. Journal of Biomaterials Applications. 2023 Mar;37(8):1355–1375.

[7] Omidian H. AI-powered breakthroughs in material science and biomedical polymers. Journal of Bioactive & Compatible Polymers. 2024;0(0):161–174.

[8] Gharibshahian M, Torkashvand M, Bavisi M, Aldaghi N and Alizadeh A. Recent advances in artificial intelligence strategies for tissue engineering and regenerative medicine. Skin Research and Technology. 2024 Sep;30(9):e70016.
[9] Serrano DR, Luciano FC, Anaya BJ, Ongoren B, Kara A, Molina G, Ramirez BI, Sánchez-Guirales SA, Simon JA, Tomietto G and Rapti C. Artificial intelligence (AI) applications in drug discovery and drug delivery: Revolutionizing personalized medicine. Pharmaceutics. 2024 Oct 14;16(10):1328.
[10] Yan C and Li G. The rise of machine learning in polymer discovery. Advanced Intelligent Systems. 2023 Apr;5(4):2200243.
[11] Juan Y, Dai Y, Yang Y and Zhang J. Accelerating materials discovery using machine learning. Journal of Material Science Technology. 2021;79:178–190.
[12] Zhou H, Fang Y and Gao H. Using active learning for the computational design of polymer molecular weight distributions. ACS Engineering Au. 2023;4(2):231–240.
[13] Pyzer-Knapp EO, Pitera PWJ, Takeda S, Laino T, Sanders DP, Sexton J, Smith JR and Curioni A. Accelerating materials discovery using artificial intelligence, high-performance computing and robotics. Npj Comput Mater. 2022;8(1):84.
[14] Deneault JR, Chang J, Myung J, Hooper D, Armstrong A, Pitt M and Maruyama B. Toward autonomous additive manufacturing: Bayesian optimization on a 3D printer. MRS Bulletin. 2021;46:566–575.
[15] Reiser P, Neubert M, Eberhard A, Torresi L, Zhou C, Shao C, Metni H, Van Hoesel C, Schopmans H, Sommer T and Friederich P. Graph neural networks for materials science and chemistry. Communications Materials. 2022 Nov 26;3(1):93.
[16] Agarwal R, Singh J and Gupta V. Predicting the compressive strength of additively manufactured PLA-based orthopedic bone screws: A machine learning framework. Polymer Composites. 2022 Aug;43(8):5663–5674.
[17] Himanen L, Geurts A, Foster AS and Rinke P. Data-driven materials science: Status, challenges, and perspectives. Advanced Science. 2019 Nov;6(21):1900808.
[18] Xie C, Qiu H, Liu L, You Y, Li H, Li Y, Sun Z, Lin J and An L. Machine learning approaches in polymer science: progress and fundamental for a new paradigm. SmartMat. 2025;6:e1320.
[19] Dong Y, Yang T, Xing Y, Du J and Meng Q. Data-driven modeling methods and techniques for pharmaceutical processes. Processes. 2023 Jul 13;11(7):2096.
[20] Bhattamisra SK, Banerjee P, Gupta P, Mayuren J, Patra S and Candasamy M. Artificial intelligence in pharmaceutical and healthcare research. Big Data and Cognitive Computing. 2023 Jan 11;7(1):10.
[21] Farizhandi AA, Alishiri M and Lau R. Machine learning approach for carrier surface design in carrier-based dry powder inhalation. Computers & Chemical Engineering. 2021 Aug 1;151:107367.
[22] Agarwal R, Singh J and Gupta V. Application of machine learning in rotary ultrasonic-assisted orthopedic bone drilling: A biomechanical pull out in vitro study. Proceedings of the Institution of Mechanical Engineers Part C Journal of Mechanical Engineering Science. 2024 Dec;238(23):10911–10921.
[23] Zeb A, Qureshi OS, Kim HS, Kim MS, Kang JH, Park JS and Kim JK. High payload itraconazole-incorporated lipid nanoparticles with modulated release property for oral and parenteral administration. Journal of Pharmacy and Pharmacology. 2017 Aug;69(8):955–966.
[24] Khong J, Wang P, Gan TR, Ng J, Anh TT, Blasiak A, Kee T and Ho D. The role of artificial intelligence in scaling nanomedicine toward broad clinical impact. Nanoparticles for Biomedical Applications. 2020 Jan 1;1:385–407.
[25] Liu D, Yang F, Xiong F and Gu N. The smart drug delivery system and its clinical potential. Theranostics. 2016;6(9):1306.
[26] Ginsburg GS and Phillips KA. Precision medicine: From science to value. Health Affairs. 2018 May 1;37(5):694–701.

[27] Rajitha A, Kansal L, Raj G, Kalra R, Dhamija K and Abdul-Zahra DS. Biomaterials and artificial intelligence: predictive modeling and design. In: E3S Web of Conferences. 2024; (Vol. 505, (pp. 01003). EDP Sciences.
[28] N. Musthafa HS, Walker J, Rahman T, Bjørkum A, Mustafa K and Velauthapillai D. In- silico prediction of mechanical behaviour of uniform gyroid scaffolds affected by its design parameters for bone tissue engineering applications. Computation. 2023 Sep 12;11(9):181.
[29] Rojek I, Mikołajewski D, Dostatni E and Macko M. AI-optimized technological aspects of the material used in 3D printing processes for selected medical applications. Materials. 2020 Nov 29;13(23):5437.
[30] Conev A, Litsa EE, Perez MR, Diba M, Mikos AG and Kavraki LE. Machine learning- guided three-dimensional printing of tissue engineering scaffolds. Tissue Engin (Part A). 2020 Dec 1;26(23–24):1359–1368.
[31] Elsen R and Nayak S. Artificial intelligence-based 3D printing strategies for bone scaffold fabrication and its application in preclinical and clinical investigations. ACS Biomaterials Science & Engineering. 2024 Jan 22;10(2).
[32] Chen H, Zhang B and Huang J. Recent advances and applications of artificial intelligence in 3D bioprinting. Biophysics Reviews. 2024 Sep 1;5(3):031301.
[33] Al-Antari MA. Artificial intelligence for medical diagnostics – Existing and future AI technology. Diagnostics. 2023 Feb 12;13(4):688.
[34] Kumar Y, Koul A, Singla R and Ijaz MF. Artificial intelligence in disease diagnosis: A systematic literature review, synthesizing framework and future research agenda. Journal of Ambient Intelligence and Humanized Computing. 2023 Jul;14(7):8459–8486.
[35] Kumar R, Sharma A, Sharma P and Thaman R. Role of artificial intelligence in diagnosis and treatment of various medical diseases in patients. AMEI's Current Trends in Diagnosis & Treatment. 2021 Dec 8;5(2):92–98.
[36] Al-antari MA. Advancements in artificial intelligence for medical computer-aided diagnosis. Diagnostics. 2024 Jun 15;14(12):1265.
[37] Chassagnon G, Vakalopoulou M, Paragios N and Revel MP. Artificial intelligence applications for thoracic imaging. European Journal of Radiology. 2020 Feb 1;123:108774.
[38] Liu B, Chi W, Li X, et al. Evolving the pulmonary nodules diagnosis from classical approaches to deep learning-aided decision support: Three decades' development course and future prospect. Journal of Cancer Research and Clinical Oncology. 2020;146:153–185.
[39] Nair A, Bartlett EC, Walsh SL, Wells AU, Navani N, Hardavella G, Bhalla S, Calandriello L, Devaraj A, Goo JM and Klein JS. Variable radiological lung nodule evaluation leads to divergent management recommendations. European Respiratory Journal. 2018 Dec 20;52(6): 1801359.
[40] Agarwal R, Singh J and Gupta V. An intelligent approach to predict thermal injuries during orthopaedic bone drilling using machine learning. Journal of the Brazilian Society of Mechanical Sciences and Engineering. 2022 Aug;44(8):320.
[41] Agarwal R, Singh J and Gupta V. A data-driven ensemble machine learning approach for predicting the mechanical strength of 3D printed orthopaedic bone screws. Proceedings of the Institution of Mechanical Engineers, Part E: Journal of Process Mechanical Engineering. 2023 Nov;23:09544089231211235.
[42] Johnson KB, Wei WQ, Weeraratne D, Frisse ME, Misulis K, Rhee K, Zhao J and Snowdon JL. Precision medicine, AI, and the future of personalized health care. Clinical and Translational Science. 2021 Jan;14(1):August 29. 2024.
[43] Mabrouk M, Das DB, Salem ZA and Beherei HH. Nanomaterials for biomedical applications: Production, characterisations, recent trends and difficulties. Molecules. 2021 Feb 18;26(4):1077.
[44] Gokcekuyu Y, Ekinci F, Guzel MS, Acici K, Aydin S and Asuroglu T. artificial intelligence in biomaterials: A comprehensive review. Applied Sciences. 2024 Jul 28;14(15):6590.

[45] Srivastava S, Varshney B, Sharma VP and Ali B. Applications of artificial intelligence in polymer manufacturing. Application of Artificial Intelligence in New Materials Discovery. 2023 Jul 5;147:105–122.
[46] Lad S. AI-driven automation in custom manufacturing: enhancing precision and efficiency in automotive components production. Journal of Artificial Intelligence General Science (JAIGS) ISSN: 3006-4023. 2024 Aug 18;5(1):372–380.
[47] Li D, Ru Y, Chen Z, Dong C, Dong Y and Liu J. Accelerating the design and development of polymeric materials via deep learning: Current status and future challenges. APL Machine Learning. 2023 Jun 1;1(2): 021501.
[48] Wei R and Mahmood A. Recent advances in variational autoencoders with representation learning for biomedical informatics: A survey. IEEE Access. 2020 Dec 31;9:4939–4956.
[49] Gracias A and Broklyn P. Development of sustainable polymers using AI-driven molecular modeling polymer. Polymer Dec 2024;1:105–112.
[50] Idumah CI. Recent advancements in self-healing polymers, polymer blends, and nanocomposites. Polymers and Polymer Composites. 2021 May;29(4):246–258.
[51] Kroeze RJ, Helder MN, Govaert LE and Smit TH. Biodegradable polymers in bone tissue engineering. Materials (Basel). 2009 Jul 24;2(3):833–856. doi: 10.3390/ma2030833. PMCID: PMC5445745.
[52] Dufour T. From basics to frontiers: A comprehensive review of plasma-modified and plasma-synthesized polymer films. Polymers. 2023 Aug 30;15(17):3607.
[53] Brunsveld L, Folmer BJ, Meijer EW and Sijbesma RP. Supramolecular polymers. Chemical Reviews. 2001 Dec 12;101(12):4071–4098.
[54] Agarwal R, Singh J and Gupta V. Prediction of temperature elevation in rotary ultrasonic bone drilling using machine learning models: An in-vitro experimental study. Medical Engineering & Physics. 2022 Dec 1;110:103869.
[55] Srivastava S, Varshney B, Sharma VP and Ali B. Applications of artificial intelligence in polymer manufacturing. Application of Artificial Intelligence in New Materials Discovery. 2023 Jul 5;147:105–122.
[56] Karasu T, Çalışır F, Pişkin S, Özgür E, Armutcu C, Çorman ME and Uzun L. An intriguing future is approaching: Artificial intelligence meets molecularly imprinted polymers. Journal of Pharmaceutical and Biomedical Analysis Open. 2024 Sep;5:100041.
[57] Jayaraman A and Olsen B. Convergence of artificial intelligence, machine learning, cheminformatics, and polymer science in macromolecules. Macromolecules. 2024 Aug 9;57(16):7685–7688.
[58] Agarwal R, Gupta V, Singh J and Jain V. Prediction of surface roughness and cutting force induced during rotary ultrasonic bone drilling via statistical and machine learning algorithms. Proceedings of the Institution of Mechanical Engineers Part C Journal of Mechanical Engineering Science. 2022 Dec;236(23):11123–11135.
[59] Gokcekuyu Y, Ekinci F, Guzel MS, Acici K, Aydin S and Asuroglu T. Artificial Intelligence in Biomaterials: A Comprehensive Review. Applied Sciences. 2024;14(15):6590.
[60] Adhikari C. Polymer nanoparticles-preparations, applications and future insights: A concise review. Polym-Plast Technol Engineer. 2021 Dec 12;60(18):1996–2024.

Kanaka Durga Devi Nelluri, S.K. Abdul Rahaman*, Minakshi Kommoju, and Bhavana Alaparthi

Chapter 9
Predictive precision: AI-enhanced measurement of mechanical properties in nanostructured polymer composites

Abstract: Nanostructured polymer composites have revolutionized materials science with fabulous mechanical properties and diverse applications. However, assessing the measurements and predictions of such properties remain grossly complex because of the interactions at the nanoscale. Therefore, this chapter discusses the transformative role artificial intelligence (AI) has in improving the measurement or predictive accuracy of mechanical properties in nanostructured polymer composites. It incorporates the latest in AI algorithms together with experimental and simulation-based methodologies to characterize key properties such as tensile strength, elasticity, and impact resistance precisely. By training machine learning models on a large dataset, which encompasses a range of material composition, manufacturing parameters, and testing conditions, important property–structure relationships can be identified. The chapter also focuses on the use of AI in optimizing material design and reducing experimental overheads by predictive modeling. It demonstrates the possibility of successful application of AI-driven insights for tailored properties to advanced aerospace, biomedicine, and energy storage applications. Finally, issues like data trustworthiness, computational expense, and interpretability are taken into account when providing future opportunities and combining AI with new technologies like quantum computing and molecular simulations. This chapter describes the widespread nature of the potential application of AI in formulating and analyzing the next generation of polymer composites.

***Corresponding author: S.K. Abdul Rahaman,** Dean and Professor, Department of Pharmacy, School of Medical and Allied Sciences, Galgotias University, Greater Noida, Uttar Pradesh, India, e-mail: rahamandr123@gmail.com
Kanaka Durga Devi Nelluri, Minakshi Kommoju, Bhavana Alaparthi, KVSR Siddhartha College of Pharmaceutical Sciences, Vijayawada, Andhra Pradesh, India

https://doi.org/10.1515/9783111681870-009

Keywords: Nanostructured polymer composites, artificial intelligence (AI), mechanical properties, predictive modeling, machine learning, material design, property–structure relationships, advanced applications

9.1 Introduction

The rapid pace of advancement of additive manufacturing technologies, especially polymer composites, has revolutionized the design and production of materials. Among such innovations, artificial intelligence (AI) in the predictive modeling of mechanical properties stands as a transforming approach that makes material formulations and processing parameters exactly optimized. This chapter covers the crossroads of AI with nanostructured polymer composites (NPCs) to understand how techniques enhanced with AI improve predictability, hence furthering high-performance material development [1–3].

Material performance largely depends on tensile strength, elasticity, toughness, and resistance to fatigue, mechanical properties that all depend on several factors including type of nanofiller, their distribution, type of matrix, and the process conditions applied to NPCs. The classical methods of prediction of these types, such as empirical models and finite element analysis, fail to properly account for all these complex interactions. AI, specifically, is machine learning and deep learning that is able to look into massive quantities of data, discover the patterns, and generate probabilistic predictions that include nonlinear effects. AI models are very adaptive, and they are continuously improving on new data. These models get better with time and optimize the hyperparameters so that development becomes smoother and is not based on trial and error.. The best formulation and processing, guided by predictions from AI, give the desired properties for specific applications, such as composites with specific strength at lightweight in aerospace and a mechanical integrity of biocompatible materials in biomedical engineering. Even so, challenges lie ahead. With high–quality data and availability for training robust models, complex algorithms become hard to interpret. So, overcoming the hurdles will open the way for using AI for exploitation of full potential in predictive modeling to drive innovations in NPCs for diverse industries [4–6].

9.1.1 Nanostructured polymer composites

Polymer composites are the most significant step forward in the history of materials science. NPC materials utilize some unique properties from nanomaterials combined with polymers, such as flexibility. Materials within this family have initiated wide interest primarily due to their enhanced mechanical, thermal, and electrical properties and usability in a variety of applications from aerospace to biomedical engineering.

This overview covers the basic concepts of NPCs – what they are, their properties, and ways of preparation and application – with an emphasis on the incorporation of nanomaterials as one of the ways to improve the functionalities of polymer matrices.

9.1.1.1 Composition of polymers

NPCs tend to contain a polymer matrix with a very broad range of nanofillers such as nanoparticles, nanofibers, or nanotubes. Carbon-based fillers, mainly carbon nanotubes (CNTs) and graphene, along with some inorganic particles such as silica, clay, and metal oxides, dominate the list of commonly used fillers. The choice of the nanofiller will considerably be influenced by the properties of the composite due to the high surface area-to-volume ratio that nanomaterials possess with resulting better interaction with the polymer matrix, yielding superior mechanical and thermal properties. The nanofillers are dispersed in the continuous phase, which is the polymer matrix. From these, amongst the thermoplastics are polyethylene, polypropylene, and polystyrene while amongst the thermosetting resins are epoxy and phenolic. The selection of the polymer matrix is vital since it needs to provide the required mechanical properties while being compatible with the nanofillers for uniform dispersion. Nanofillers could be classified based on origin and structure. For instance, CNTs possess superior mechanical strength and electrical conductivity, and their uses are indeed in lightweight materials with high conductivity requirements. Inorganic nanoparticles such as silica and clay are used to enhance the thermal stability and barrier properties. The overall properties of the composite will be greatly enhanced with the nanofillers incorporated into the polymer matrix [7].

9.1.1.2 Properties of nanostructured polymer composites

NPCs with added nanomaterials exhibit better properties than conventional polymers. The most significant improvements are in the following aspects:

NPCs have improved mechanical properties such as tensile strength, stiffness, and toughness. The nanofillers can serve as reinforcement for the polymer matrix; therefore, the stronger material will be derived. For instance, in the case of CNTs, the tensile strength of polymers can be severalfold. Thus, they will be able to work in high-performance conditions. The thermal stability of nano-structured polymer composites is generally more pronounced compared to that of the pure polymers. Nanofillers often help in acting as a barrier towards thermal activity thereby lowering the rates of heat transfer while also increasing thermal degradation temperatures in a composite material. This would make them applicable where the involved material would work in elevated temperature environments. Polymer composites can provide very high enhancement in electrical conductivity if filled with conductive nanofillers like CNTs or graphene. Actually, one of the significant applications for the barrier properties is in electronics, sensors, and electromagnetic shielding. Advanced technologies also lie open by their ability to control the electrical property of such composites. NPCs generally exhibit higher barrier properties towards gases and liquids. Nanofillers can offer a tortuous path for the molecules, which facilitates a decrease in the permeability of the composite. This is highly advantageous for packaging applications where barrier performance forms a dominant factor.

9.1.1.3 Fabrication techniques

In the fabrication process, there exist different techniques for NPC along with merits and demerits. Solution mixing is the process in which a suitable solvent dissolves the polymer, and nanofillers are added to the solution. This method ensures the uniform dispersion of nanofillers, but extra steps may be required to remove the solvent [9, p. 1]. It is the technique of melting blending polymer and nanofillers. This method is extensively used for thermoplastic polymers, which gives good dispersion of nanofillers. The process is not very effective for thermosetting resins. In situ polymerization is used for the polymerization of monomers in the presence of nanofillers. The method may provide excellent bonding between the nanofillers and polymer interface, leading to enhanced properties. Electrospinning is used to prepare nanofibers from polymer solutions. Depending on the incorporation of nanofillers into the electrospun fibers, it is possible to prepare nanocomposite mats. Electrospinning nanofibers are applied in filtration as well as tissue engineering.

9.1.1.4 Applications of nanostructured polymer composites

NPCs possess exclusive properties, and as such, applications are seen in a very broad number of applications. In aerospace and automotive, fuel efficiency and performance are enhanced by lightweight, high-strength materials. The NPCs have been increasingly used in structural parts, interior components, and lightweight panels. The NPCs are also used in biomedical devices, drug delivery systems, and tissue engineering scaffolds. These are eco-friendly, and the mechanical behavior of these can be controlled for their applications in medicine. Their electrical conductivity does great service for electronic devices, sensors, and conductive coatings. Their lightness also further aids flexible and portable electronic systems. It aims at improving the properties of the barriers used in packaging through applying NPCs with higher shelf lives to food products for environmental protection. NPCs marked the vanguard in materials science as a whole with promised great performances with regard to potential versatile applications. Materials, in which the polymer matrix is blended with well-chosen nanofillers, turn out to have improved mechanical, thermal, electrical, and barrier properties than the conventional composites. There is much research that continues to emerge, opening the door for developing new applications as well as improvement of the current technology [8, 9].

9.1.2 Importance of mechanical properties in materials science

Mechanical parcels are abecedarian characteristics defined in terms of how kits behave when subjected to usable forces and conditions in the environment. Such parcels are essential in material wisdom, determining kits' choice, design, and operation in all colorful spheres of diligence that include aerospace, automotive, construction, and biomedical engineering. To masterminds and scientists, knowledge of mechanical parcels is important to predict materials. Basic mechanical properties are the characteristics that define the behavior of a material under stress. Tensile strength is defined as the ultimate quantum of tensile (pulling) stress that a material can resist to failure. The property is very vital for accoutrements used in structural operations. It determines cargo-carrying factors. High tensile strength is fundamental for accoutrements utilized in construction, islands, and aircraft, where consummate safety and structural integrity have to be adhered to. Yield strength is that stress at which a material begins to deform plastically. Beyond this value, the cargo will no longer return to its original shape once it is unloaded. Knowing yield strength is important so that masterminds can ascertain whether fittings have sufficient strength for resistance to function loads without causing permanent deformation – that is, safe for safety-critical applications. Rigidity is the capacity of a material to deform plastically before breaking. This is a very useful attribute in applications in which parts experience dynamic loads, as in the automotive crash factors where energy absorption will save lives. Hardness is the measurement of resistance to localized plastic distortion, for instance, indentation. This is a very vital property for wear and tear-resistant accoutrements, such as cutting tools and ministry factors. The hardness of the accoutrements makes them retain shape and functionality even in abrasive conditions, thereby increasing the product's lifetime. Durability is the ability to be able to withstand the absorption by energy without breaking. Toughness is an elastic deformation capacity; it is a very crucial property for equipment, especially for shock-resistant applications such as defensive equipment and structural parts of buildings. The equipment has an ability to withstand the unexpected forces without breaking thus providing safety and reliability. Fatigue strength is the surface stress which a material is able to withstand for a given number of cycles without fracture. This is the characteristic, though critical, of accoutrements that face cyclic loading in the system; these include the rotating ministry and structural elements. Understanding the fatigue strength of materials enables designers to develop components that can withstand repeated stress without failure. Mechanical properties play a crucial role in material selection. [10, 11].

9.1.2.1 Importance in design processes

Mechanical parcels are critically involved in the design processes of engineering factors. Designers apply these parcels to:
1. Perform stress analysis
2. Optimize material operations
3. Formulation of design criteria
4. Guide manufacturing processes
5. Performance evaluation

This data is pivotal for:
1. Quality control
2. Failure analysis
3. Research and development

9.1.3 Role of AI in modern material characterization

Having the possibility of understanding and analyzing a range of physical and chemical properties in materials science has made material characterization important. Data on the composition and properties of materials has been provided through extensive and significant production from X-ray diffraction, scanning electron microscopy, and transmission electron microscopy though frequently involving human time to interpret the high amounts of their complicated data. AI and ML enabled automation in data analysis, improvement of accuracy, and the accelerated discovery of new materials in material characterization. Below is a summary of applications, advantages, challenges, and future directions of AI in contemporary material characterization.

9.1.3.1 AI techniques in material characterization

AI is a technique that has been applied in and encompasses a vast array of techniques in the characterization of material; these incorporate a machine learning algorithm, deep models, and myriad data mining approaches. This technology can analyze great amounts of dataset generated with several characterization methods while providing meaningful insight and patterns, perhaps which cannot unveil through simple analysis.

9.1.3.1.1 Machine learning algorithm

It primarily depends on ML algorithms, which vary from support vector machines (SVMs), and decision trees to random forests. Algorithms are trained on all the data that are present and correspond to material characteristics – their input features. For instance, one should predict the best material based on the models of ML about the mechanical properties, which can be related best to chemical composition. That is, they will guide in developing the best material for the intended use.

9.1.3.1.2 Deep learning algorithms

Deep learning is the ML subcategory in which one can train the several-layered neural networks to study the complex data. In order to work on analyzing the data coming out from SEM and TEM, convolutional neural networks (CNNs) are very effective. They are further trained over the labeled datasets so as to classify the particle morphology, size, and distribution autonomously. This thus saves large amounts of time, which one would have spent if they were analyzing them manually.

9.1.3.1.3 Data mining techniques

Data mining can be utilized for extracting informative material from the volumes of data produced from characterization techniques. Trends, correlations, and anomalies in data are identified, which lends insight into the behavior and performance of the material. For example, clustering algorithms group similar materials by their properties, thus discovering new materials with desired characteristics [12, 13].

9.1.3.2 AI in material characterization applications

Material characterization with the involvement of AI has been utilized in multiple industries and has simultaneously increased the level of efficiency involved in material exploration. Applications of AI are found in surface analysis techniques like XPS (X-ray photoelectron spectroscopy) and AFM (atomic force microscopy). AI calculates the XPS spectrum automatically, giving an extremely accurate representation in regard to the composition and chemical state of the surface. Similarly, algorithms of AI, used in the evaluation of images from AFM, can be applied for the calculation of the surface roughness and morphology responsible for material performance. Scientists report techniques of deep learning model classification for the first time, applicable to a crystal structure derived from X-ray diffraction (XRD) data. Models are trained in known crystal structures, and then algorithms can automatically classify a material by its diffraction pattern. That will increase the speed of new materials identification, but also make easier the search for novel crystal phases.AI can be used in predictive modeling in material characterization in relating material properties with the structural features. ML models, for instance, can be utilized with the data gained from microstructural characterization methods to predict mechanical properties. Predictive capability has enabled the development of materials with engineered properties that meet requirements for application and thus avoid the trial-and-error method. This will allow the combination of AI and high-throughput characterization methods to analyze large samples much more quickly. Automated systems can be fitted with AI algorithms that will process data streaming from several characterization methods at the same time. The approach could greatly accelerate material discovery, which is highly useful for fields like energy storage and catalysis where materials need to be screened as quickly as possible.

9.1.3.3 Advantages of AI in material characterization

The benefits of the utilization of AI in material characterization are that big datasets will be processed and analyzed quickly through the algorithms used in AI as compared to the traditional techniques. Data interpretation time would reduce considerably. More attention can then be devoted towards the design of experiment and material preparation than to analysis. Artificial methods minimize the human error rates in data interpretation and allow characterization of the material. Improved designs of ML models ensure that very little patterns and data linking patterns are under observation. This in turn delivers probable results unseen by human analysts. These big and complicated sets of data could be called the path through which data-driven revelations are given to the AI; it enables learners to know about the behavior of the material. Trends and interrelationships within the data guide researchers in designing and optimizing material systems. Applying AI to material characterization hastens the finding of new materials automatically by doing analyses. This is a very critical aspect whereby fast identification of promising candidates can be done in drug discovery and energy materials.

9.1.3.4 Challenges in applying AI to material characterization

Applying AI to the characterization of materials comes with numerous challenges though the benefits are many; some of the challenges are the amounts and qualities of the data available have great impact on the AI systems. Most available datasets are relatively small in size and noisy or biased. This limits the performance in ML models. It would be necessary in the successful applications of AI if high-quality as well as diverse sets of datasets have been used. Most AI models are extremely difficult, especially deep learning, and to explain precisely why a specific model arrived at a certain type of decision is really tough. It means understanding the mechanisms driving the AI model from the inside out so the predictions can become trustworthy and appropriate for material characterization. This usually makes it a bit more complicated to integrate AI into the existing material characterization workflow. One of the challenges that researchers, including compatibility of new instrumentation with existing data management systems, experience is the incorporation of novel technology with established practice [14, 15].

9.2 Fundamentals of nanostructured polymer composites

NPCs represent the class of advanced materials, where polymers are mixed with nanoscale fillers. These materials enhance the properties and functions of the polymer greatly. It is the remarkable properties of nanomaterials such as high surface area, strength, and electrical conductivity, which make it a better alternative to traditional materials. NPCs can offer enhancement in the mechanical, thermal, and barrier properties of the polymer matrix. In this chapter, the general overview of NPCs in terms of composition, properties, fabrication processes, and applications is presented.

9.2.1 Nanostructured polymer composite structure

The continuous phase is the polymer matrix holding the nanofillers. Common polymer matrices are thermoplastics such as polyethylene, polypropylene, and polystyrene; also included are thermosetting resins: epoxy and phenolic, among others. Selection of the polymer matrix is an important step. On the one hand, it must offer the desired properties from mechanical performance and must show compatibility with the nanofillers in order to enable uniform dispersion of nanofillers. The classification of nanofillers is based on origin and structure. The most frequently used nanofillers include carbon-based nanomaterials; graphene exhibits very good mechanical strength and electrical conductivity. These could greatly improve the mechanical and electrical characteristics of the composite when added to the polymer matrix. Inorganic fillers are most often silica, clay, and metal oxides. These can enhance thermal stability, barrier properties, and mechanical strength. For instance, silica nanoparticles increase thermal stability of polymer composites. Interest in the use of sustainable alternatives to synthetic fillers, such as natural nanofillers is growing; interest in natural fibers and biopolymers is also growing. These can have good mechanical properties along with green capability (see Table 9.1) [16–19].

9.2.2 Mechanical properties: key metrics

Mechanical properties are simple measurements that describe how materials respond to coupled powers and natural conditions. In the context of nanoparticle-based poly-

Table 9.1: Overview of nanostructured polymer composites: composition, properties, and applications.

Aspect	Details
Matrix	Polymer (thermoplastics or thermosets)
Nanofillers	– Carbon-based (e.g., CNTs, graphene) – Inorganic (e.g., silica and metal oxides) – Natural (e.g., natural fibers and biopolymers)
Properties	– Mechanical: Increased strength and stiffness – Thermal: Improved heat resistance – Electrical: Enhanced conductivity (e.g., CNTs and graphene) – Barrier: Better resistance to gases and liquids
Fabrication methods	– Solution mixing, in situ polymerization, melt blending, and electrospinning
Applications	– Aerospace, automotive, biomedical, electronics, and packaging

mer composites, these properties are fundamental to optimize fabric performance in various applications. The chapter addresses the key mechanical characteristics of polymer nanocomposites reinforced with nanoparticles and highlights the following: ductile quality, surrender quality, ductility, hardness, durability, and weariness resistance. Pieces of knowledge are derived from various sources that analyze the basic and mechanical properties of such advanced materials.

Tensile quality is the highest stretching that a textile can endure while being stretched or pulled some time back before rupture. It is a critical property of materials used in simple applications. For nanoparticle-based polymer composites, the integration of nanoparticles like CNTs or silica can significantly enhance flexible strength. CNT composites lead to the positing that due to their flexibility, the composites of polymers reinforced with CNTs may be even more flexible than the pure polymer network. For instance, poly(methyl methacrylate) (PMMA) composites containing CNTs can be more than 36% ductile than the unreinforced matrix [22, pp. 1, 6]. Silica nanocomposites and in addition, the dilution of silica nanoparticles into polymer networks have been found to improve the flexible nature due to successful stack exchange between the framework and the nanoparticles. This property is especially critical in applications where materials are subjected to tall loads. Yield quality is the extension at which a material starts to plastically distort. After passing this point, the fabric will not revert to its original shape when the stack is evacuated. For example, the surrender qualities of CNT-strengthened composites are altogether higher compared to those of unadulterated polymer, whose values can reach up to 35 MPa for certain formulations. Interfacial retention between the polymer lattice and the nanoparticles is crucial to establish the abdicate quality. Improvements in interfacial bonding result in better stress transfer, leading to higher tensile strength and yield strength. Ductility is a material's ability to deform plastically for a certain period of time before it breaks. It is a key property for materials that need to absorb energy on deformation, for example, car applications. Agglomeration of nanoparticles could potentially affect the ductility of polymer composites. Though it is indicated that ductility loss occurs due to the hardness of nanoparticles, other results indicate that the optimal composition of nanoparticles enhances ductility through more homogenized stretch distributions. For instance, CNT-reinforced PMMA composites have demonstrated improved ductility, with abdicate strains as high as 0.9%. Hardness is a measure of the material's resistance to localized plastic distortion (e.g., space). It is a basic property for applications requiring wear resistance. Nanoparticles, such as silica or CNTs, will significantly improve the hardness of polymer composites. For instance, it is estimated that the use of silica nanoparticles will result in the improvement of the hardness level of polymer composites by 20–30%.

The improvement in hardness finds immense application in coatings, for which wear resistance is vital. Toughness is the capacity of a material to absorb energy and plastically deform without breaking. It is a key property for applications requiring impact resistance. Nanoparticle reinforcement improves the toughness of polymer composites by increasing the energy absorption in deformation. For instance, composites reinforced with CNT have been reported to exhibit improved toughness due to the energy dissipation capability of CNTs at fracture. Strength can be estimated by performing affect tests, which measure the amount of vitality remaining in the material just before failure. Research studies have found that the strength of CNT-reinforced polymers is likely to be much higher than that of unreinforced polymers. Fatigue resistance is the capacity of a material to withstand cyclic loading without cracking. It is a key property for all materials for applications that required repeated tensile loading, for example, car and aeronautical components. Nanoparticles can enhance the resistance of polymer composites to weariness. Composites reinforced with carbon nanotubes (CNTs) have been reported to exhibit significantly improved fatigue life, which can be attributed to the highly ordered alignment and superior energy transport properties of the CNTs. Typically, weakness resistance is checked using cyclic loading tests where the number of cycles to failure is recorded. The greater resistance to weakness in nanocomposites may cause a longer life of benefits as well as greater reliability (see Table 9.2) [20–23].

9.2.3 Challenges in conventional measurement techniques

In characterization of the mechanical, thermal, and structural properties of polymer nanocomposites, standard measurement methods have been prominent. The main process in the context of desired properties is the homogeneity of nanoparticles in the polymer matrix. However, these pose serious challenges in their applications and hinder the proper assessment and further development of advanced materials. Some of the challenges faced in conventional measurement techniques in polymer nanocom-

Table 9.2: Key mechanical properties of nanoparticle-based polymer composites.

Mechanical property	Details
Tensile strength	– Increased by up to 36% with CNTs (e.g., PMMA composites) – Silica nanoparticles also improve tensile strength via effective load transfer
Yield strength	– Enhanced by nanofillers like CNTs, reaching up to 35 MPa – Interfacial bonding plays a key role in improving stress transfer
Ductility	– Can improve or decrease depending on nanoparticle type and distribution – CNT-reinforced PMMA composites show improved ductility (yield strains up to 0.9%)
Hardness	– Increased by 20–30% with silica and CNTs – Important for wear-resistant applications (e.g., coatings)
Toughness	– Enhanced by CNTs that improve energy dissipation during fracture – Increased toughness in CNT-reinforced composites compared to unreinforced polymers
Fatigue resistance	– Improved with nanoparticles like CNTs due to enhanced load transfer and energy dissipation – Higher fatigue life in CNT-reinforced composites under cyclic loading

posites are discussed on the basis of insights from the recent literature on additive manufacturing, healthcare applications, and ordered polymer composite materials.

9.2.3.1 Drawbacks of conventional methods

9.2.3.1.1 Sample preparation

Sample preparation is the primary limitation of conventional measurement methods. Introductions of variability in handling and preparation can easily be made during the testing of polymer nanocomposites. The main process in the context of desired properties is the homogeneity of nanoparticles in the polymer matrix. Poor dispersibility may cause an inclination for agglomeration, and thus the mechanical and thermal measurements get distorted. Agglomerated nanoparticles tend to introduce defects within the composite. This may mean that they give unreliable readings for tests such as tensile strength and toughness. In the most fundamental ways, the problems in this approach are, therefore, SEM and TEM: preparation of sample tends to modify the dispersion in nanoparticles. Conventional methods fail to provide a solution, especially when the features are at the nanoscale. Methods such as XRD and AFM may lack enough resolution to determine the distribution and orientation of nanoparticles in the polymer matrix. The ability to distinguish nanoscale characteristics defines the polymer matrix–nanoparticles interaction. For example, XRD is not capable of differentiating between tightly spaced peaks, which are normally signatures of different phases or nanocrystal orientations and nonuniqueness that sometimes leads to very ambiguous interpretations in the process. The traditional approaches to the measurements might suffer from a lack of reproducibility. Such inconsistency will jeopardize data reliability and complicate making comparisons between the results of different studies. The majority of the conventional testing methods rely on trained operators to interpret the outcome of results. This allows room for human errors, and mechanical test result interpretation is different from operator to operator because of the varied experiences and understanding of the behavior of material.

9.2.3.1.2 Environmental sensitivity

Conventional measurement techniques are sensitive to environmental conditions like temperature and humidity. The properties of polymer nanocomposites strictly depend on these factors, which can influence the measuring outcome. For example, differential scanning calorimetry is a type of thermal analytical technique whose outcome could be susceptible to the thermal history of the sample. When the sample is not well-conditioned, then transitions can be detected to the extent that it does not precisely portray the true characteristics of the material.

9.2.3.1.3 Data interpretation difficulties

The data from the conventional measurement methods is complex and multidimensional for interpretation. For example, the data of mechanical testing normally needs advanced analysis to tease out meaningful insights regarding the performance of the material involved. The methods like DMA bring data at numerous frequencies and temperatures, which are sometimes not easy to understand unless complex modeling is used. It may lead to wrong conclusions sometimes and obstructs the formation of predictive models of material response. To avoid such consistency between different labs, a standardized set of procedures regarding the measurement for the properties of polymer nanocomposites could easily be established. The different methodologies of measurement as well as samples and conditions taken may make comparison complicated. The lack of standards for universally accepted values of tensile strength, modulus, and stability to heat can introduce enormous differences between the reported values. This variability makes it very challenging to progress in the field by establishing benchmarks for new material development. At the same time, the reliability of these values makes comparison tricky for novelties.

9.2.3.1.4 Increase in alternatives and solutions

Advanced techniques for characterization are used by scientists to improve the normal measurement techniques posed by conventional measures. This will be a way of transcending some of the problems in classical techniques, as it involves AI combined with ML into the analysis. These algorithms can look at complex data sets, establish patterns, and make predictions as to how the material may react compared to past data. Efforts are made to acquire standardization protocols in the measurement of polymer nanocomposites. Methods also involve AFM and scanning tunneling microscopy (STM) among many other synchrotron radiations, where higher resolution or even higher sensibility in determining features at a nanoscale is offered. Such high-resolution methods could possibly offer improved resolutions in terms of the morphological and dispersion characteristics of nanoparticles in a polymer matrix for better analysis of material properties. Overcoming these limitations will be important in furthering this area and optimizing polymer nanocomposites for the variety of applications envisioned. Sophisticated characterization methods, AI-based data analysis, and standardization could now possibly resolve these issues, thereby resulting in the creation of new materials with next-generation properties and functionalities. Further breakthroughs for polymer nanocomposites in several applications will require further research into evolving measurements [24–26].

9.3 Artificial intelligence in materials science

AI is transforming materials science by enhancing how materials are designed and characterized and optimized. AI techniques, especially ML and deep learning, enable researchers to analyze great amounts of data that come from experiments and simulations. For instance, AI can forecast material properties from composition and processing conditions, cutting down on time and expense associated with conventional trial-and-error processes. Development of the AI algorithms in material characterization allows the analysis of complex data from spectroscopy and microscopy to be automated. The accuracy and reproducibility will improve through such an application, which will benefit highly in advanced material development such as nanocomposites and biomaterials where precision is a major need for property control. New materials can be found by AI through patterns and correlations discovered in data that might have gone unnoticed for a human researcher. It fast-tracks innovation in developing novel materials to meet applications in the realms of healthcare, energy, and electronics. With AI, materials science promises to revolutionize the field, allowing materials to be designed with specific properties and functionalities.

9.3.1 Overview of AI and machine learning in materials science

AI and ML have emerged as the revolutionary technologies in material wisdom, which greatly enhanced the processes of material discovery, characterization, and optimization. This has exerted an impact on huge datasets and complex algorithms that reveal patterns, predict parcels, and automate the design of new accoutrements. This review has synthesized insights from colorful sources, culminating in the principles, operations, and future directions of AI and ML in materials science. AI is capable of performing tasks that demand human intelligence. Among them, literacy, logic, problem-solving, and understanding natural language stand out. AI can create machines that can mimic human cognitive functions and, thereby, human acclimatization and learning through experience to execute activities autonomously. ML is a field of AI, actually creating algorithms that enable computers to learn from and forecast data. Unlike traditional programming, which operates by giving clear instructions to a computer, ML algorithms improve with time as they are exposed to more data. This feature is very valuable in material wisdom, where intricate relationships between material portions and processing conditions often reside. ML may be classified into three fundamental classes: supervised learning trains a model in a dataset wherein the input affair dyads are known. The model learns how to collude the inputs to the labor and then give some prophetic suggestions on any data it was never exposed to earlier; unsupervised literacy, which is the training process of a model on an unlabeled dataset. Thus the algorithm detects such patterns and structure in the provided data without explicit direction. This method is effective for reduced dimensionality clustering; latent literacy, via which an agent learns to make decisions by taking actions that maximize a reward signal. often used within optimization problems as well as the simulations.

9.3.1.1 Operations of AI and ML in materials science

AI and ML are revolutionizing the discovery of new accoutrements with predictions of the parcels based on the composition and processing conditions. The traditional discovery of materials follows styles that typically consume much time and money in terms of just trial-and-error procedures. On the other hand, ML algorithms will be able to analyze datasets to pinpoint some potential candidates for a particular operation. Application researchers use ML to model the mechanical loads of composite parts to be quickly applied to optimize the expressions. The application of AI- and ML-based techniques is decreasing in the characterization of accoutrements. It automates complex data analysis, usually generated from colorful characterization styles. For example, ML algorithms rely on data that is generated by other methods, like spectroscopy and microscopy, which are fine-tuned for delicacy and reproducibility. Visualization of deep learning models has been applied to break down images obtained through scanning electron microscopy (SEM), which has allowed for automated bracketing of flyspeck morphology and size. Predictive modeling with AI and ML allows the experimenter to read material. Under different conditions, using literal data, models can be designed that provide predictive tools for the design and processing of materials. Materials processed under optimized AI- and ML-assisted conditions result in effi-

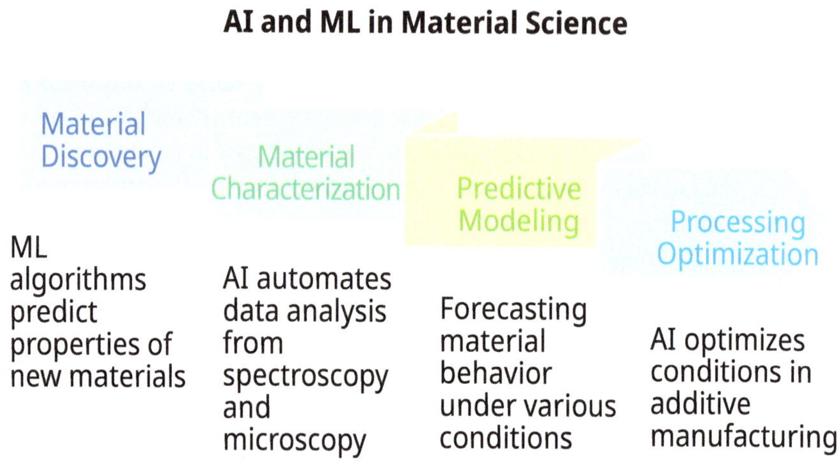

Figure 9.1: AI and ML in material science.

cient manufacturing with less loss. ML will find the optimality of processing conditions of certain issues by relationship testing between processing parameters as well as among material parcels for desired conditions (see Figure 9.1) [27–29].

9.3.1.2 Challenges and limitations

There are numerous challenges and limitations of the combination of AI and ML in materials science:

A) Data quality and integrity
B) Interpretability of models
C) Integration with existing workflows

9.3.2 Applications of AI in materials research

AI is in many ways integrated into materials research, which is dramatically changing the area of discovering, characterizing, and optimizing materials. AI methods, especially ML, allowed mass data analysis, such as predictive material properties but also streamlining experiments. The following section explains the various applications of AI in materials research and insights derived from different sources regarding its impact on different fields.

9.3.2.1 Accelerating materials discovery

Aiding AI allows for the high-throughput screening of materials because the process automatically processes large datasets of experiments and simulations. This type of ability can quickly allow researchers to discover which promising candidates could be used for specific applications by significantly reducing material discovery time. For instance, millions of materials can be screened for new catalyst designs, and the best material with the highest activity at the appropriate selectivity level for a specific reaction can be identified. AI predictive modeling can predict properties of materials when the composition is known along with the conditions used for processing. Such models can be trained with existing datasets that would prove handy in the coming future as more predictive tools are designed by researchers about how new materials will behave in various conditions. ML models allow the mechanical properties of polymer composites to be predicted, and here they can be used to identify the optimal formulation within a very short time [30].

9.3.2.2 Material characterization

AI methods are helpful in characterization by automating the analysis procedure whereby they work on very sophisticated data resulting from most techniques used in characterization. They mostly exclude the chance of an error and offer chances for reproducibility with the relatively short times needed to complete the analysis process of the data. Application of the deep learning networks is being done recently so as to automatically classify the morphology and size of particles in SEM. AI can take these unstructured sources of information, research papers, and patents to build knowledge graphs that allow easy access and retrieval of data and even its analysis. The knowledge graph enables the researchers to ask questions regarding the relationships between materials, properties, and processing conditions. The IBM Deep Search platform follows NLP algorithms that detect material-related entities in documents. It generates a knowledge graph that could be termed an information quest: complexities of querying for information [31].

9.3.2.3 Optimization of material properties

AI can optimize design experiments in a way that it might identify parameters, which are very likely to impact the outcome, thus performing the experiments to achieve some specific outcome. The process does enhance efficiency during material development. For example, Bayesian optimization methods can rank conditions of the experiment in respect to the achieved results so that researchers may concentrate on promising research lines. AI algorithms can optimize multiple material properties in one computation so that it can handle the tangled trade-offs encountered during designing materials most of the time. This is particularly helpful in applications wherein conflict requirements must be satisfied at the same time by the materials used. Energy storage materials can apply AI to optimize properties such as conductivity, stability, and capacity so that the final material meets all performance criteria [32].

9.3.2.4 AI in simulation and modeling

AI can be used to develop ML potentials, which may simulate the behavior of the materials very well. The simulation can be carried out at only a fraction of the computational effort for some other traditional approaches. The potentials will predict atom and molecule interactions; therefore, complex material systems can be explored. In recent times, ML potentials have been utilized to predict material properties in a solid state and even explore new compounds and phases. AI accelerates the simulation workflow because it allows the choice of auto and optimized computational protocols, which accelerate the workflow of a researcher. For instance, he can look for a larger design space and produce data in lesser time. Real-time feedback will allow AI-driven simulation frameworks to continually make dynamic adjustments to the parameters in order to enhance the efficiency of the computational experiments [33].

9.3.2.5 Applications in specific domains

AI has emerged as irreplaceable in energy materials, including application in batteries, fuel cells, and solar cells, with prediction of material and optimization of its composition that accelerates discoveries for energy performance materials.

A finding that illustrates electrode material utilization towards the development of a new Li–ion battery (LIB) of enhanced energy density and cycling life is demonstrated. Biomedicine field is a place where AI finds an application for developing and improving material designs related to drug delivery, tissue engineering, and medicinal device development. It can predict biocompatibility and efficiency for novel biomedical material development. AI-based strategies have been used in the fabrication of polymeric nanocomposites with the aim of enhancing drug delivery, which should yield expected results and thus, would optimize therapeutic response and lower the possibility of adverse reactions.AI is also used in the designing of structural materials such that their mechanical properties are optimized, and they function optimally considering various loading conditions. It has been used on a very vast scale in the aerospace and automobile industries because it all depends on the performance of materials. An ML algorithm is utilized to forecast the mechanical behavior of composite materials so that lightweight and durable structures are designed.

9.3.2.6 Challenges and future directions

A) Data quality and availability
B) Explainability of AI models
C) Improving the integration with existing workflows

9.3.3 Benefits of AI-enhanced measurement in materials science

This is revolutionary in the sense that it introduces an enhancement of accuracy, speed, and range in the analysis and characterization of materials, with ML and data analytics being used in complex data for automating analyses and gaining greater insights into material properties in measurement techniques augmented by AI. This chapter details the benefits of AI-enhanced measurement with information from the following sources.

9.3.3.1 Increased accuracy and precision

It is where the AI algorithm analyzes enormous data sets derived from measurement technologies such as spectroscopy, microscopy, and mechanical tests. Here, state-of-the-art statistical modeling and identification of patterns are deployed to uncover hidden trends or interactions that others may even fail to recognize by using other forms of analysis. Application of AI in recognizing phase has accuracy level comparable to humans since AI detects patterns based on diffraction peaks that may provide evidence that perhaps some phases or compounds exist. AI measurement system will eliminate or eradicate errors caused by human inaccuracy. For auto-generation of measurements data analysis results are entirely consistent across the entire experiment mainly because there is very minimal exposure to such variations that result from different experimenter's variations. Deep learning-based computer vision can classify the microscopic features of the material with high accuracy and also avoids subjective bias that is associated with manual inspection.

9.3.3.2 Efficiency

AI can automate repetitive measuring processes and, therefore save much time on data collection and processing. In this automation, researchers are able to allocate themselves to complex tasks that further speed up the pace of the research process. In high-throughput screening of materials, AI can automate measuring mechanical properties for a number of samples; it makes rapid evaluation of candidate materials that would be appropriate for specific applications. AI-based measurement systems allow real-time processing of data and provide immediate feedback to the researchers, thus enabling dynamic adjustment of experimental conditions based on real-time analysis for optimum measurement. In additive manufacturing, AI can monitor the printing process and adjust parameters in real time so that material properties are optimum with less chance of defects.

9.3.3.3 Predictive capabilities

AI can construct predictive models predicting material properties that could be a function of composition and conditions of processing and microstructural features. Such models will enable researchers to predict material performance effects of design variation: this will help in decision-making. ML models can predict tensile strength of polymer composites based on composition and parameters of processing such that one can optimize formulations without having to do physical testing beforehand. AI predicts new materials with required properties that are likely to expedite the discovery of a novel material; exploration of enormous design spaces is one of the key applications of AI, and once-promising candidates can be determined for further research. AI will predict the activities of the materials of different chemical composition according to the findings at much more speed than this current method of research.

2.3.3.4 Inclusive data integration

AI allows multisource measurement technique analyses of data so that the overall material property can be understood in deep detail since it forms a better whole, thereby providing an entire explanation of the overall material behavior. Inclusion of AI in the integrated outcome of mechanical testing, thermal analysis, and microstructural characterization would unravel the different property–interrelation patterns of an individual performance that can evolve during a generalized material response. Knowledge extraction from raw unstructured information like patents and research papers by AI methodologies, particularly NLP, have been shown to enable a researcher in making a knowledge base for guiding future studies. For instance, AI can scan scientific literature to identify trends in material research and the latest developments with potential applications for researchers.

9.3.3.5 Cost-effectiveness

The cost of materials research will significantly decrease because AI enhances the efficiency of measurement procedures, which would mean fewer experiments to produce reliable results. Cost-effectiveness is highly useful for industries characterized by tight budgets and very high materials costs. New alloys may use AI to optimize the composition and conditions of processing in a way that incurs minimum physical testing and waste generation. This measurement, augmented by AI, will determine the best research directions for resource allocation. Resources will be allocated to best materials with higher potential to ensure high return on investment in R&D. With AI focused on materials that are likely to provide high performance, resource allocation by research groups would be effective, and advancement in material technology would be speeded up.

9.3.3.6 Collaboration and knowledge sharing

Through this application of AI in the generation of visualizations, it would very well make very complicated results of measurement visually accessible to researchers so they could convey the information to a big number of audiences. Conditions of processing versus properties of material will become visually intuitive using AI-generated visualizations. It can rapidly propagate to the scientific as well as industrial world for further diffusion. In the future, AI along with perfect synergy with materials science will not only design and optimize materials but also do so with unprecedented precision and efficiency [34, 35].

9.4 AI-driven measurement techniques

AI-driven measurement techniques are revolutionizing the accuracy, efficiency, and scope of characterization in materials science. ML algorithms and data analytics process complex datasets generated by a wide range of measurement methods, including spectroscopy, microscopy, and mechanical testing. One of the critical benefits of measuring by AI is that it automatically does the data interpretation. Compared to traditional analyses, AI applies sophisticated statistical measures to assess patterns and correlation among data, so shorter time is taken to analyze data, which provides far better estimates of the material properties compared to traditional analyses. For example, deep learning models automate classification of microstructural features of the materials, meaning there is elimination of human error and consistency results. Another strength of AI-based methods is real-time data processing. This indicates that, at the time of experiments, immediate feedback will be provided to researchers. This also helps in the dynamical adjustment of experimental conditions so that measurement processes and general efficiency can be optimized. An important point is assimilation of data from a variety of sources and general view of material properties given by AI. The multimodal analysis of data will be useful for enhancing comprehension of how the factors affect the behavior of the material, thereby enhancing the decision-making process in material design and optimization.

9.4.1 Data collection and preprocessing in material engineering

Two major tasks of material engineering are collecting data and preprocessing when working with the application of AI and ML. This section explains the data gathering methodologies, tells why the need to preprocess data arises, and discusses the practices used to ready the data for use within AI for the application of materials science (see Table 9.3).

9.4.1.1 Data collection

There are multiple sources available, which provide the needed data for material engineering, and appropriately, one would term such data as being the information in general, most of which have been generated by conducting experiments in laboratories on mechanical tests, thermal analyses, and the measurement by spectroscopic methods. This experimental data would be much more detailed and explores the properties besides various behaviors in the case of different materials. Huge datasets of material properties, atomic resolution, etc. are generated using computational methods like DFT and MD simulations. Such data might include electronic structures, vibrational frequencies, and thermodynamic properties, which are extracted from the simulations. There have been several public databases created for exposing structure of materials and properties into the public realm. Some of the important ones include Materials Project, JARVIS–DFT, Open Quantum Materials Database, and such others that are storing curated data sets for data used during training AI models. From research articles in sciences, patents, and technical reports, material properties and range of material performance may be found. The unstructured sources, after recovery, may be further processed with the application of NLP techniques.

9.4.1.2 Diversity in data

In addition to developing robust models of AI, collection of different kinds of data ensures that diverse kinds of new and unseen material could be generalized from the data. This range is important in order to capture the very intricate relationships of the processing and the structure properties in performance, which form the fundamental principle of material engineering.

9.4.1.3 Data preprocessing

Data preprocessing is one of the steps that involve cleaning, transforming, and organizing the raw data to put it in its analysis form. The performance of AI models will depend on data quality; thus the need to preprocess, in search of correct results. Data cleaning is the removal of errors or inconsistencies in a dataset. AI models may be misled by missing values, outliers, and duplicate entries. Some common techniques that are used in data cleaning are imputation of missing values and the detection of outliers where anomalous data points may be removed or adjusted. Normalization and standardization methods scale normalize such that each feature has equal contributions to analysis. Standardization modifies data in that all the values have mean = zero and standard deviation = 1, but normalization is simply to rescale data into the range say, 0 through 1. These are very important if features have more than one size or unit.

An appropriate subset of a given database that possesses the best models and reduces model complexity likely increases performance. Techniques like correlation analysis and recursive feature elimination can be applied to tree-based methodologies to determine a subset of important variables, which affect material properties to the highest extent. Dimensionality reduction: if a feature set becomes too large in size, there is the threat of overfitting and increases in the number of computations necessary. Principal component analysis (PCA) finds a special mention here. These, along with a few other methods of dimension reduction, serve as its aid further ahead. t-SNE falls within this category. Artificial data augmentation techniques can be used if the dataset is small. Interpolation or noise adding, etc. Generative models such as GANs can be used for synthesizing data points.

9.4.1.4 Challenges in data preprocessing

Though data preprocessing is a crucial process, it brings with it challenges. Data collection and preprocessing are most vital to the use of AI and ML in materials science. With diversified data sources and proper implementation of strong preprocessing techniques, quality and relevance will be enhanced in data used for training AI models, making predictions more accurate or giving incisive insights into material properties and behaviors. With further developments in the field, improved methods of data acquisition and preprocessing will be the bedrock on which successful AI implementation in materials science will rest, driving innovation and discovery in new materials [36].

Table 9.3: AI-driven measurement techniques and data collection/preprocessing in material engineering.

Aspect	Details
AI–driven measurement techniques	Enhance accuracy, efficiency, and scope of material characterization using machine learning and data analytics.
Advantages	– Automates data analysis, reducing time and human error. – Real-time data processing for dynamic adjustments. – Integrates data from multiple sources.
Applications	– Classification of microstructural features. – Material behavior prediction. – Optimization of experimental processes.
Data collection sources	– Experimental data. – Simulation data. – Public databases. – Literature data.
Importance of diverse data	Ensures AI models generalize well to new materials and captures complex relationships between processing, structure, and performance.
Data preprocessing	– Data cleaning. – Normalization and standardization. – Feature selection. – Dimensionality reduction. – Data augmentation.
Challenges in data preprocessing	– Data quality. – Complexity of data. – Integration of the diverse data sources.

9.4.2 Machine learning models for mechanical property prediction

ML has proven to be an effective tool for the prediction of material mechanical properties, greatly improving the efficiency and accuracy of material discovery and design. Through the use of large datasets and advanced algorithms, ML models are able to discern intricate relationships between material composition, structure, and mechanical behavior. This section discusses different ML models applied to predicting mechanical properties, their methodologies, applications, and strengths [37].

9.4.2.1 Types of machine learning models

The most popular method in ML for the prediction of mechanical properties is supervised learning. Here, models are trained on labeled data, with the input features being material descriptors and the output being the target mechanical property.

9.4.2.1.1 Regression models

Regression models are used when the target variable is continuous, such as tensile strength or elastic modulus. Common regression techniques include:

Linear regression is a basic method that creates a linear relationship between input features and the target variable. Although straightforward, it might not be able to capture intricate interactions in the data. Gradient boosting regression ensemble method constructs several weak learners (usually decision trees) sequentially, with each subsequent model adjusting the mistakes of the preceding ones. It has been used effectively to predict mechanical properties of different materials, such as metals and polymers. Random forest regression is another ensemble technique that builds several decision trees and takes the average of their predictions. It performs best with high-dimensional data and has been predominantly used for predicting characteristics such as yield strength and hardness.

9.4.2.1.2 Classification models

When the target variable is categorical in nature, as in the case of classifying a material to be ductile or brittle, classification models are used. They involve SVMs that can handle binary problems efficiently and be adapted to work for multi-class situations. They operate through discovering the best hyperplane separating distinct classes within feature space. Neural networks such as deep learning models, in particular feedforward neural networks, may be applied both in classification and regression operations. They have the ability to learn nonlinear complicated relationships and have been implemented for the purpose of predicting mechanical properties from microstructural characteristics.

9.4.2.1.3 Unsupervised learning models

Unsupervised learning methods are utilized to find patterns in data that do not have labeled outputs. Such models can prove to be useful for exploratory data analysis and feature extraction. Clustering algorithms methods like *k*-means clustering can classify materials that possess comparable mechanical properties, which can help determine trends and relationships in the dataset. PCA is a method of dimensionality reduction that may be utilized to visualize high-dimensional data and determine the most important features that affect mechanical properties [38–40].

9.4.3 Regression techniques in machine learning for mechanical property prediction

Regression techniques are the backbone of ML for predicting continuous outcomes, such as mechanical properties of materials. These techniques allow researchers to model complex relationships between input features (e.g., material composition and processing conditions) and target properties (e.g., tensile strength and hardness). This section discusses various regression techniques used in materials science, particularly focusing on symbolic regression and Gaussian process regression (GPR), as highlighted in the sources provided.

9.4.3.1 Symbolic regression

Symbolic regression is the kind of regression analysis that involves searching for a mathematical expression describing the best fitting of a certain dataset. Different from traditional regression, which assumes the existence of a predetermined functional form, symbolic regression uses algorithms, such as genetic programming, to evolve models from the data itself. With this flexibility, it can then discover complex relationships that may otherwise not be detected by predefined models. Symbolic regression has so far shown promising potential in materials science for the modeling of mechanical properties. For instance, it can be employed to develop expressions linking material composition and processing parameters to mechanical properties like yield strength or elastic modulus. It is particularly valuable to generate models that are interpretable, which gives insight into the underlying physical mechanisms governing material behavior. Symbolic regression came up with transformation kinetics in materials and showed the power of revealing relations that ordinary regression methods fail to capture.

9.4.3.2 Gaussian process regression

GPR is a nonparametric Bayesian regression technique that offers a probabilistic model for modeling input–output relationships. In GPR, a Gaussian process is defined over the function space, allowing for the estimation of not only the mean prediction but also the uncertainty associated with that prediction. GPR is particularly useful in materials science for predicting mechanical properties due to its ability to quantify uncertainty. This is important in scenarios where data is noisy or scarce, as it enables researchers to make decisions based on the confidence of the predictions. The study highlighted the use of GPR in predicting properties of materials and molecules, emphasizing its effectiveness in capturing complex relationships while providing uncertainty estimates.

9.4.3.3 Comparison of regression techniques

9.4.3.3.1 Symbolic regression versus Gaussian process regression

While both symbolic regression and GPR are powerful tools for predicting mechanical properties, they have distinct characteristics that make them suitable for different scenarios. Symbolic regression is superior in producing understandable mathematical expressions, which makes it simpler for researchers to comprehend the relationships between variables. In contrast, GPR provides probabilistic predictions but may not yield easily interpretable models. GPR inherently provides uncertainty estimates, which is beneficial in applications where data quality is variable. Symbolic regression does not directly provide uncertainty estimates, although it can be combined with other techniques to assess model reliability. Symbolic regression can be more effective in scenarios with limited data, as it can discover relationships without requiring a predefined model. GPR, while flexible, may require more data to accurately capture complex relationships.

Regression techniques, particularly symbolic regression and GPR, play a crucial role in predicting mechanical properties in materials science. Symbolic regression offers interpretability and flexibility, allowing for the discovery of new relationships, while GPR provides uncertainty quantification and adaptability to complex datasets. As materials science evolves, the incorporation of these regression methods into research pipelines will further develop the capacity to forecast material behavior; ultimately stimulating innovation and discovery in material design and engineering. Continued development of these techniques will further enable researchers to address challenges faced in mechanical property prediction, resulting in more effective and efficient material development processes [41, 42].

9.4.4 Deep learning architectures in materials science

Deep learning has proven to be the most revolutionary method in materials science, enabling researchers to simulate complex relationships between material properties and underlying structures. Several deep learning architectures have been developed to meet specific challenges in materials research – from image classification to property prediction. This section focuses on the primary deep learning architectures used in materials science, along with their functionalities, applications, and advantages [43].

9.4.4.1 Convolutional neural networks (CNNs)

CNNs are a class of deep learning models that have been designed specifically for handling structured grid data, that is, images. CNNs use convolutional layers to learn features from input data automatically, making them extremely efficient for spatial hierarchy based tasks. CNNs have seen extensive use in materials science for tasks like image classification that identifies the SEM images of various materials for identifying different phases and defects. This CNN, using transfer learning, is trained by Modarres et al. for distinguishing multiple material systems and it demonstrated its success to reach accuracy [43, p. 1]. CNNs can extract quantitative information from microstructural images, including grain size, shape, and distribution. This would help establish a correlation with the mechanical properties of materials.

9.4.4.2 Recurrent neural networks (RNNs)

Since it is designed with sequential data processing, recurrent neural networks (RNNs) are one of the finest architectures for times series or text data. RNN's hidden state represents information from all previous time steps making them successful in modeling many temporal dependencies. Some applications of the RNN algorithm in materials science include: predicting time-dependent properties like RNNs can also model the aging of polymers or degradation due to stress, and so on with time. NLP uses RNNs for textual analysis in materials science, like data extraction from the research paper and patents.

9.4.4.3 Graph neural networks (GNNs)

GNNs were designed to act on data having a graph structure; hence, those networks are especially optimized to account for substances in atomic and molecular terms. GNN captures the interaction between nodes (atoms) and the edges (bonds) between them, representing complex interactions, and is therefore better suited for complex models. GNN's popularity among the materials science community has been infectious for predicting material properties. GNNs are able to predict a variety of material properties from their atomic structure, such as formation energies, electronic bandgaps, and mechanical properties. Modeling complex structures, they are optimally applied to materials with complicated geometries, such as polymers and nanomaterials, where traditional means would collapse.

9.4.4.4 Generative adversarial networks (GANs)

Generative adversarial networks (GANs) are two neural networks – a generator and a discriminator that are competing against each other. The generator generates synthetic data, and the discriminator checks if the generated data is authentic or not. It is through adversarial training that GANs can generate high-quality data. GANs have been applied in materials science for data augmentation by which GANs can develop artificial datasets in supplementing low quantities of experimental data that the machine learning model becomes less brittle. Microstructure generation: they generate high-quality real-life microstructure, which enhances designing and optimization of material simulations.

9.4.4.5 Deep reinforcement learning (DRL)

Deep reinforcement learning (DRL) combines reinforcement learning and deep learning techniques, allowing agents to learn optimal strategies by trial and error. DRL is particularly useful for decision-making tasks in complex environments. DRL has been applied in materials science for, inverse design by which DRL can optimize material properties by exploring the chemical space and identifying suitable compounds with desired characteristics. Process optimization is able to optimize manufacturing processes, such as additive manufacturing, by learning the optimal parameters for the desired material properties.

Deep learning architectures, including CNNs, RNNs, GNNs, GANs, and DRL, are revolutionizing materials science by enabling accurate predictions and insights into material properties and behaviors. Each architecture offers unique advantages tailored to specific challenges in the field, from image analysis to property prediction and process optimization [44, 45].

9.4.5 Integration of experimental and simulation data in materials science

This is the merging of experimental and simulation data: a critical characteristic of modern materials science that gives researchers the power to exploit strengths from both camps in enhancing the design and the prediction of materials. It allows a thorough understanding of the materials by interlocking empirical observation with theoretical models to lead into more accurate predictions and innovative solutions for materials. This section explains the merging methods, advantages, and challenges of experimental and simulation data. For this purpose, references will be taken from the given sources. The integration of experimental and simulation data allows researchers to validate theoretical models against real-world observations. Verification of these theoretical models is necessary to confirm that simulations are indeed accurately simulating the behavior of materials under different conditions. By comparing the results of simulation with experimental data, researchers refine their models and improve predictive capabilities. That is integrating data from both experiments and simulations, which can really speed up the process of developing materials. Using simulations to predict how materials will behave will actually help narrow down potential candidates to be studied experimentally. This results in saving a lot of time and resources needed to find new materials. One of the approaches such as those for the Materials Genome Initiative might work by streamlining the cycle of material development [46].

9.4.5.1 Methodologies for data integration

Effective data management systems are crucial for integrating experimental and simulation data. These systems facilitate the storage, retrieval, and analysis of large datasets generated from both sources. For instance, the Sci Server platform, as it has been conceived when considering the Materials in Extreme Dynamic Environments (MEDE) Collaborative Research Alliance, provides a comprehensive framework for managing and analyzing materials data. It allows researchers to upload, store, and share data seamlessly, enabling collaborative efforts across different research groups. Automating workflows that involve both experimental and simulation data can enhance efficiency and reduce human error. By establishing standardized protocols for data collection, processing, and analysis, researchers can ensure consistency and reliability in their results. For example, the integration of automated data ingestion processes within platforms like Sci Server allows for the immediate upload and processing of experimental data, facilitating real-time comparisons with simulation results.

ML techniques can be at the center of the integration of experimental and simulation data. By training models on combined datasets, researchers can uncover complex relationships between material properties and processing conditions. This makes it possible to develop predictive models that can guide material design and optimization. For instance, ML algorithms can be used to correlate microstructural features obtained from simulations with mechanical properties measured experimentally, leading to more accurate predictions.

9.4.5.2 Applications of integrated data

Several case studies successfully integrate experimental and simulation data for materials research. For instance, MEDE used experimental characterization of the material as well as large-scale simulations of dynamic loading on the material in studying boron carbide. Their results, including comparison of results from simulation data with the obtained experimental data, helped refine models and make improvements in prediction. While data integration across various sources does improve the reliability of a predictive model, it is possible to validate and calibrate a simulation framework by integrating experimental data into the former approach. This can be relevant in many applications where materials are submitted to extreme conditions, for instance, in aerospace or defense applications where failure mechanisms must be understood so as to anticipate material behavior well beyond acceptable material performance.

9.4.5.3 Challenges in data integration

One of the primary challenges in integrating experimental and simulation data is ensuring compatibility between different data formats and structures. Materials research often involves diverse datasets generated from various sources, each with its own format and organization. Developing standardized protocols for data representation and storage is essential for facilitating seamless integration. The quality and reliability of both experimental and simulation data are crucial for successful integration. Experimental data may be affected by measurement errors, while simulation data may rely on assumptions and approximations that can introduce uncertainties. Establishing rigorous validation processes to assess the quality of both data types is essential for ensuring the integrity of integrated datasets. Integrating experimental and simulation data often requires collaboration between researchers from different disciplines, such as materials science, computer science, and engineering. Effective communication and collaboration are essential for overcoming disciplinary barriers and ensuring that integrated approaches are successful. Formation of interdisciplinary groups with the ability to bridge theory and practice is critical in advancing materials research [47, 48].

9.5 Case studies

9.5.1 AI-optimized nanocomposites for aerospace applications

The aerospace industry requires materials that are light, strong, and capable of withstanding extreme conditions. CNT-reinforced nanocomposites are now serious players given their advanced mechanical, thermal, and electrical properties. Application of AI techniques for optimization and design of such materials has also advanced them for their application in aerospace technology. It discusses several case studies that show how AI-optimized nanocomposites have found applications in the aerospace sector and, in doing so, explore relevant information from the sources.

9.5.1.1 Carbon nanotube-based polymer nanocomposites

Polymer nanocomposites of CNTs have gained significant attention for use in aerospace because they possess a high strength-to-weight ratio, increased thermal stability, and improved electromagnetic shielding capabilities. They are well-suited for structural components, aircraft interiors, and satellite systems.

CNT-reinforced polymer nanocomposites' mechanical performance was investigated in a research study. The researchers employed AI techniques to maximize the dispersion of CNTs in the polymer matrix, something they considered key to achieving uniform characteristics and maximum performance. By utilizing ML models, they effectively predicted the optimum CNT loading and processing path that would achieve the best mechanical properties. The optimized nanocomposites were found to exhibit significantly increased tensile strength and stiffness compared to the traditional polymer composites. The study presented the necessity for AI-based prediction of how processing conditions impacted the mechanical characteristics of nanocomposites in order to design materials efficiently.

9.5.1.2 Damage detection in composite materials

Damage detection in composite materials is extremely critical for ensuring the reliability and safety of aerospace structures. AI techniques, particularly ML algorithms, have been employed to develop advanced damage detection systems that can analyze data from various sensors embedded in composite materials.

A study reviewed various AI techniques for damage detection in composite materials used in aerospace applications. The study concentrated on combining AI with traditional nondestructive testing (NDT) methods to enhance damage detection efficiency and accuracy. Application of AI algorithms, such as SVMs and neural networks, enabled effective damage type identification from sensor signals. Combining AI with traditional NDT methods yielded a robust damage detection system that was capable of detecting micro-level damages undetectable by the traditional methods.

9.5.1.3 Electromagnetic interference shielding

Electromagnetic interference shielding is crucial in aerospace application to protect delicate electronic equipment against external electromagnetic fields. Nanocomposites doped with CNTs demonstrated potential for effectively shielding EMI while retaining a lightweight nature.

In one study the electromagnetic shielding property of CNT/polymer composites was examined. AI techniques were employed by the authors to design the formulation of such composites with high EMI shielding property. The study found that addition of CNTs dramatically enhanced the EMI shielding performance of the polymer matrix. Optimization by AI was useful in identifying the best CNT loading and dispersion strategies to provide composites that satisfy the very high EMI shielding requirements of aerospace applications.

9.5.1.4 Thermal management in aerospace structures

Effective thermal management is essential in aerospace applications to ensure the performance and longevity of materials exposed to extreme temperatures. Nanocomposites with enhanced thermal conductivity can play a vital role in thermal management systems.

Experiments were conducted for the thermal conductivity of CNT-reinforced polymer nanocomposites. AI techniques were applied to model the influence of CNT concentration, dispersion, and thermal conductivity. AI models were able to predict the thermal conductivity of the nanocomposites based on varying CNT concentrations and dispersion methods with high accuracy. Optimized nanocomposites possessed significantly improved thermal conductivity, making them suitable for thermal management applications in aerospace systems.

9.5.1.5 Structural health monitoring

Structural health monitoring (SHM) is crucial for maintaining the integrity of aerospace structures. AI techniques have been integrated into SHM systems to analyze data from sensors and predict potential failures.

In a study the use of CNT-modified adhesive films for SHM in composite structures was explored. The researchers employed AI algorithms to analyze sensor data and detect changes in the adhesive properties over time. The AI-based SHM system was able to accurately predict the onset of adhesive failure, allowing for timely maintenance and repairs. The integration of CNTs in the adhesive films enhanced their mechanical properties, contributing to the overall durability of the bonded joints in aerospace applications.

Applications of AI techniques in nanocomposite design and optimization for aeronautical applications have led to significant advancements in material performance as well as functionality. Case studies reflect the successful utilization of AI in the enhancement of mechanical properties, better damage detection, EMI shielding optimization, and efficient thermal management. As the aerospace industry continues to progress, the ongoing development of AI-optimized nanocomposites will become crucial in meeting the stringent requirements of cutting-edge aerospace use, thereby propelling innovation and improving safety in the industry. The convergence of materials science and AI will precipitate the development of the next-generation high-performance materials intended for aerospace engineering [49–51].

9.5.2 AI-driven innovations in biomedical materials

Application of AI in the biomedical materials sector has ushered in a significant impetus in material developing, designing, and deploying within the healthcare field. AI-driven innovations are transforming traditional approaches to materials science, enabling faster discovery, enhanced performance, and improved patient outcomes. This section explores the key innovations in biomedical materials driven by AI, drawing insights from the provided sources.

9.5.2.1 Autonomous experimentation systems

Autonomous experimentation (AE) systems, or self-driving labs or material acceleration platforms, are an excellent development in biomedical research capability. AE systems utilize AI and ML algorithms to design, perform, and analyze experiments automatically, reducing the time required for material discovery and optimization considerably.

AE systems have been applied especially in biomedical materials design, such as drug delivery vehicles, tissue engineering scaffolds, and responsive biomaterials. For instance, the ability of AE systems allows researchers to synthesize and characterize numerous materials in a short time, to determine the ones with optimal properties for a specific biomedical application. The production of nanoparticles for targeted drug delivery has been facilitated by AE systems that can quickly test various formulations and conditions to provide optimal performance. AE systems can accomplish in days what would take months or years conventionally, thus enabling quicker innovation cycles for biomedical materials. Through automation of experiments, AE systems generate a lot of data that can be used to train ML algorithms, making the predictability of material properties better.

9.5.2.1.1 Machine learning in material design

ML techniques are increasingly being applied to the design of biomedical materials. By analyzing existing data regarding composition and performance of material, ML algorithms will identify patterns and relationships applied for creating new materials. ML is capable of forecasting mechanical, thermal, and biological properties of new material based on composition and structure. It is particularly handy when it comes to developing biocompatible materials for the purpose of implant and prosthetics. AI-powered computer algorithms can optimize drug formulations by modeling the effect of composition changes on release characteristics and potency, yielding improved treatments. ML can reduce to a significant extent the time taken to identify material candidates with promise, allowing scientists to focus on the most hopeful possibilities.

By reducing extensive trial-and-error experimentation, ML can reduce the cost of material development.

9.5.2.1.2 AI in nanomedicine

Nanomedicine, which involves the use of nanotechnology for medical applications, has seen substantial advancements through AI integration. AI tools are being used to design, synthesize, and evaluate nanomaterials for various therapeutic and diagnostic purposes. AI algorithms can also be employed to engineer nanoparticles in such a manner that their targeting ability to particular cells or tissues is optimized, thereby enhancing the effectiveness of treatments with minimal side effects. AI can predict the toxicity of nanomaterials, helping researchers design safer materials for clinical use. AI-driven approaches enable the development of personalized nanomedicine solutions tailored to individual patient needs, enhancing treatment outcomes. The ability to quickly synthesize and test a variety of nanomaterials accelerates the development of innovative therapeutic agents.

9.5.2.2 AI-enhanced drug discovery

AI is revolutionizing the process of drug discovery by enabling scientists to search through vast chemical spaces and identify promising candidates for drugs more effectively than can be achieved by traditional techniques. AI can search through large databases of chemical compounds to identify those that have the highest potential for therapeutic activity. AI is able to predict the success of drug candidates on clinical trials and thus prioritize those that will be most likely to succeed. By leveraging AI to identify promising candidates early in the process, the likelihood of success in clinical trials can be significantly improved. AI-driven drug discovery can reduce the costs associated with R&D [52, 53].

9.5.3 Advancements in energy storage materials

Improving energy storage efficiency and sustainability has become very important because more and more consumers are turning toward renewable energy sources while also demanding greater use of EVs. This has prompted the researchers to delve deep into the progress of nanomaterials energy storage materials, which have exhibited extraordinary potential for enhancing the performance efficiency and sustainability of energy storage devices. The recently developed advancement in energy storage materials, that is, LIBs, sodium-ion batteries (SIBs), supercapacitors, and zinc-air batteries, is being reported in this section as per the references.

9.5.3.1 Lithium-ion batteries (LIBs)

LIBs are the most common energy storage systems, mainly applied in portable devices and electric vehicles. Progress in nanomaterials over the past few years has greatly improved their performance by addressing issues of energy density, cycle life, and safety. Silicon is a promising anode material since it possesses a high theoretical capacity of 4,200 mAh/g. High volume expansion of silicon upon lithiation and delithiation is an issue. Recent studies have shown that silicon nanowires can store these changes to exact capacities of around 3,100 mAh/g after 10 cycles. The addition of graphene in LIBs has resulted in an improved electrical conductivity and mechanical stability. Graphene-based anodes demonstrate a higher rate capability for charge/discharge cycles and cycling stability, making them applicable to high-performance applications.

9.5.3.2 Sodium-ion batteries (SIBs)

SIBs are also being looked at as an inexpensive alternative to LIBs, particularly for large-scale energy storage applications. Its high sodium content is a strong potential candidate as an alternative energy source. Recently, titanium dioxide nanotubes are reported to be effective anodes for SIBs. Its unique structure facilitates the large ionic radius of sodium with reversible capacities of around 300 mAh/g and stable cycling characteristics. Carbonaceous materials such as hard carbon and graphene have also been studied for enhancing the electrochemical properties of SIBs. They display increased conductivity and structural stability, which lead to increased cycling stability and capacity retention.

9.5.3.3 Supercapacitors

Supercapacitors, or ultracapacitors, are an energy storage device, which may display extremely high power density and high-rate charge/discharge capability. Use in applications where very high-speed energy delivery is possible makes such supercapacitors appropriate for use in electric vehicles, renewable energy systems, and so on. Its superior electrical conductivity and high surface area make it an appropriate material for supercapacitor electrodes. Graphene-based supercapacitors have been shown in recent studies to possess high specific capacitance and improved cycle stability to be appropriate for high-performance applications.

The addition of CNTs to supercapacitor electrodes has further improved their performance. CNTs form a conductive network that enhances charge transport and increases the overall capacitance of the device.

9.5.3.4 Zinc-air batteries

Zinc-air batteries are metal-air batteries of a type having an anode as zinc and cathode as oxygen in the air. They have high energy density and are viewed as a very promising large-scale energy storage candidate. Recent studies have considered using cobalt oxide nanoparticles to enhance the performance of zinc-air batteries. Such nanoparticles enable the delivery of reducing agents to oxygen molecules, which significantly enhances energy efficiency and cycle stability. Nanostructured electrodes have been constructed to enhance electrochemical reaction kinetics in zinc-air batteries, leading to improved performance and longer cycle life [54, 55].

9.6 Challenges and limitations

In materials science, data quality and availability present significant challenges that can hinder research and development. One major limitation is the inconsistency in data formats and standards across different studies and databases. This inconsistency makes it difficult to integrate and compare data, leading to potential errors in analysis and interpretation. Another challenge is the scarcity of high-quality experimental data, particularly for novel materials. Many emerging materials lack comprehensive datasets, which can impede the validation of theoretical models and simulations. Additionally, proprietary data from industry can limit access to critical information, creating barriers for academic researchers. Data availability is also affected by the rapid pace of technological advancement. As new materials and methods are developed, existing databases may become outdated, and researchers may struggle to find relevant and current data. Furthermore, the lack of robust data-sharing practices within the scientific community can lead to duplicated efforts and wasted resources.

To address these challenges, there is a growing need for standardized data formats, improved data-sharing platforms, and collaborative efforts to create comprehensive databases that encompass a wide range of materials and their properties. This would enhance the reliability and accessibility of data in materials science, ultimately accelerating innovation and discovery.

9.6.1 Data quality and availability

The availability and quality of data issues and limitations in materials science are the most important issues that have considerable impacts on research and development in the field. Based on the related documents, we can elaborate on these issues in detail.

9.6.1.1 Data scarcity

One of the most critical issues in materials science is the scarcity of high-quality data availability. Many emerging materials, particularly those developed through innovative methods, lack comprehensive datasets. This scarcity can hinder the validation of theoretical models and simulations, which are essential for predicting material behavior and performance. The limited availability of experimental data can lead to reliance on outdated or incomplete datasets, which may not accurately represent the properties of new materials.

9.6.1.2 Inconsistency in data formats

Another significant limitation is the inconsistency in data formats and standards across various studies and databases. Different study groups use varying data collection and reporting techniques, which leads to inconsistencies making comparison and integration more challenging. This inconsistency can result in errors during analysis and interpretation, ultimately affecting the reliability of research findings.

9.6.1.3 Proprietary data and access barriers

Proprietary data from industry can also pose a challenge. Many companies conduct extensive research and development but do not share their findings publicly due to competitive reasons. This lack of access to critical information can create barriers for academic researchers who rely on comprehensive datasets to validate their work. The absence of collaboration between industry and academia further exacerbates this issue, as valuable insights and data remain siloed within organizations.

9.6.1.4 Rapid technological advancements

The rapid pace of technological advancement in materials science can lead to data becoming outdated quickly. As new materials and methods are developed, existing databases may not keep up with the latest findings, making it difficult for researchers to find relevant and current data. This situation can result in duplicated efforts and wasted resources, as researchers may unknowingly work on problems that have already been addressed.

9.6.1.5 Lack of standardized data sharing practices

The other significant concern is the nonexistence of proper data sharing practices in science. Although lots of initiatives about open data already exist, plenty of researchers would not share them due to misuse prospects of intellectual properties,, or, for that matter, lack of awareness about the rewards of data sharing. This hesitation can lead to a patchwork landscape where valuable data is not accessible to those who could benefit from it.

9.6.1.6 Data quality issues

Data quality is a major issue in materials science. Poor-quality data can be due to experimental errors, poor measurement techniques, and biases in data collection. Such issues may result in wrong conclusions and hinder the development of reliable predictive models. Ensuring high data quality requires rigorous validation processes and adherence to best practices in data collection and reporting.

9.6.1.7 The role of machine learning

ML is now the tool making this area of materials science powerful – because it allows its users to consider processing large datasets and pattern identification beyond direct human eyes. However, the efficiency of such an ML algorithm as with theory relies on data quality and amount employed to train. With sparse or low-quality underlying data, prediction produced by ML models will be flawed. Consequently, data quality and availability become of utmost importance in maximizing the potential provided by ML in material discovery.

9.6.1.8 Collaborative efforts and solutions

To overcome these challenges, there is a growing need for collaborative efforts among researchers, industry, and funding agencies. Initiatives aimed at creating standardized data formats and promoting data-sharing platforms can facilitate better integration and accessibility of data. In addition, cultivating a culture of openness and collaboration within the scientific community will make it easier to connect academia with industry, enabling more extensive and richer datasets and improved research outcomes.

The challenges and limitations of data quality and availability in materials science are quite multifold and therefore need concerted efforts to address the issue. In improving data sharing practices, standardizing data formats, and facilitating collaboration between stakeholders, the materials science community would enhance the reliability and accessibility of data. This will, in turn accelerate innovation and discovery into new materials and technologies that will make a difference in society as a whole [56, 57].

9.6.2 Computational costs

The challenges and limitations associated with computational costs in computational materials science are significant and multifaceted. Drawing insights from the connected documents, we can explore these challenges in detail.

9.6.2.1 High computational demands

Computational materials science often involves complex simulations that require substantial computational resources. The need for high-performance computing (HPC) arises from the intricate nature of materials and their properties, which can be influenced by various factors such as atomic structure, electronic interactions, and environmental conditions. As the complexity of the models increases, so does the computational cost, making it challenging for researchers to conduct extensive simulations within reasonable timeframes.

9.6.2.2 Multiscale modeling challenges

One of the primary challenges in computational materials science is the need for multiscale modeling, which integrates different levels of detail – from atomic to macroscopic scales. While multiscale approaches are essential for accurately predicting material behavior, they also introduce significant computational costs. High-fidelity simulations at the atomic level, such as those based on DFT, can be computationally intensive and time-consuming. Coupling these with larger-scale models that describe macroscopic behavior requires careful balancing of accuracy and computational efficiency.

9.6.2.3 Limitations of current algorithms

The algorithms used in computational materials science, while powerful, often have limitations that can lead to increased computational costs. For instance, traditional methods may struggle with systems that exhibit complex behaviors, such as phase transitions or nonequilibrium dynamics. As a result, researchers may need to employ more sophisticated and computationally demanding techniques, which can further escalate costs.

9.6.2.4 Resource availability and accessibility

Access to HPC resources is another critical factor influencing computational costs. Many researchers, particularly those in academia, may face limitations in accessing the necessary computational power to conduct their simulations. This can lead to delays in research progress and may restrict the scope of investigations. Furthermore, the costs associated with using HPC facilities can be prohibitive, especially for smaller research groups or institutions.

9.6.2.5 Energy consumption and environmental impact

Computational costs in materials science are both financial and environmental. Running such large-scale simulations requires much energy, thus putting into question whether such computations will be sustainable or not. There is an increase in demand for computational resources and it becomes paramount to address environmental issues surrounding the activities as more energy-efficient solutions in computing can be used to mitigate such energy consumption.

9.6.2.6 Balancing accuracy and efficiency

Even more difficult is the appropriate balance between computational effectiveness and accuracy in computational materials science. High accuracy techniques, such as DFT, allow researchers to get detailed insight into material properties but are often applicable to small systems because of their computational requirements. Slower techniques are more likely to be less accurate with a commensurate decrease in predictability. This can make material design difficult, as appropriate computational techniques are often obscure and harder to find.

9.6.2.7 The role of machine learning

ML has great potential to reduce some of the computationally costly parts of materials science. Large datasets allow ML algorithms to make rapid predictions of material properties, which can reduce the number of simulations. However, the training data density and quality determine the performance of the ML models. If the data is of poor quality or it is sparse data, then the ML model may produce untrusted predictions and thereby cause further complications in the research process.

9.6.2.8 Future directions and solutions

There are multiple ways to address the challenges of computational costs in materials science. Development of more efficient algorithms that ensure accuracy while reducing the computational burden is needed. Advances in parallel computing, cloud computing, and GPU computing would enhance accessibility and efficiency for computing calculations. Access to HPC, data, and methodology sharing also foster collaboration between academia and industry, thus narrowing the gap that has been a huge divide between theory and practical research, finally giving a material discovery process much needed efficiency with fewer costs, which is attributed to Heine [58, 59, p. 2].

9.6.3 Interpretability of AI models

Interpretability of AI models in materials science is a significant area of concern, especially as the field increasingly relies on complex algorithms for materials discovery and property prediction. Following the context-relevant documents, we can probe into issues and challenges facing interpretability in the case of AI models in this regard.

The most critical problem to achieving interpretability is that many AI models are intrinsically complex, especially deep learning architectures. The former models tend to be many-layered and possess thousands of parameters, so the logic by which they generate a specific prediction cannot be grasped easily. For example, though deep neural networks may yield good predictions in the determination of material properties, the way a decision is reached in the networks remains unclear. This transparency shortfall will make it difficult for researchers to have faith in the model's predictions and in comprehending the physical laws governing material behavior. The representation of input features is another critical factor influencing interpretability. In most cases, AI models depend on high-dimensional feature spaces that do not have a clear physical interpretation. For instance, conventional machine learning models tend to rely much on feature engineering, which makes use of domain knowledge to represent data in meaningful ways. This method does introduce bias and the relationship between input features and predicted outputs could be lost. Hence, even when the model is accurate, the contribution made by each individual feature towards the ultimate prediction cannot be appropriately determined.

In materials science, domain-specific issues stem from the nature of the materials data itself. For example, material properties are often governed by many factors such as atomic structure, composition, and processing conditions. This complexity can make it difficult to isolate the effect of individual variables in a model. Second, nonlinear relations, which are rather typical for materials data, may complicate the interpretation of model predictions since slight changes in input values may cause significant differences in predicted properties.

The growing demand for explainable AI in materials science underscores the necessity of developing models that are not only efficient but also explanatory of their decision-making. Scientists increasingly need tools and techniques that can enhance the interpretability of AI models. This includes developing hybrid models that combine traditional physics-based approaches with data-driven approaches, allowing for more transparency in understanding material behavior.

The explainability of AI models also has practical and ethical implications. In materials science, stakes can be critical – that is, in creating new energy materials – the ability to comprehend the thought process behind model predictions is crucial for guaranteeing safety and effectiveness. Also, regulatory bodies can ask for explanations of AI-based decisions, particularly in industries where materials are used in mission-critical applications. Therefore, improving interpretations is not only a technical matter but also a demand of responsible innovation.

In short, interpretability of AI models in materials science is a multifaceted challenge that requires continuous attention and innovation. By resolving the complexities entailed with model structure, feature description, and accuracy–interpretability trade-offs, scientists can render AI-based predictions more transparent and reliable. Ultimately, increasing interpretability will not only advance the field of materials science but also contribute to the ethical and responsible application of AI technologies in scientific research [60, 61].

9.7 Future perspectives

9.7.1 Integration with emerging technologies

The integration of emerging technologies in materials science is a critical part of the field's advancement, especially in terms of AI and electrochemical devices. The future perspectives regarding this integration are explored in this section, considering the potential benefits, challenges, and opportunities that are ahead. AI is shaping materials science due to the assessment of huge volumes of data, optimization of the properties of materials, and acceleration in material discovery. Pattern recognition and connection between variables often not easily derivable by routine methods are obtained through AI algorithms. In particular, in the case of electrochemical devices, the performance is affected by very many factors among which are: composition, structure and processing conditions. Implementation of AI in the field of materials science has the benefit of discovering new material. ML can predict properties of new compounds from existing data. Experimental verification is now avoided due to time and effort saved. For example, AI assists in discovering potential candidates for materials in batteries according to their stability and electrochemical behavior. This predictive capability also accelerates the development of new materials but streamlines existing ones for particular uses. The future of materials science will increasingly rely on interdisciplinary collaboration.

The convergence of AI, materials science, and engineering will require the expertise of professionals from various fields, including computer science, physics, and chemistry. Collaborations can lead to the development of hybrid models that combine data-driven approaches with traditional physics-based methods, providing a more comprehensive understanding of material behavior. While AI is increasingly used in the field of materials science, ethical issues also find immense importance in the context. Questions arise whether AI-based decision-making processes are accountable, transparent, and less biased.

Researchers should ensure that AI models are interpretable, that is, they can be understood and confirmed by human experts. In addition, ethics should be established to govern the use of AI in materials research, especially for applications that have a bearing on public health and safety. AI and electrochemical devices are likely to open vast areas of future applications. For instance, AI can be used to optimize the design of electrodes and electrolytes, thereby enhancing the performance and efficiency of energy storage systems like batteries and supercapacitors. Educational institutions should include AI and data science in their materials science curricula, thus preparing a new generation of scientists who are proficient in both domains. Industry partnerships can also offer practical training opportunities, thus ensuring that professionals are well-prepared to navigate the complexities of modern materials research [62, 63].

9.7.2 Quantum computing and molecular simulations

The future of quantum computing in molecular simulations presents a revolutionary prospect to the discipline of computational materials science. As suggested in the referenced papers, the integration of quantum computing with molecular dynamics and other simulation techniques has the capability to revolutionize our understanding and manipulate molecular systems. This debate will reflect on the possibilities for quantum computation in molecular calculations in the years to come, and its promise, limitations, and ramifications for science and industry alike. Quantum computing offers an alternative philosophy for computation compared with classical computing. Quantum computing relies on quantum mechanics, such as entanglement and superposition, to do things that are not possible using standard computers. This capability is particularly valuable for molecule simulations, where the complexities of quantum systems will often lead to exponential scaling problems in traditional computational methods. Quantum algorithms like quantum phase estimation and variational quantum eigensolver rely on identifying more efficient paths to solve quantum mechanical issues. One of the largest advantages of quantum computing with regards to molecule simulations is the ability to be more accurate.

Traditional methods, such as DFT, normally rely on approximations that have room for error, especially in high electron correlation systems. Quantum computers are capable of providing exact electronic Schrödinger equation solutions, and hence accurate prediction of the behavior and properties of molecules. This capability is fundamental to be used in drug development, materials development, and catalysis in scenarios where accurately describing molecular interactions is necessary. The use of quantum computing for molecular simulations necessitates interdisciplinary work among physicists, computer scientists, chemists, and engineers.

This is required in order to develop new quantum algorithms for specific molecular problems and optimize quantum hardware for practical purposes. With the advancement of quantum technology, it will be significant to facilitate collaboration between academia and industry to translate theoretical advances into useful applications. Despite the encouraging prospects, there are still some challenges to be addressed for the complete realization of quantum computing in molecular simulations. Quantum ML algorithms can improve data analysis and pattern recognition in molecular simulations, allowing for improved exploration of chemical space. Quantum information can be used to learn and create models using the power of quantum computing, which can lead to new discoveries and findings in materials science and chemistry. Quantum computing industry applications are highly significant. The more advanced quantum computing's algorithms get and are released, the faster sectors like the pharmaceutical industry, materials science, and energy can advance with the acceleration of the research and development process (see Figure 9.2).

9.7.3 Pathways to industry adoption

The adoption of AI-enhanced measurement techniques for NPCs in industrial settings requires a strategic approach to overcome technical, economic, and organizational challenges. First, developing user-friendly AI platforms tailored to the needs of material engineers is essential. These platforms must integrate seamlessly with existing measurement and production systems, minimizing disruptions while maximizing efficiency. Collaboration between academia, industry, and technology providers is crucial to translating AI research into practical, scalable solutions. Standardization plays a key role in industry adoption. Establishing universally accepted protocols for AI-driven measurements will build trust and consistency in results. Industries must also invest in data quality and management, as AI models rely heavily on robust and representative datasets. Creating shared databases of material properties can accelerate innovation while maintaining competitive advantages.

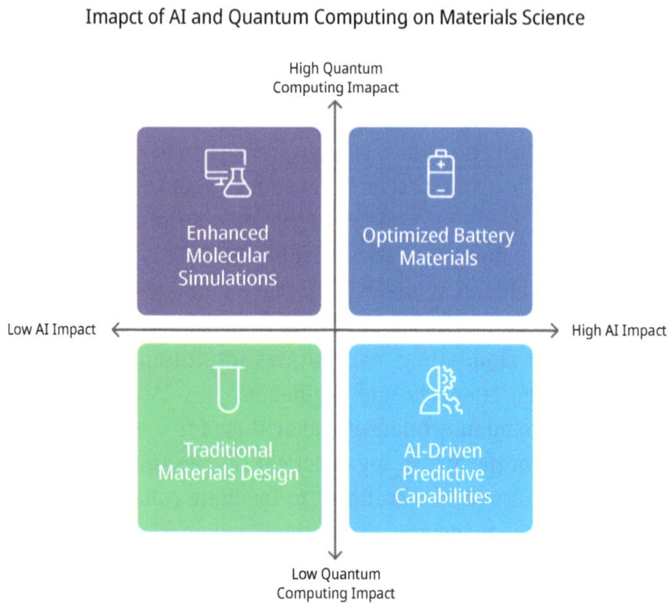

Figure 9.2: Impact of AI and quantum computing on materials science.

Economic considerations, such as the cost of implementing AI infrastructure and training personnel, must be addressed. Government incentives, subsidies, and partnerships can support industries during this transition phase. Moreover, showcasing successful case studies demonstrating the tangible benefits of AI in reducing costs, improving efficiency, and enhancing product performance will encourage wider acceptance. Finally, fostering a culture of innovation within organizations is key. This involves training engineers and scientists in AI technologies and promoting interdisciplinary collaboration. As AI continues to prove its value in research and development, the integration of these tools in commercial manufacturing will enable industries to leverage NPCs' full potential, driving technological advancements and competitive growth [64, 65].

9.8 Conclusion

This chapter showcases the revolutionary potential of AI to revolutionize measurement and prediction of mechanical properties of NPCs. With its ability to bypass the limitation of traditional measurement methods, AI has also become incredibly powerful with the ability to process humongous data sets, visualize simulation and experiment data, and accurately predict material performance. Use of machine learning models, including regression models and deep neural network architectures, has uncovered complex patterns and interactions to enable a greater understanding of the properties of the material. Case studies also unveil the practical impact of AI by showing how nanocomposites can be engineered with it for aerospace, biomedical devices, and energy storage systems. AI integration into materials science not only improves the characterization process but also facilitates quick development of tailored materials for fulfilling a particular industrial requirement. This development is highly advantageous to companies employing high-performance materials, with more pressure for innovation and the ability to tackle complicated engineering challenges. The combination of AI, quantum computing, and molecular simulation can also give rise to more extensive research opportunities. With increasingly enhanced quality data and diminishing computation bottlenecks, AI applications in nanocomposite science in the future will become increasingly more powerful, bridging the gap between theory and application closer. With the aid of AI, scientists and engineers can realize the greatest potential of NPCs and initiate advances in materials science as well as inspire innovation in many fields.

List of abbreviations

AI	Artificial intelligence
ML	Machine learning
CNN	Convolutional neural network
RNN	Recurrent neural network
GNN	Graph neural network
GAN	Generative adversarial network
DRL	Deep reinforcement learning
AE	Autonomous experimentation (self-driving lab)
SHM	Structural health monitoring
CNT(s)	Carbon nanotube(s)
PMMA	Poly(methyl methacrylate)
SEM	Scanning electron microscopy
TEM	Transmission electron microscopy
XRD	X-ray diffraction
AFM	Atomic force microscopy
XPS	X-ray photoelectron spectroscopy
DMA	Dynamic mechanical analysis
NDT	Nondestructive testing
EMI	Electromagnetic interference
LIB(s)	Lithium-ion battery/ies
SIB(s)	Sodium-ion battery/ies
DFT	Density functional theory
MD	Molecular dynamics
PCA	Principal component analysis
t-SNE	t-Distributed stochastic neighbor embedding
SVM	Support vector machine
GPR	Gaussian process regression

References

[1] Rooney K, Dong Y, Basak AK and Pramanik A. Prediction of mechanical properties of 3D printed particle–reinforced resin composites. Journal of Composites Science, 2024;8(10):416.
[2] Dong H, Lin J, Tao Y, Jia Y, Sun L, Li WJ and Sun H. AI-enhanced biomedical micro/nanorobots in microfluidics. LOC, 2024;24(5):1419–1440.
[3] Chen H, Zheng Y, Li J, Li L and Wang X. AI for nanomaterials development in clean energy and carbon capture, utilization and storage (CCUS). ACS Nano, 2023;17(11):9763–9792.
[4] Wang C, He T, Zhou H, Zhang Z and Lee C. Artificial intelligence enhanced sensors – Enabling technologies to next-generation healthcare and biomedical platform. Bioelectronic Medicine, 2023;9(1).
[5] Rafiee R and Shahzadi R. Predicting mechanical properties of nanoclay/polymer composites using stochastic approach. Composites Part B: Engineering, 2018;152:31–42.
[6] Tafreshi OA, Saadatnia Z, Ghaffari-Mosanenzadeh S, Okhovatian S, Park CB and Naguib HE. Machine learning-based model for predicting the material properties of nanostructured aerogels. SPE Polymers, 2023;4(1):24–37.
[7] Tofighy MA and Mohammadi T. Carbon nanotubes–polymer nanocomposite membranes for pervaporation. In: Polymer Nanocomposite Membranes for Pervaporation. 2020; (pp. 105–133). Elsevier.
[8] Sundarram S, Kim Y-H and Li W. Preparation and characterization of poly(ether imide) nanocomposites and nanocomposite foams. In: Manufacturing of Nanocomposites with Engineering Plastics. 2015; (pp. 61–85). Elsevier.
[9] Rasool A, Rizwan M, Ur Rehman Qureshi A, Rasheed T and Bilal M. Thermo–responsive functionalized polymeric nanocomposites. In: Smart Polymer Nanocomposites. 2023; (pp. 219–240). Elsevier.
[10] Liu K and Wu J. Mechanical properties of two-dimensional materials and heterostructures. Journal of Materials Research, 2016;31(7):832–844.
[11] Astakhov VP. Mechanical properties of engineering materials: Relevance in design and manufacturing. In Materials Forming, Machining and Tribology. 2018; (pp. 3–41). Springer International Publishing.
[12] Goswami L, Deka MK and Roy M. Artificial intelligence in material engineering: A review on applications of artificial intelligence in material engineering. Advanced Engineering Materials, 2023;25(13).
[13] Lau ML, Burleigh A, Terry J and Long M. Materials characterization: Can artificial intelligence be used to address reproducibility challenges? Journal of Vacuum Science & Technology. A, Vacuum, Surfaces, and Films: An Official Journal of the American Vacuum Society, 2023;41(6).
[14] Batool M, Sanumi O and Jankovic J. Application of artificial intelligence in the materials science, with a special focus on fuel cells and electrolyzers. Energy and AI, 2024;18(100424):100424.
[15] Sha W, Guo Y, Yuan Q, Tang S, Zhang X, Lu S, Guo X, Cao Y-C and Cheng S. Artificial intelligence to power the future of materials science and engineering. Advanced Intelligent Systems (Weinheim an der Bergstrasse, Germany), 2020;2(4).
[16] Assad H, Sharma S, Kaya S, Sharma PK and Kumar A. Overview and fundamentals of polymer nanocomposites. In: Nanocomposites–Advanced Materials for Energy and Environmental Aspects. 2023; (pp. 41–66). Elsevier.
[17] Hazarika D and Karak N. Fundamentals of polymeric nanostructured materials. In: Advances in Polymeric Nanomaterials for Biomedical Applications. 2021; (pp. 1–40). Elsevier.
[18] Karak N. Fundamentals of nanomaterials and polymer nanocomposites. In: Nanomaterials and Polymer Nanocomposites. 2019; (pp. 1–45). Elsevier.

[19] Moharana S, Sahu BB, Nayak AK and Tiwari SK. Eds. Polymer Composites: Fundamentals and Applications. 2024, Springer Nature Singapore.
[20] Kara Y, Acar V and Seydibeyoglu MO. Mechanical properties of nanoparticle–based polymer composites. In: Nanoparticle–Based Polymer Composites. 2022; (pp. 95–108). Elsevier.
[21] Tjong SC. Structural and mechanical properties of polymer nanocomposites. Materials Science & Engineering. R, Reports: A Review Journal, 2006;53(3–4):73–197.
[22] Arash B, Wang Q and Varadan VK. Mechanical properties of carbon nanotube/polymer composites. Scientific Reports, 2014;4(1).
[23] Chen T-H, Wang I-H, Lee Y-R and Hsieh T-H. Mechanical property of polymer composites reinforced with nanomaterials. Polymers and Polymer Composites, 2021;29(6):696–704.
[24] Al Rashid A, Khan SA, Al-Ghamdi G and Koç M. Additive manufacturing of polymer nanocomposites: Needs and challenges in materials, processes, and applications. Journal of Materials Research and Technology, 2021;14:910–941.
[25] Kumar S, Sarita N, Dilbaghi M, Tankeshwar N, K. and Kim K-H. Recent advances and remaining challenges for polymeric nanocomposites in healthcare applications. Progress in Polymer Science, 2018;80:1–38.
[26] Wang Y, Desroches GJ and Macfarlane RJ. Ordered polymer composite materials: Challenges and opportunities. Nanoscale, 2021;13(2):426–443.
[27] Adetunla A, Akinlabi E, Jen TC and Ajibade S-S. Harnessing the power of artificial intelligence in materials science: An overview. 2024 International Conference on Science, Engineering and Business for Driving Sustainable Development Goals (SEB4SDG), 2024;1–6.
[28] Mobarak MH, Mimona MA, Islam MA, Hossain N, Zohura FT, Imtiaz I and Rimon MIH. Scope of machine learning in materials research – A review. Applied Surface Science Advances, 2023;18 (100523):100523.
[29] Brimo N, Serdaroğlu DÇ and Muhittin H. Artificial intelligence and machine learning for material characterizations and mechanical properties. In: Tissue Repair and Reconstruction. 2024; (pp. 55–72). Springer Nature Singapore.
[30] Batool M, Sanumi O and Jankovic J. Application of artificial intelligence in the materials science, with a special focus on fuel cells and electrolyzers. Energy and AI, 2024;18(100424):100424.
[31] Papadimitriou I, Gialampoukidis I, Vrochidis S and Kompatsiaris I. AI methods in materials design, discovery and manufacturing: A review. Computational Materials Science, 2024;235(112793):112793.
[32] Pyzer-Knapp EO, Pitera JW, Staar PWJ, Takeda S, Laino T, Sanders DP, Sexton J, Smith JR and Curioni A. Accelerating materials discovery using artificial intelligence, high performance computing and robotics. Npj Computational Materials, 2022;8(1).
[33] Goswami L, Deka MK and Roy M. Artificial intelligence in material engineering: A review on applications of artificial intelligence in material engineering. Advanced Engineering Materials, 2023;25(13).
[34] Parvin BG and Parvin LG. Enhancing mechanical properties of materials using artificial intelligence. Figshare, 2024.
[35] Yu B, Zhang L, Ye X, Wang Y, Ying H, Zhu W, Xiang W and Wu X. State–of–the–art review on various applications of machine learning techniques in materials science and engineering. Chemical Engineering Science, 2024;121147:121147.
[36] Maharana K, Mondal S and Nemade B. A review: Data pre–processing and data augmentation techniques. Global Transitions Proceedings, 2022;3(1):91–99.
[37] Allen AEA and Tkatchenko A. Machine learning of material properties: Predictive and interpretable multilinear models. Science Advances, 2022;8(18).
[38] Chibani S and Coudert F-X. Machine learning approaches for the prediction of materials properties. APL Materials, 2020;8(8).

[39] Akbari P, Zamani M and Mostafaei A. Machine learning prediction of mechanical properties in metal additive manufacturing. Additive Manufacturing, 2024;91(104320):104320.
[40] Jacobs R, Schultz LE, Scourtas A, Schmidt KJ, Price-Skelly O, Engler W, Foster I, Blaiszik B, Voyles PM and Morgan D. Machine learning materials properties with accurate predictions, uncertainty estimates, domain guidance, and persistent online accessibility. 2024.
[41] Wang Y, Wagner N and Rondinelli JM. Symbolic regression in materials science. MRS Communications, 2019;9(3):793–805.
[42] Deringer VL, Bartók AP, Bernstein N, Wilkins DM, Ceriotti M and Csányi G. Gaussian process regression for materials and molecules. Chemical Reviews, 2021;121(16):10073–10141.
[43] Choudhary K, DeCost B, Chen C, Jain A, Tavazza F, Cohn R, Park CW, Choudhary A, Agrawal A, Billinge SJL, Holm E, Ong SP and Wolverton C. Recent advances and applications of deep learning methods in materials science. Npj Computational Materials, 2022;8(1).
[44] Mienye ID and Swart TG. A comprehensive review of deep learning: Architectures, recent advances, and applications. Information (Basel), 2024;15(12):755.
[45] Yu B, Zhang L, Ye X, Wang Y, Ying H, Zhu W, Xiang W and Wu X. State-of-the-art review on various applications of machine learning techniques in materials science and engineering. Chemical Engineering Science, 2024;121147:121147.
[46] Enoki M and Horiya T. Materials integration – Data-driven approach to materials design using simulation and database. In: Springer Series in Materials Science. 2023; (pp. 105–126). Springer Nature Singapore.
[47] Carey NS, Budavári T, Daphalapurkar N and Ramesh KT. Data integration for materials research. Integrating Materials and Manufacturing Innovation, 2016;5(1):143–153.
[48] Yu B, Zhang L, Ye X, Wang Y, Ying H, Zhu W, Xiang W and Wu X. State-of-the-art review on various applications of machine learning techniques in materials science and engineering. Chemical Engineering Science, 2024;121147:121147.
[49] Das M, Sahu S and Parhi DR. Composite materials and their damage detection using AI techniques for aerospace application: A brief review. Materials Today: Proceedings, 2021;44:955–960.
[50] Joshi M and Chatterjee U. Polymer nanocomposite. In: Advanced Composite Materials for Aerospace Engineering. 2016; (pp. 241–264). Elsevier.
[51] Gadtya AS, Mahaling RN, Rout L and Moharana S. Carbon nanotube-based polymer nanocomposites for aerospace applications. In: Engineering Materials. 2024; (pp. 435–459). Springer Nature Singapore.
[52] Omidian H. AI-powered breakthroughs in material science and biomedical polymers. Journal of Bioactive & Compatible Polymers, 2024;39.
[53] Da Silva RGL. The advancement of artificial intelligence in biomedical research and health innovation: Challenges and opportunities in emerging economies. Globalization & Health, 2024;20(1).
[54] Khan MK, Raza M, Shahbaz M, Farooq U and Akram MU. Recent advancement in energy storage technologies and their applications. Journal of Energy Storage, 2024;92(112112):112112.
[55] Santhosh BM, Shivarudrappa HP, Vasantha Kumar BC, Prashantha Kumar PN, Naik SD, JammapraKallappa P and Divya T. Case studies: Nanomaterials in specific energy storage devices. In Nanostructure Science and Technology. 2024; (pp. 123–138). Springer Nature Switzerland.
[56] Nandy A, Duan C and Kulik HJ. Audacity of huge: Overcoming challenges of data scarcity and data quality for machine learning in computational materials discovery. Current Opinion in Chemical Engineering, 2022;36(100778):100778.
[57] Himanen L, Geurts A, Foster AS and Rinke P. Data-driven materials science: Status, challenges, and perspectives. Advanced Science (Weinheim, Baden-Wurttemberg, Germany), 2019;6(21).
[58] De Borst R. Challenges in computational materials science: Multiple scales, multi-physics and evolving discontinuities. Computational Materials Science, 2008;43(1):1–15.

[59] Heine T. Grand challenges in computational materials science: From description to prediction at all scales. Frontiers Materials, 2014;1.

[60] Liu Y, Yang Z, Yu Z, Liu Z, Liu D, Lin H, Li M, Ma S, Avdeev M and Shi S. Generative artificial intelligence and its applications in materials science: Current situation and future perspectives. Journal of Materiomics, 2023;9(4):798–816.

[61] Gupta V, Liao W-K, Choudhary A and Agrawal A. Evolution of artificial intelligence for application in contemporary materials science. MRS Communications, 2023;13(5):754–763.

[62] Phogat P, Sharma S, Jha R and Singh S. Future perspectives and challenges. In: Engineering Materials. 2024; (pp. 307–325). Springer Nature Singapore.

[63] Maqsood A, Chen C and Jacobsson TJ. The future of material scientists in an age of artificial intelligence. Advanced Science (Weinheim, Baden–Wurttemberg, Germany), 2024;11(19).

[64] Lappala A. The next revolution in computational simulations: Harnessing AI and quantum computing in molecular dynamics. Current Opinion in Structural Biology, 2024;89(102919):102919.

[65] Liu H, Low GH, Steiger DS, Häner T, Reiher M and Troyer M. Prospects of quantum computing for molecular sciences. Materials Theory, 2022;6(1).

Jatinder Singh Dhaliwal*, Pravesh Kumar, Rajesh M. Badhe, Alex C. Pulikottil, and Alok Sharma

Chapter 10
Artificial intelligence in polymer molding and extrusion

Abstract: Polymer injection molding and extrusion are critical production techniques for manufacturing plastic parts, which play a vital role in industries like industrial goods, automotive, electronics, healthcare, sports, construction, and consumer products. Producing high-quality molded products requires precise control over factors such as injection pressure, speed, and mold design. Traditionally, defect detection in molding depends on operator's experience, making quality control less efficient. Likewise, polymer extrusion is a vital technique for manufacturing plastic products like pipes, films, fibers, and profiles, involving stages such as plasticization and shaping. However, maintaining high-quality standards in both molding and extrusion presents a challenge, as accurate and automated defect detection remains difficult. Conventional methods are labor-intensive and heavily reliant on human judgment.

Research has demonstrated that intelligent manufacturing can address these challenges by optimizing process parameters for consistent and efficient molding and extrusion. Hybrid approach combining artificial neural networks , design of experiments , and genetic algorithms can meet multi-objective models that balance customer and production needs. Artificial intelligence (AI)-based data mining techniques can optimize engineering processes using minimal data, eliminating the need for complex, problem-specific strategies. Moreover, advanced deep learning models such as convolutional neural networks and recurrent neural networks have demonstrated exceptional capabilities in analyzing polymer microstructures and sequential data, uncovering insights that conventional methods often fail to detect. This chapter explores groundbreaking research leveraging AI in polymer extrusion and molding, focusing on enhancing defect-free production, developing advanced materials, optimizing mold design, and improving overall manufacturing efficiency.

Keywords: Artificial intelligence, machine learning, polymer extrusion, and injection molding

***Corresponding author: Jatinder Singh Dhaliwal**, Indian Oil Corporation Ltd., R&D Centre, Faridabad 121007, Haryana, India
Pravesh Kumar, Rajesh M. Badhe, Alex C. Pulikottil, Alok Sharma, Indian Oil Corporation Ltd., R&D Centre, Faridabad 121007, Haryana, India

https://doi.org/10.1515/9783111681870-010

10.1 Introduction

10.1.1 Artificial intelligence

Artificial intelligence (AI) has become a pivotal research focus in the twenty-first century, impacting a wide range of disciplines such as education, marketing, finance, economics, engineering, science, medicine, business, accounting, the stock market, and law [1–7]. Despite its reputation as a contemporary breakthrough, the concept of AI dates to 1955, when Stanford Professor John McCarthy coined the term and described it as "the science and engineering of making intelligent machines" [8]. Since then, AI has evolved dramatically, with machine learning (ML) driving transformative changes in industries, governments, and society [9]. Additionally, AI plays a key role in advancing global sustainability efforts by optimizing energy consumption, logistics, supply chain operations, and waste management. Several branches of AI such as natural language processing, image recognition, ML, and data mining have become foundational to contemporary technological development. With ongoing innovations and increasing global interest, AI continues to be at the forefront of scientific exploration and advancement. Extensive research has been done to develop programs that enable machines demonstrating intelligent behavior, such as playing chess. The current focus in AI is on developing systems that can learn in ways that resemble human learning processes. As a dynamic and continually evolving discipline, AI includes multiple subfields such as ML and deep learning. Central to this is ML, which empowers computers to improve their predictive capabilities and problem-solving skills using algorithms that ultimately aim to replicate elements of human cognitive behavior [10]. Rather than representing a single entity, AI functions as a broad umbrella term, encompassing a diverse and continuously advancing range of modeling capabilities. These capabilities evolve over time, integrating new techniques and methodologies to enhance problem-solving, decision-making, and automation. Figure 10.1 [12] provides a visual depiction of the dynamic evolution of AI, highlighting its ongoing growth driven by progress in ML, deep learning, natural language processing, and computer vision.

10.1.2 Machine learning

ML is a distinct subset of AI focused on enabling computer systems to enhance their perception, comprehension, decision-making, and actions by learning from experience or data. Unlike traditional programming, which depends on explicitly defined rules and instructions, ML allows computers to autonomously recognize patterns, analyze vast datasets, and make predictions or decisions without direct human intervention. Instead of following predefined logic, ML systems learn from data, adapting their behavior based on observed trends and outcomes. ML is an inherently interdis-

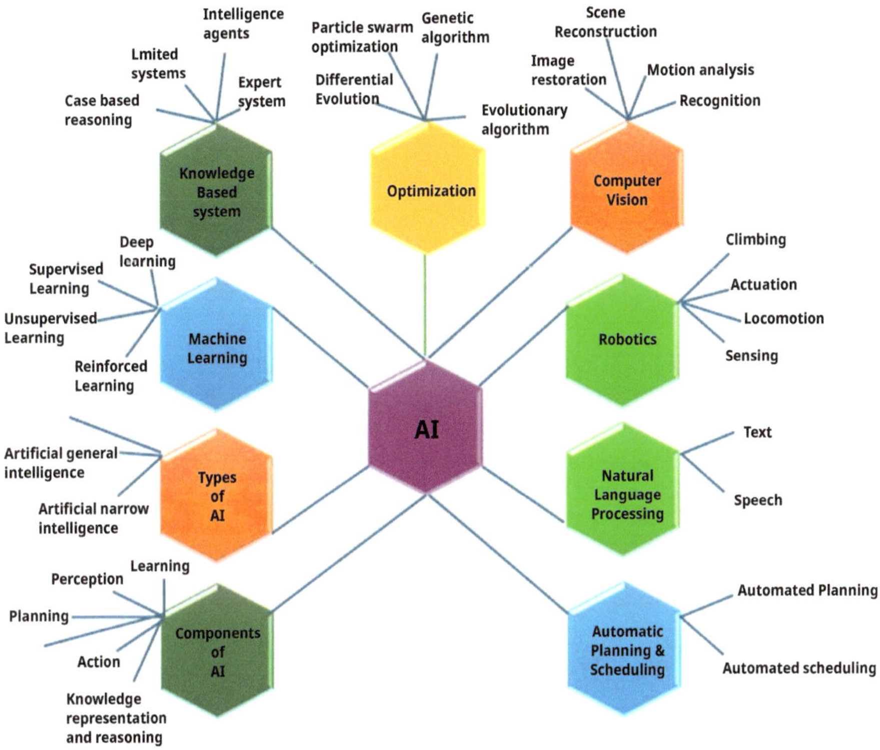

Figure 10.1: AI components, types, and subfields [3, 12].

ciplinary field, combining concepts from computer science, statistics, mathematics, psychology, neuroscience, economics, and control theory. By leveraging insights from these diverse areas, it establishes a strong framework for creating intelligent systems that can tackle complex challenges across multiple industries.

At its core, ML is the scientific study of algorithms and statistical techniques that enable computers to perform complex tasks efficiently. By identifying the underlying patterns in data, these models support intelligent, data-informed decision-making processes. Through continuous learning and refinement, ML algorithms enhance their accuracy and efficiency over time. This iterative improvement allows them to adapt to evolving datasets, optimize performance, and foster innovation across diverse fields such as healthcare, finance, autonomous systems, and AI research.

Through iterative learning processes, these models refine their accuracy over time, continuously enhancing performance without direct human intervention. ML techniques are widely applied across industries, from predictive analytics in finance and healthcare diagnostics to speech recognition, autonomous systems, and personalized recommendation engines. With the increasing demand for intelligent automation, ML continues to advance, driving the future of AI-powered innovations. As a distinct field within AI, ML algorithms develop mathematical models from sample data, known as training data, enabling them to make predictions or decisions without requiring explicit programming [12]. Figure 10.2 shows an overview of some of the most accepted ML methods.

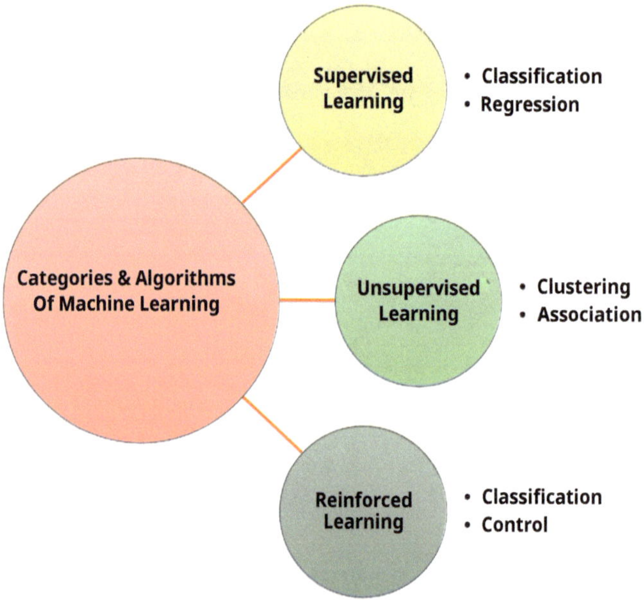

Figure 10.2: Overview of some of the most accepted ML methods [13].

The era of explicitly programming machines to solve problems has given way to ML, where computers independently uncover patterns within datasets to generate solutions. By analyzing hidden trends, ML enables predictive insights, allowing for the anticipation and prevention of potential issues before they arise. AI has grown into a diverse and expansive field within computer science, dedicated to creating systems that mimic and enhance human cognitive functions. Intelligent technologies are now capable of executing intricate tasks traditionally dependent on human intelligence, including decision-making, problem-solving, language understanding, pattern recognition, and visual analysis. AI draws from a diverse range of disciplines such as mathematics, data science, neuroscience, and linguistics to create advanced algorithms that learn from data, adapt to new inputs, and progressively enhance their performance. This progression has led to the emergence of core AI domains, such as ML, deep learning, natural language processing, and computer vision, all of which are instrumental in driving automation, enabling smart decision-making, and solving complex real-world problems.

AI-powered technologies are transforming industries worldwide, spanning sectors such as healthcare, finance, manufacturing, and autonomous systems. From advancing medical diagnostics and fraud detection to driving autonomous vehicles and personalizing virtual assistants, AI is redefining how humans interact with digital environments – boosting efficiency, enhancing precision, and fostering innovation. These intelligent systems perform tasks such as problem-solving, visual recognition, reasoning, decision-making, language processing, and pattern identification. AI utilizes sophisticated algorithms, statistical models, and large datasets to replicate cognitive processes, analyze complex issues, identify hidden patterns, and continuously adapt to new data. With the help of ML, deep learning, and natural language processing, AI is advancing the boundaries of automation and intelligent decision-making. Whether in financial forecasting, self-driving cars, groundbreaking medical innovations, or intelligent digital assistants, AI is reshaping how both businesses and individuals engage with technology and data.

As AI progresses, its capacity to replicate, enhance, and expand human cognitive functions is driving the development of more advanced and adaptive solutions across various industries. Meanwhile, ML stands as a specialized branch of AI, dedicated to empowering computer systems to "learn" from data by identifying patterns and relationships using mathematical models. Instead of depending on explicit programming, ML algorithms employ statistical methods to analyze and interpret data, allowing them to predict outcomes, identify trends, and adjust to new information over time. The principles of ML are strongly grounded in traditional statistical concepts, as depicted in Figure 10.3. ML, often termed "statistical learning" [14, 15], is deeply rooted in statistical principles such as regression analysis, probability theory, and optimization techniques. This strong connection highlights the intersection between classical statistics and modern ML, as both aim to derive meaningful insights from data. ML employs various methodologies tailored to different problem domains, including supervised, unsupervised, and reinforcement learning.

Supervised learning employs labeled datasets to train models, enabling them to predict outcomes based on established input–output relationships. In contrast, unsupervised learning analyzes unlabeled data, revealing hidden structures and patterns without predefined categories. Reinforcement learning takes a distinctive approach, where models improve through a trial-and-error process, earning rewards for optimal actions and facing penalties for suboptimal ones. By integrating these diverse learning methods, ML has transformed industries such as healthcare, finance, automation, and cybersecurity. Its ability to process large datasets, identify complex patterns, and continuously enhance performance has made it fundamental to modern AI applications. Additionally, ML bridges the gap between theoretical statistical principles and real-world decision-making, driving innovation, efficiency, and progress across a range of fields.

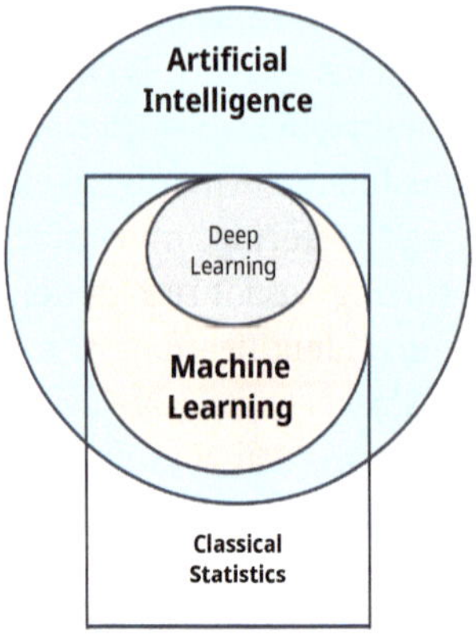

Figure 10.3: The relationship between deep learning, classical statistics, artificial intelligence, and machine learning [16].

10.1.3 Polymer molding and extrusion

The extrusion process traces its origins to the metallurgical industry, with Joseph Bramah patenting the first extrusion method in 1797. His design used a piston-driven mechanism to produce seamless lead pipes. Modern advancements in extrusion technology and machine design have largely resulted from research within the plastics industry. In the food sector, extrusion has been in use since the 1930s and has gained continuous popularity due to its efficiency, consistency, and sustainability [17]. Extrusion has become a fundamental manufacturing process across various industries that require advanced solutions to meet evolving and complex demands. It is particularly crucial in the plastic, food, and pharmaceutical sectors, where precise material processing and consistent product quality are essential. In polymer processing, extrusion involves the use of specialized metal extruders equipped with either single or twin screws, which can operate in a corotating or counterrotating manner. These extruders are designed with highly efficient screw elements that significantly enhance material flow, shear, and mixing. They play a vital role in several key processes, including polymer compounding, the homogeneous blending of fillers and polymers, and facilitating in situ chemical reactions that modify or enhance material properties. The characteristics and performance of extruded polymers are influenced by multiple factors, including screw design, processing conditions, material composition, and temperature control, as illustrated in Figure 10.4. Optimizing these parameters is essential for achieving desired mechanical properties, uniformity, and functionality in the final product. With continuous advancements in extrusion technology, industries can develop high-performance materials tailored to specific applications, ensuring efficiency, sustainability, and innovation in modern manufacturing.

Injection molding is a commonly used method for manufacturing versatile polymer components and structural parts in industries such as automotive, electronics, packaging, and aerospace. The ultimate properties of injection molded products are primarily influenced by factors such as the development of crystalline structures, the

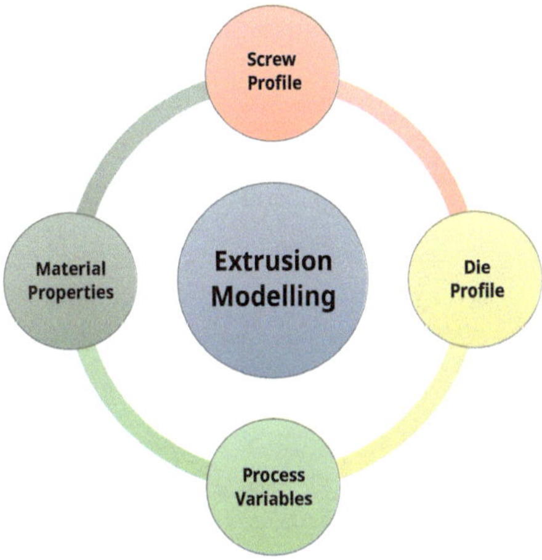

Figure 10.4: Factors involved in polymer extrusion [18].

distribution and alignment of functional fillers, and the shear and extensional flow fields created during the molding process. Additionally, the cooling phase plays a crucial role in shaping the final characteristics of the manufactured components [19]. A core element of the injection molding system is the injection unit, which is composed of three primary sections: the plasticizing unit, the injection unit, and the pressure maintenance unit. Their functions are as follows:

a) **Plasticizing:** The raw material is uniformly melted and plasticized through the combined action of the screw mechanism and heating elements.
b) **Injection:** Plastic material in the molten state is injected into the mold cavity at a controlled pressure and speed, facilitated by the movement of the screw.
c) **Pressure maintenance:** Once the cavity is filled, the screw remains stationary to counteract shrinkage during cooling, ensuring uniform product quality and preventing material backflow [16].

10.2 Artificial intelligence in polymer extrusion

Advancements in AI – particularly in ML and deep learning – are bringing notable benefits to the polymer extrusion industry. These advanced technologies are playing a key role in solving persistent challenges, enhancing productivity, and fostering innovation. Figure 10.5 highlights how AI can be effectively utilized across different stages of the polymer extrusion process.

Munir et al. [20] demonstrated that AI-driven real-time monitoring systems can significantly improve the control of extrusion processes, particularly in optimizing film thickness. These systems utilize real-time data to adjust parameters, ensuring consistent product quality. By adopting these advanced techniques, manufacturers

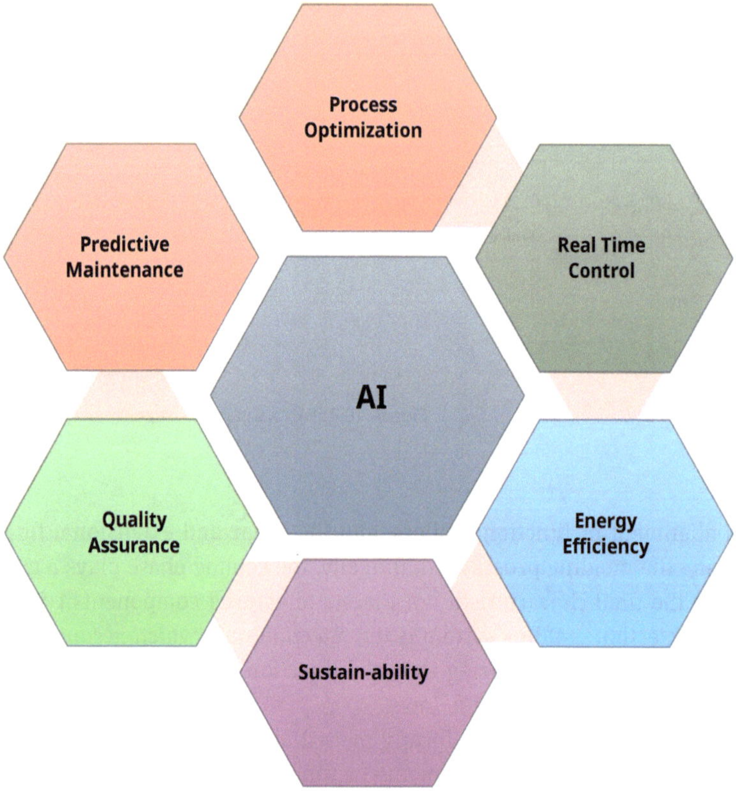

Figure 10.5: AI applications in polymer extrusion.

can significantly reduce defects and minimize material waste during production. In a separate investigation, Munir et al. [21] focused on real-time monitoring of polylactic acid (PLA) degradation during the extrusion process. By leveraging sensor data and machine settings, they developed predictive models to estimate the polymer's molecular weight and mechanical characteristics. The study found that the combination of recursive feature elimination (RFE) and the random forest (RF) algorithm delivered the highest accuracy, efficiency, and interpretability. Among the key variables influencing PLA degradation and mechanical behavior were pressure and temperature, especially at the extrusion exit. Based on these findings, the researchers recommended the RFE-RF model as a reliable quality control strategy for managing the processing of heat-sensitive polymers like PLA.

Additionally, Çınar et al. [22] examined the application of ML for predictive maintenance within the context of sustainable smart manufacturing, specifically targeting polymer extrusion processes. Their study highlighted the effectiveness of advanced algorithms in analyzing sensor data to identify significant patterns that signal potential equipment failures. By leveraging predictive models, manufacturers can anticipate equipment malfunctions, enabling timely maintenance interventions. This forward-thinking strategy reduces operational expenses, minimizes unplanned downtime, and improves overall efficiency in industrial environments. The integration of ML in manufacturing not only enhances process reliability but also contributes to sustainability by optimizing resource utilization and reducing waste. Abeykoon et al. [23] investigated AI's potential to reduce energy consumption in extrusion without compromising product quality. Their research provided a detailed analysis of energy usage, pinpointed areas of energy loss, and proposed strategies to improve process efficiency.

AI is also accelerating the development of new polymer formulations and product customization. By predicting the behavior of new materials during extrusion, AI empowers industries such as automotive and aerospace to push the boundaries of high-performance polymers. The technology is transforming the speed and scope of innovation, enabling companies to work faster while fostering greater creativity and flexibility. Park et al. [24] illustrated this impact by applying AI not only to the design of new materials but also to their processing. This is particularly valuable for advanced polymers and composites, which require precise shaping during extrusion and other manufacturing processes.

10.2.1 Optimization in extrusion through artificial intelligence: concepts and use of DAMICORE and FS-OPA

Optimizing polymer processing requires developing a multidimensional model – referred to as the response surface – that maps the relationship between selected input variables and the resulting process outputs. This approach aims to pinpoint the optimal peak or extreme values within this model to maximize efficiency and overall performance. In modern manufacturing, computer simulations play a crucial role in designing polymer processing systems by utilizing mathematical models to replicate and analyze production processes. These simulations allow for precise predictions of process flow by considering factors such as material properties, machine geometry, and operating conditions. For instance, extrusion processes can be assessed based on factors such as flow rate (throughput), pressure and temperature profiles, energy usage, and mixing effectiveness. By utilizing input data, these critical output parameters can be accurately predicted and optimized for improved performance. However, optimizing extrusion is a complex, multi-objective challenge due to the vast number of interrelated parameters, their nonlinear relationships, and conflicting optimization criteria, such as maximizing output while minimizing energy consumption. This complexity makes it difficult to identify a global optimum without getting trapped in local optima [25].

The earliest efforts to optimize the extrusion process using computer simulations were pioneered by Tadmor and Klein [26], who applied regression analysis and response surface methodology to improve screw design through the development of the EXTRUD program [27]. This was followed by Maddock and Smith [28], who showcased extruder design capabilities using computer-generated simulations. Helmy and Parnaby [29] further advanced the field by creating optimization software that employed a steady-state hill-climbing algorithm to enhance melt extruder design. As research progressed, more advanced optimization approaches began to emerge. Potente et al. [30, 31] introduced a screw optimization technique that combined design of experiments with multiple regression, using the REX extrusion model [32–34]. Likewise, Thibodeau and Lafleur [35, 36] utilized the STATISTICA software to identify optimal response surfaces based on an extrusion model developed at École Polytechnique de Montréal [37, 38]. Wilczyński et al. [39] also adopted statistical methods via the SSEM program [40]. Despite these advances, a major limitation of traditional statistical approaches was the need for a large number of simulations to construct a detailed response surface. This not only increased computational demands but also raised the risk of converging on local optima rather than achieving true global optimization.

The efficiency and overall performance of a single-screw extruder are influenced by a combination of key factors. These include the properties of the processed material – such as its physical, thermal, and rheological characteristics – the geometric design of the system, particularly the screw configuration, and the operational settings, including barrel temperature and screw speed. An extruder's effectiveness is generally assessed using indicators such as mass throughput, melt temperature, the screw length needed for full material melting, energy consumption, mixing performance, and the extent of viscous heat generation.

To optimize the extrusion process, it is essential to fine-tune screw geometry and operating conditions to maximize output and mixing while minimizing inefficiencies. António Gaspar-Cunha et al. [41] investigated data-driven optimization techniques for single-screw extruders, focusing on multi-objective optimization problems, which require balancing multiple competing factors such as material behavior and screw configurations. Traditionally, these optimization processes rely heavily on expert knowledge; however, data-driven approaches aim to reduce manual intervention by utilizing advanced data analysis and computational methods. By combining data analysis with optimization techniques, the decision-making process becomes more efficient and precise. Data-driven optimization replaces conventional objective function calculations with metamodels or surrogate models that use existing datasets to estimate system behavior. These models enable efficient prediction and optimization of performance metrics. Common techniques include polynomial regression [42], kriging [43], artificial neural networks (ANNs) [44], radial basis function networks [45], and swarm optimization algorithms [46]. AI further enhances this process by allowing computers to independently identify optimal solutions through the analysis of historical data [47]. A variety of AI-based metamodels have been developed for optimization purposes, such as linear and nonlinear regression, incremental dynamic model decomposition [48, 49], support vector machines (SVMs) [50], ANNs [51], decision trees, and the Code2Vect model [52]. However, a significant challenge in applying ML to extrusion lies in the potential impact of unaccounted variables, which can affect the accuracy and generalizability of predictive models. The objective is to leverage the DAMICORE framework, which utilizes distance estimation through compression algorithms known as normalized compression distance (NCD), to aid in investigations where data availability is limited, and external factors may influence the system [53, 54]. Simultaneously, feature sensitivity optimization using phylogram analysis will be utilized to determine the most influential features in real-world scenarios, with a focus on feature interactions and their impact on a specific objective or target [53]. Research has shown that phylogram analysis (OPA) is highly effective in addressing complex combinatorial single- and multi-objective optimization challenges [54, 55]. The FS-OPA approach is built upon the DAMICORE methodology, which facilitates the direct processing of raw data, enables adaptive learning, and generates predictive models for optimization applications.

In practice, the methodology involves storing data for each analyzed object, applying DAMICORE to compute NCD-based distances between data pairs, and generating a distance matrix. Neighbor-joining is then used to construct a phylogram, and finally, feature nodes (FNs) identify clades (or communities). When applied to single-screw extrusion, this approach generates phylograms that provide two levels of learning. The data mining process is designed to analyze and optimize polymer extrusion in a single screw extruder. The dataset utilized for this computational analysis is limited in availability, as it was randomly collected and covers various aspects of the extrusion process, including operational parameters, system geometry, and optimization targets. By implementing the DAMICORE-based learning approach, the study uncovers relationships between decision variables and optimization goals, revealing critical interactions. This analysis provides key insights into the significance of specific design parameters in achieving optimal performance.

10.2.2 Scale-up and optimization for polymer extrusion

xProcess scaling involves adjusting the size of a process according to the defined criteria while striving to preserve critical parameters that mirror those of the original setup. In the context of extrusion scale-up, this means determining the most suitable geometry and operational settings for a larger extruder so that it delivers performance equivalent to that of a smaller, benchmark extruder. This practice is essential in engineering, as it enables the design of industrial-scale extruders using insights gained from laboratory-scale experiments. Over time, various methodologies for extrusion scale-up have been developed and refined. The fundamental principles of this process have been extensively covered in key literature by experts such as Rauwendaal [56], Hensen et al. [57], McKelvey [58], Stevens and Covas [59], Campbell and Spalding [60], and Chen et al. [61]. In addition to these foundational texts, numerous research studies have explored the intricacies of extrusion scale-up. Pearson [62] was the pioneer in performing a thorough and detailed analysis of extrusion scale-up. His study focused on critical aspects of the process, such as the transportation of solids, melting behavior, and the dynamics of melt flow. He determined that successful scale-up could be achieved by maintaining a consistent screw helix angle and barrel temperature, while ensuring that key dimensionless numbers – Graetz, Brinkman, and Nahme – remained constant across different sections of the screw. This method offered several advantages, including balanced solids transport, efficient melting and flow, stable specific energy consumption, and a power–law relationship among the primary scaling factors. Later, Rauwendaal [56, 63] conducted a comparative analysis of various scale-up strategies, uncovering that many resulted in imbalances in solids and melt flow rates or excessive viscous dissipation, thereby highlighting their limitations. Building upon these findings, Covas and Gaspar-Cunha [64] proposed a multi-objective optimization approach for extrusion scale-up, which aimed to identify the best geometry and operating conditions for the target extruder to ensure its key performance metrics closely matched those of the reference extruder. In essence, the objective of the scale-up process is to minimize discrepancies in critical operational parameters between the two machines. Typically, the reference extruder's specifications are known, and the challenge lies in determining the optimal parameters for the target extruder. This methodology has also been applied to the optimization of both single- and twin-screw extrusion systems [65–72]. Implementing this scale-up strategy requires specific steps, as illustrated in Figure 10.6.

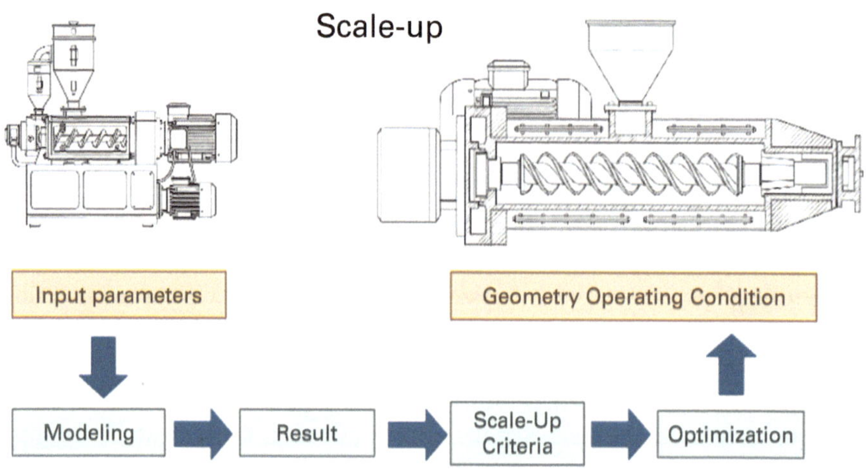

Figure 10.6: The modeling and optimization approach for scaling up polymer extrusion involves defining key parameters: D1 represents the reference screw diameter, while D2 denotes the target screw diameter.

Berzin et al. [73] investigated the difficulties involved in scaling up twin-screw reactive extrusion. Their research revealed that conventional scale-up techniques, which depend on diameter ratios, are less effective when complex processes, such as chemical reactions, occur during extrusion. To overcome this challenge, they proposed that mathematical formulas of varying accuracy could be used to estimate critical parameters and identify suitable scaling exponents, resulting in more dependable scale-up guidelines.

10.3 Artificial intelligence in polymer molding: leveraging artificial intelligence for precision injection molding and achieving zero-defect manufacturing

In today's highly competitive market, maintaining high-quality standards is essential for companies to stand out. Minimizing errors in production processes is crucial, as it leads to lower costs, reduced downtime, and increased overall efficiency. One of the most effective approaches to achieving this is through AI, which has been shown to greatly improve system performance and productivity [74, 75]. Injection molding is an extensively utilized manufacturing method for mass producing polymer components. The primary goal of injection molding is to manufacture consistent, repeatable parts with a uniform appearance. However, the dynamic nature of the process, where parameters continuously fluctuate, can lead to defects, resulting in unexpected costs. Common defects in plastic injection molding (PIM) include unfilled sections, burrs, burn marks, short shots, warping, and flow lines [76, 77]. To mitigate these issues, implementing a monitoring and predictive system that detects defects based on process variations is highly beneficial. An effective and reliable classification system that can identify and alert operators to potential defects is the ideal solution for ensuring quality and minimizing production losses.

Silva et al. [78] developed an advanced intelligent system designed to manage the entire process, from data collection to real-time defect classification. Their approach utilized the European Zero-Defect Manufacturing Platform (ZDMP), which harnesses AI-driven technologies to streamline production workflows and minimize defects. ZDMP operates through a comprehensive framework that incorporates various tools, such as product life cycle assessment, diagnostic and predictive analytics, production control enhancements, quality assurance strategies, and inspection techniques, all aimed at promoting sustainable manufacturing. These tools enable quick process adjustments through both forward and feedback control mechanisms. As part of their work, the researchers proposed an architecture that integrates the RAILES Smart Manufacturing Execution System (MES) with components of the ZDMP platform. RAILES is a state-of-the-art MES that supports real-time monitoring of manufacturing

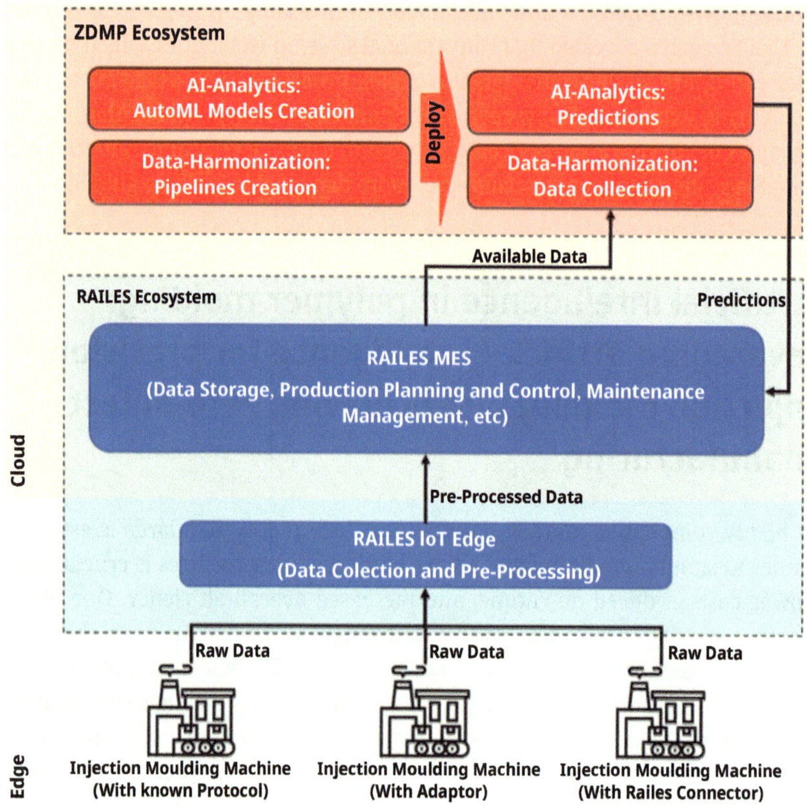

Figure 10.7: Generic system architecture [78].

operations. As a next-generation industrial software solution, RAILES provides manufacturers with real-time, data-driven insights to improve decision-making on the production floor. The proposed architecture seeks to seamlessly incorporate RAILES into the ZDMP ecosystem, enhancing its capabilities by utilizing additional features from the ZDMP platform. The general system architecture is depicted in Figure 10.7.

By seamlessly integrating all tools into a unified ecosystem using the RAILES platform and ZDMP, substantial improvements were achieved in shop floor monitoring and operational reliability. This integration led to notable improvements in key performance indicators (KPIs), including a boost of up to 12% in overall equipment effectiveness, a 9% reduction in process downtime, and a significant decline in defective part production. Many researchers have investigated various approaches for quality assurance and defect detection in plastic molding processes. Christian et al. [79] introduced a hybrid methodology that integrates ML techniques with data obtained from both real-world experiments and simulations. Their goal was to improve product quality predictions using limited experimental data to optimize system parameters.

However, a key challenge remains in bridging the gap between simulated and experimental data to develop a more effective learning process. Ogorodnyk et al. [80] explored the use of ANN, particularly multilayer perceptron (MLP), in conjunction with decision tree algorithms like J48, to enhance the production process of high-density polyethylene dog bone specimens. However, selecting the most relevant features remained a challenging task. Struchtrup Alexander et al. [81] examined various feature selection techniques, including filter, wrapper, and embedding methods. Their study compared multiple feature sets and evaluated their effectiveness using seven supervised ML algorithms. Although these methods showed promising results, their adoption in industrial settings remains limited, despite the availability of comparable commercial solutions. Park et al. [82] focused on real-time monitoring and control by developing a learning algorithm that adjusts machine input parameters based on sensor data. This proportional correction system was designed to improve the quality of the final product. While traditional quality control methods are available, AI-driven approaches provide substantial benefits over conventional methods, making them a more efficient solution for contemporary manufacturing.

Park et al. [83] developed an advanced system for real-time monitoring, diagnostics, and autonomous quality control in injection molding, aiming to enhance process stability and product consistency. The system gathers vast amounts of data from millions of injection molding cycles using cavity pressure and temperature sensors. Each dataset is assigned a unique cycle ID, linking it to key process parameters, real-time sensor readings, and corresponding product quality classifications. This extensive database allows for in-depth analysis and the creation of reliable reference models. By analyzing the data, the researchers identified correlations between machine input parameters, sensor outputs, and the final product quality. Using these insights, they designed a compensatory algorithm that dynamically adjusts machine input parameters in real time, ensuring optimal processing conditions and consistent product quality. The system continuously compares real-time sensor data with predefined reference values, and when discrepancies are detected, it automatically fine-tunes the process parameters to address any inconsistencies, reducing defects and maintaining uniform quality. This AI-driven approach significantly cuts down on manual adjustments, boosts production efficiency, and improves the overall reliability of injection molding processes. Figure 10.8 illustrates a conceptual flow diagram of the control system's implementation, which includes two temperature sensors and two pressure sensors placed within the cavity of a car door module. These sensors provide continuous feedback on cavity conditions, enabling precise adjustments to optimize the injection molding process.

Hail Jung et al. [84] conducted a study demonstrating that autoencoder-based models outperform tree-based models in defect classification. Tree-based models are typically built using two main techniques: bagging and boosting. Bagging focuses on effectively organizing the data before building the model, while boosting emphasizes the importance of sensor values and iteratively adjusts residuals to improve the model's accuracy. As shown in Table 10.1, gradient boosting achieved the highest perfor-

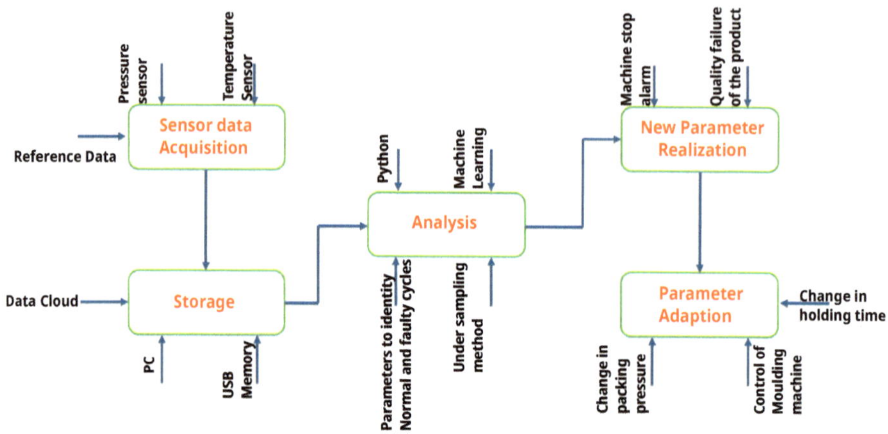

Figure 10.8: Flow diagram for the control system [83].

Table 10.1: Showing the model results.

	Accuracy	Precision	Recall	F_1 score
Panel A: regression-based models				
	Accuracy	**Precision**	**Recall**	**F_1 score**
Logistic regression	0.8449	.0833	0.8947	0.1521
Support vector machine	0.86642	0.961	0.9210	0.1741
Panel B: tree-based models				
	Accuracy	**Precision**	**Recall**	**F_1 score**
Random forest	0.9918	0.7647	0.6841	0.7222
Gradient booting	0.9862	0.5576	0.7638	0.6444
XGBoost	0.989366	0.6761	0.6052	0.6388
CatBoost	0.9905	0.6923	0.7105	0.7012
LightGBM	0.9914	0.7575	0.6578	0.7042
Panel C: autoencoder models				
	Accuracy	**Precision**	**Recall**	**F_1 score**
Autoencoder	0.9959	0.9469	1.0000	0.9727

mance score of 0.7638, indicating that the model correctly classified 76% of defective products. However, only 55% of the predictions identifying products as defective were actually accurate, highlighting a limitation in the model's precision.

Selvaraj Senthil et al. [85] conducted an extensive review of ML models applied in injection molding machines. Their study explored various factors, including the types of algorithms employed, classifications of AI models, depth layer configurations, training and testing durations, datasets, frameworks, programming languages, user interface designs, and the strengths and weaknesses of each approach. In the realm of ML, supervised learning (Figure 10.9) refers to training a model using labeled input data, allowing it to make highly accurate predictions. This process is like a teacher guiding a student through problem-solving until they achieve precision. Once trained, the model is tested on a separate dataset – typically a subset of the training data in different proportions – to assess its reliability. The approach outlined in [86] yielded the most effective results by precisely predicting the optimal time to replace machine components, thereby reducing damage costs and enhancing operational efficiency. This was accomplished by training the model with multiple supervised learning algorithms and comparing its performance with other research findings. Unsupervised learning (Figure 10.10), on the other hand, is used in scenarios where labeled data is unavailable, requiring the model to identify hidden patterns within the dataset. Without direct supervision, the model autonomously detects similarities and groups related to data points through clustering. This learning process mimics the way humans naturally identify patterns and organize information. The approach outlined in [87] achieved remarkable success by implementing unsupervised cluster analysis for real-time monitoring of mold wear. This method incorporated a sophisticated early warning system designed to detect wear patterns in advance, preventing potential operational failures and improving overall efficiency through proactive maintenance strategies.

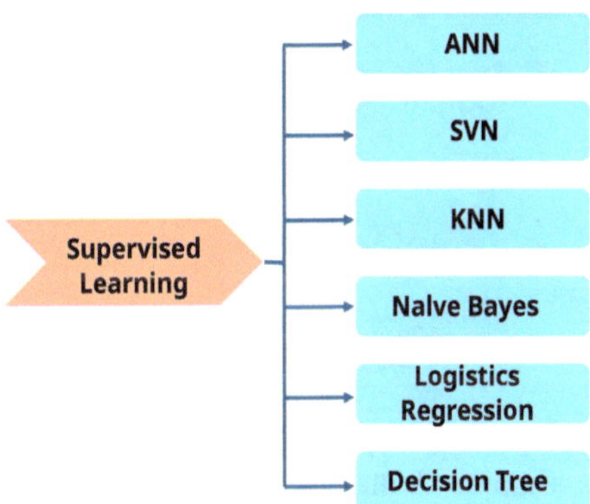

Figure 10.9: Supervised learning.

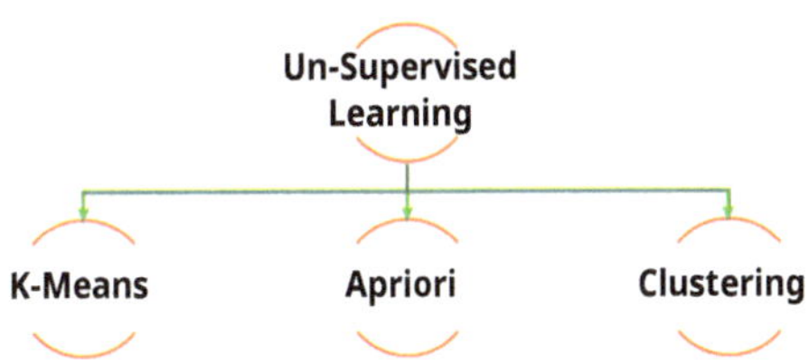

Figure 10.10: Unsupervised learning.

Reinforcement learning (Figure 10.11) is a subset of ML in which a system enhances its dataset classification by learning through trial and error. The system receives rewards for correct outputs and penalties for incorrect ones, enabling it to continuously improve its decision-making process. By utilizing this approach, the model learns to optimize task performance by maximizing rewards and minimizing penalties, initially relying on random data trials before progressively improving its accuracy. Shiraishi et al. [88] applied reinforcement learning to predict operating and molding conditions within a short timeframe, achieving efficient molding with reduced power consumption compared to findings from other research studies.

Based on their research, several key conclusions were drawn:

1. ML algorithms can be effectively applied throughout the injection molding process, from the initial molding stage to the final steps.
2. Regression equations derived from experimental designs can reduce the need for extensive training datasets in ANNs.
3. By minimizing repetitive tasks, ML addresses the limitations of simulations, significantly lowering computational power requirements, resource consumption, and processing time. This automation reduces dependence on expert interpretation, facilitating smoother implementation.
4. The most utilized simulation tools in these studies were MATLAB and C-MOULD.

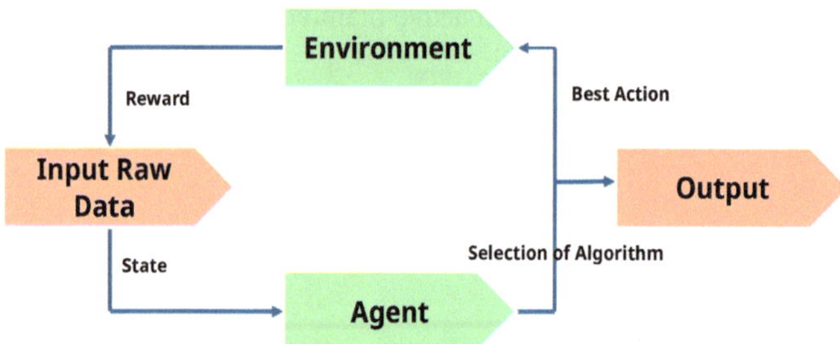

Figure 10.11: Reinforcement learning.

Ensuring consistent product quality in the PIM process requires the automatic detection of operational defects, which remains a major challenge. Ekta et al. [89] proposed an ML framework aimed at identifying complex faults during PIM. Their approach involved the development of a novel ML-driven gating system (MLGS-PIM), which utilizes ANNs and SVMs to enhance fault detection accuracy and improve overall process reliability. By integrating these two models, the system improves product quality under various conditions. This innovative ML-driven method optimizes the injection molding process by effectively managing simulation data and measurements, while predicting quality characteristics based on multiple parameter values. The study focused on three thermoplastic materials – polystyrene, acrylonitrile butadiene styrene (ABS), and polyvinyl chloride (PVC) – and employed gating systems with configurations of 3, 4, and 5 gates. The evaluation was based on KPIs such as sum rate, bit error rate, and convergence analysis. The results demonstrated that the proposed MLGS-PIM framework significantly outperformed traditional meth-

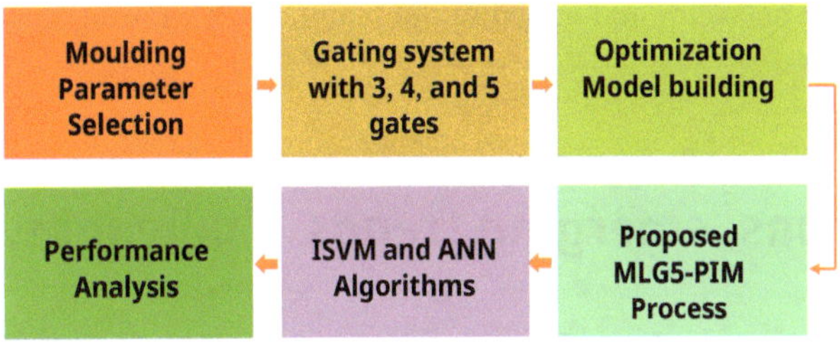

Figure 10.12: Workflow of the MLGS-PIM method.

ods like AntLion optimization and PSO-MSQPA, leading to notable gains in efficiency and accuracy. Figure 10.12 illustrates the MLGS-PIM workflow.

Numerous researchers have explored innovative methods to enhance the accuracy and quality of injection-molded components. Ogorodnyk et al. [90] examined the application of ANN, particularly MLP, in combination with decision tree algorithms such as J48, to develop predictive models aimed at optimizing the production process of high-density polyethylene dog bone specimens. However, a significant challenge in this approach remains the selection of the most relevant features, which is crucial for improving model accuracy and performance. To tackle this challenge, Struchtrup Alexander et al. [91] conducted a comprehensive evaluation of various feature selection techniques, including filter, wrapper, and embedding methods. Their study evaluated the performance of various feature sets by applying seven supervised ML algorithms to derive meaningful insights from the data. While these techniques demonstrated significant potential, their adoption in industrial settings remains limited, despite the increasing accessibility of AI-driven solutions.

Building upon their previous research, Struchtrup Alexander et al. [92] designed a comprehensive framework to automate data preprocessing, enhancing the accuracy of part quality prediction. This framework enhances the analysis of injection molding data by streamlining feature selection and optimizing model performance. By improving the efficiency and reliability of quality assessments, this approach has the potential to facilitate broader integration of ML in manufacturing, ultimately leading to smarter, data-driven production processes.

The choice of ensemble hyperparameters was found to significantly influence overall performance. Despite considerable advancements, fully unlocking AI's potential in polymer extrusion remains a challenge. A key obstacle is the lack of high-quality, domain-specific datasets required to train AI models effectively. Without robust data, models struggle to generalize across different polymers and extrusion techniques. Another critical issue is the need for interpretable AI models that provide clear justifications for their predictions and decisions. Munir et al. [93] emphasized the critical role of collaboration among materials scientists, data scientists, and engineers in overcoming these challenges. By fostering interdisciplinary cooperation, these efforts can help eliminate obstacles and expedite the seamless integration of AI into polymer extrusion processes.

10.4 Future directions, emerging trends, challenges, and solutions

The foremost challenge in any ML application is determining whether the appropriate data is available, an issue that is particularly significant in extrusion and molding due to the numerous variables involved. Despite these complexities, there is strong optimism about AI's ability to enhance extrusion efficiency and sustainability, particularly by reducing material waste and energy consumption [94]. Key recommendations include acquiring high-quality data, ensuring model interpretability, and promoting interdisciplinary collaboration [20]. However, certain factors, including ambient temperature, humidity, and the inherent properties of raw materials, are difficult to regulate. These variables, categorized as uncontrollable data, cannot be directly adjusted but significantly influence the process. Recognizing their impact is crucial for developing robust and reliable AI models [95]. The chapter insights take a forward-thinking approach to the future of polymer extrusion and molding, exploring emerging trends and potential breakthroughs that could shape the plastics industry. This underscores the revolutionary role of AI in driving advancements in process optimization, sustainability, and energy efficiency, three critical areas for enhancing polymer extrusion. To guide future advancements, a detailed research roadmap outlines key questions and areas requiring further investigation. By identifying significant gaps and opportunities for innovation, this framework supports the structured and systematic development of AI-driven polymer extrusion, paving the way for continued progress in this evolving field. Future research in AI for polymer extrusion and molding is expected to focus on several key areas:

- **Advanced algorithms and techniques:** The continuous advancement of sophisticated ML models, such as deep learning and reinforcement learning, is poised to significantly improve the control of complex extrusion and molding processes while enhancing predictive accuracy. These advanced techniques allow for more precise management of processing parameters, enabling manufacturers to optimize product quality, minimize material waste, and increase overall efficiency. Additionally, emerging generative AI technologies, including generative adversarial networks and variational autoencoders, hold great potential to revolutionize polymer processing. These models can be applied to create innovative polymer material designs with specific properties or to optimize extrusion parameters by simulating different processing scenarios. By leveraging these AI-driven advancements, the industry can achieve greater efficiency, sustainability, and product customization, ushering in a new era of smarter, data-driven manufacturing in polymer extrusion and molding.

- **Real-time data integration and processing:** Improving model accuracy and relevance depends on more efficient real-time data collection, integration, and processing. Progress in sensor technologies and data management systems is set to be instrumental in refining these processes, providing more accurate and dependable insights.
- **Scalability and computational efficiency:** Overcoming the challenges associated with training and deploying complex ML models is crucial for achieving scalability while maintaining cost efficiency. Future research could focus on developing more efficient computational strategies, such as edge computing and distributed systems, to enhance the performance and feasibility of large-scale execution processes.
- **Adaptability and robustness:** Developing ML models that can adapt to evolving process conditions and material characteristics is another critical research direction. Techniques such as transfer learning, meta-learning, and other adaptive methods can enhance model resilience and flexibility in dynamic extrusion environments.
- **Integration with emerging technologies:** The integration of ML models with emerging technologies, such as the Internet of Things (IoT) and digital twins, has the potential to transform process optimization and real-time monitoring. This combination fosters greater efficiency, drives innovation, and boosts system performance across a wide range of applications.

In conclusion, integrating artificial learning into polymer extrusion and molding presents substantial opportunities for improving production efficiency and quality control. However, this evolving field requires further research to address existing challenges and unlock new possibilities for optimization. Additionally, as AI becomes more fundamental to manufacturing, it is essential to tackle ethical matters. Critical issues like workforce displacement and data privacy demand proactive strategies to ensure responsible implementation. Implementing strategies for retraining and upskilling displaced workers will be essential to ensure a smooth transition into new roles, fostering a balanced and responsible integration of AI in the industry. Moreover, data security and transparency are essential, necessitating robust protection measures and clear policies on data usage to build trust and ensure responsible AI implementation.

Abbreviations

AI	Artificial intelligence
ML	Machine learning
ANNs	Artificial neural networks
GA	Genetic algorithms
CNNs	Convolutional neural networks
RNNs	Recurrent neural networks
RFE	Recursive feature elimination
RF	Random forest
PLA	Polylactic acid
SVM	Support vector machines
NCD	Normalized compression distance
OPA	Phylogram analysis
ZDMP	Zero-defect manufacturing platform
MES	Manufacturing Execution System
KPIs	Key performance indicators
ABS	Acrylonitrile butadiene styrene
PVC	Polyvinyl chloride
MLP	Multilayer perceptron
IoT	Internet of Things

References

[1] Halal WE. Artificial intelligence is almost here. Horizon. 2003;11:37–38.
[2] Masnikosa VP. The fundamental problem of an artificial intelligence realization. Kybernetes. 1998;27:71–80.
[3] Metaxiotis K, Ergazakis K, Samouilidis E and Psarras J. Decision support through knowledge management: The role of the artificial intelligence. Information Management & Computer Security. 2003;11:216–221.
[4] Raynor WJ. The international dictionary of artificial intelligence. Reference Reverend. 2000;14:1–380.
[5] Stefanuk VL and Zhozhikashvili AV. Productions and rules in artificial intelligence. Kybernetes. 2002;31:817–826.
[6] Tay DPH and Ho DKH. Artificial intelligence and the mass appraisal of residential apartments. Journal of Property Value and Investment. 1992;10:525–540.
[7] Wongpinunwatana N, Ferguson C and Bowen P. An experimental investigation of the effects of artificial intelligence systems on the training of novice auditors. Management Audit Journal. 2000;15:306–318.
[8] Andresen SL. John McCarthy: Father of AI. IEEE Intelligent Systems. 2002;17(5).
[9] Oke SA. A literature review on artificial intelligence. International Journal of Information Science and Management. 2008;19:535–570.
[10] Artificial Intelligence and the Future of Teaching and Learning. Insights and Recommendations. 2023; Washington, DC: U.S. Department of Education, Office of Educational Technology.
[11] Regona M, Yigitcanlar T, Xia B and Man RY. Opportunities and adoption challenges of AI in the construction industry: A PRISMA review. Journal of Open Innovation: Technology, Market, and Complexity. 2022;81.

[12] Geetha Gowri S, Devi R and Sethuraman K. Machine learning. International Journal of Research and Analytical Reviews. 2019;62.

[13] Ahmad J, Farman H and Jan Z. Deep learning methods and applications. In Deep Learning: Convergence to Big Data Analytics; Springer Briefs in Computer Science. 2019; (pp. 31–42) Springer: Singapore.

[14] James G, Witten D, Hastie T. et al. An Introduction to Statistical Learning. 2021; New York: Springer US.

[15] Rattan P, Penrice DD and Simonetto DA. Artificial intelligence and machine learning: What you always wanted to know but were afraid to ask. Gastro Hep Advances. 2022:70–78.

[16] Singh B, Sharma C and Sharma S. Fundamentals of Extrusion Processing In: Novel Food Processing Technologies. 2017; New Delhi: New India Publishing Agency.

[17] Hyvärinen M, Jabeen R and Kärki T. The modelling of extrusion processes for polymers – a review. Polymers. 2020;12(6):1–14.

[18] Zhou S and Hrymak AN. Injection moulding of polymers and polymer composites. Polymers. 2024;16:1796.

[19] Fu H, Xu H, Liu Y, Yang Z, Kormakov S, Wu D and Sun J. Overview of injection moulding technology for processing polymers and their composites. ES Material & Manufacturing. 2020;8:3–23.

[20] Munir N, Nugent M, Whitaker D and McAfee M. Machine learning for process monitoring and control of hot-melt extrusion: Current state of the art and future directions. Pharmaceutics. 2021;13 (9):1432.

[21] Munir N, McMorrow R, Mulrennan K, Whitaker D, McLoone S, Kellomäki M, Talvitie E, Lyyra I and McAfee M. Interpretable machine learning methods for monitoring polymer degradation in extrusion of polylactic acid. Polymers (Basel). 2023;15(17):3566.

[22] Çınar ZM, Abdussalam Nuhu A, Zeeshan Q, Korhan O, Asmael M and Safaei B. Machine learning in predictive maintenance towards sustainable smart manufacturing in Industry 4.0. Sustainability. 2020;12(19):8211.

[23] Abeykoon C, McMillan A and Nguyen BK. Energy efficiency in extrusion-related polymer processing: A review of state of the art and potential efficiency improvements. Renewable and Sustainable Energy Reviews. 2021:147.

[24] Park J, Shim Y, Lee F, Rammohan A, Goyal S, Shim M, Jeong C and Kim DS. Prediction and interpretation of polymer properties using the graph convolutional network. ACS Polymers Au. 2022;2(4):213–222.

[25] Nastaj A and Wilczynski K. Optimization and Scale-Up for Polymer Extrusion. Polymers. 2021;13:1547.

[26] Tadmor Z and Klein I. Engineering Principles of Plasticating Extrusion. 1970; New York, NY, USA: Van Nostrand Reinhold Co.

[27] Tadmor Z and Klein I. Computer Programs for Plastic Engineers. 1968; New York, NY, USA: Reinhold Book Corporation.

[28] Maddock BH and Smith DJ. Extruder design by computer printout. SPE Journal. 1972;28:12–17.

[29] Helmy HAA and Parnaby J. Computer – aided optimal melt screw design. Polymer Engineering & Science. 1976;16:437–449.

[30] Potente H, Hanhart W, Klarholz B and Schöppner V. Extruderschnecken durch Simulation optimieren. Kunststoffe. 1992;82:939–944.

[31] Potente H and Krell B. Screw optimization by means of DOE and multiple regression. In: Proceedings of the 55th Annual Technical Conferences of the Society of Plastics Engineers (ANTEC 1997). May 1997; (pp. 135–139). Toronto, ON, Canada.

[32] Potente H, Hanhart W and Schöppner V. Potential applications for computer-aided extruder design. International Polymer Processing. 1993;8:335–344.

[33] Potente H, Hanhart W and Reski T. Design and processing optimization of extruder screws. Polymer Engineering & Science. 1994;34:937–945.

[34] Potente H, Schöppner V and Ujma A. Successful simulation of wall – slipping plastics. Journal of Polymer Engineering. 1997;17:153–170.
[35] Thibodeau CA and Lafleur PG. Computer design and screw optimization. In: Proceedings of the 58th Annual Technical Conference of the Society of Plastics Engineers (ANTEC 2000). May 2000; (pp. 276–282). Orlando, FL, USA.
[36] Thibodeau CA and Lafleur PG. Computer design and screw optimization. In: Proceedings of the Polymer Processing Society 16th Annual Meeting (PPS-16). June 2000; Shanghai, China.
[37] Vincelette AR, Guerrero CS, Carreau PJ and Lafleur PG. A model for single-screw plasticating extruders. International Polymer Processing. 1989;4:232–241.
[38] Amellal K and Lafleur PG. Computer simulation of conventional and barrier screw extruders. Plastics, Rubber and Composites. 1993;19:227–239.
[39] Wilczyński K, Nastaj A and Krutysz P. Optymalizacja Procesu Wytłaczania Jedno´slimakowego Tworzyw Sztucznych. Metoda Statystyczna. Mechanik. 2003;10:618.
[40] Wilczyński K. SSEM: A computer model for a polymer single – screw extrusion. Journal of Materials Processing Technology. 2001;3:308–313.
[41] Gaspar-Cunha A, Delbem A, Costa P and Monaco F. Application of artificial intelligence techniques in the optimization of single screw polymer extrusion. In Congreso de Métodos Numéricos En Ingeniería. Las Palmas de Gran Canaria. 2022; España SEMNI.
[42] Zhou Z, Ong YS, Nguyen MH and Lim D. A study on polynomial regression and Gaussian process global surrogate model in hierarchical surrogate-assisted evolutionary algorithm. IEEE Congress on Evolutionary Computation. 2005:2832–2839.
[43] Chugh T, Jin Y, Miettinen K, Hakanen J and K. Sindhya A. Surrogate-Assisted Reference Vector Guided Evolutionary Algorithm for Computationally Expensive Many-Objective Optimization. IEEE Transactions on Evolutionary Computation. 2018;22:129–142.
[44] Jin Y and Sendhoff B. Reducing fitness evaluations using clustering techniques and neural network ensembles. In: Proc. Genet. Evol. Comput. 2004; (pp. 688–699). Conf., Seattle, WA.
[45] Regis RG. Evolutionary programming for high-dimensional constrained expensive black-box optimization using radial basis functions. IEEE Transactions on Evolutionary Computation. 2014;18:326–347.
[46] Sun C, Jin Y, Cheng R, Ding J and Zeng J. Surrogate-assisted cooperative swarm optimization of high-dimensional expensive problems. IEEE Transactions on Evolutionary Computation. 2017;21:644–660.
[47] Dimiduk DM, Holm EA and Niezgoda SR. Perspectives on the impact of machine learning, deep learning, and artificial intelligence on materials. processes, and structures engineering. Integrating Materials and Manufacturing Innovation. 2018;7:157–172.
[48] Schmid PJ. Dynamic mode decomposition of numerical and experimental data. Journal of Fluid Mechanics. 2010;656:5–28.
[49] Williams MO, Kevrekidis G and Rowley CW. A data-driven approximation of the Koopman operator: Extending dynamic mode decomposition. Journal of Nonlinear Science. 2015;25:1307–1346.
[50] Cristianini N and Taylor JS. An Introduction to Support Vector Machines: In Addition, Other Kernel-Based Learning Methods. 2000; New York: Cambridge University Press.
[51] Goodfellow I, Bengio Y and Courville A. Deep Learning. 2016; Cambridge: MIT Press.
[52] Argerich C, Ibanez R, Barasinski A and Chinesta F. Code2vect: An efficient heterogeneous data classifier and nonlinear regression technique. Comptes Rendus Mécanique. 2019;347:754–761.
[53] Kharrat FGZ, Miyoshi F,NSB, Cobre J, Mazzoncini De Azevedo-Marques J, Mazzoncini de Azevedo-marques P and Delbem ACB. Feature sensitivity criterion-based sampling strategy from the Optimization based on Phylogram Analysis (Fs-OPA) and Cox regression applied to mental disorder datasets. PLOS ONE. 2020:15.
[54] Soares A, Râbelo R and Delbem ACB. Optimization based on phylogram analysis. Expert Systems with Applications. 2017;78:32–50.

[55] Martins JP, Fonseca CM and Delbem ACB. On the performance of linkage tree genetic algorithms for the multidimensional knapsack problem. Neurocomputing. 2014;146:17–29. 2014.
[56] Rauwendaal C. Polymer Extrusion. 2014; 5th ed. Munich, Germany: Carl Hanser Verlag.
[57] Hensen F, Knappe W and Potente H. Handbuch der Kunststoff- Extrusiontechnik. 1989; Munich, Germany: Grundlagen; Carl Hanser Verlag.
[58] McKelvey JM. Polymer Processing. 1962; John Wiley & Sons Inc.
[59] Stevens MJ and Covas JA. Extruder Principles and Operation. 1995; Berlin/Heidelberg, Germany: Springer.
[60] Campbell GA and Spalding MA. Analyzing and Troubleshooting Single-Screw Extruders. 2013; Munich, Germany: Carl Hanser Verlag.
[61] Chen B, Zhu L, Zhang F and Qiu Y. Process Development and Scale-Up. In Developing Solid Oral Dosage Forms. 2017; (pp. 821–868). Amsterdam, The Netherlands: Elsevier.
[62] Pearson JRA. Scale-up of single screw extruders for polymer processing. Plastic and Rubber Processing. 1976;1:113–118.
[63] Rauwendaal C. Scale-up of single screw extruders. Polymer Engineering & Science. 1987;27:1059–1068.
[64] Covas JA and Gaspar-Cunha A. Extrusion scale-up: an optimization-based methodology. International Polymer Processing. 2009;24:67–82.
[65] Gaspar-Cunha A. Modeling and Optimization of Single Screw Extrusion. 2000; Ph.D. Thesis. Guimarães, Portugal: University of Minho.
[66] Covas JA, Cunha AG and Oliveira P. An optimization approach to practical problems in plasticating single screw extrusion. Polymer Engineering & Science. 1999;39:443–456.
[67] Covas JA and Gaspar-Cunha A. The use of an optimization approach to the design of extrusion screw. In: Proceedings of the Polymer Processing Society 16th Annual Meeting (PPS-16). June 2000; Shanghai, China.
[68] Gaspar-Cunha A and Covas JA. The design of extrusion screw: an optimization approach. International Polymer Processing. 2001;16:229–240.
[69] Gaspar-Cunha A, Covas JA and Vergnes B. An optimization methodology for setting the operating conditions in twin-screw extrusion. In Proceedings of the Polymer Processing Society 18th Annual Meeting (PPS-18). June 2002; Guimaraes, Portugal.
[70] Gaspar-Cunha A, Poulesquen A, Vergnes B and Covas JA. Optimization of processing conditions for polymer twin-screw extrusion. International Polymer Processing. 2002;17:201–213.
[71] Gaspar-Cunha A, Covas J and Vergnes B. Defining the configuration of Co-Rotating twin-screw extruders with multiobjective evolutionary algorithms. Polymer Engineering & Science. 2005;45:1159–1173.
[72] Teixeira C, Covas JA, Berzin F, Vergnes B and Gaspar-Cunha A. Application of evolutionary algorithms to the definition of the optimal twin-screw extruder configuration for starch cationization. Polymer Engineering & Science. 2011;51:330–340.
[73] Berzin F, David C and Vergnes B.Optimization and Scale-Up of Twin-Screw Reactive Extrusion: The Case of EVA Transesterification. International Polymer Processing. 2020;35:422–428.
[74] Preuveneers D and Ilie-Zudor E. The intelligent industry of the future: A survey on emerging trends, research challenges and opportunities in Industry 4.0. Journal of Ambient Intelligence and Smart Environments. 2017;9:287–298.
[75] Aminabadi SS, Tabatabai P, Steiner A, Gruber DP, Friesenbichler W, Habersohn C and Berger-Weber G. Industry 4.0 in-line AI quality control of plastic injection moulded parts. Polymers. 2022;14:3551.
[76] Rosato M and Rosato D. Injection Moulding Handbook. 2000; Berlin/Heidelberg, Germany: Springer.
[77] Bernardete R. Support Vector Machines for quality monitoring in aplastic injection moulding process. IEEE Transactions on Systems, Man, and Cybernetics, Part C. 2005;35:401–410.

[78] Silva B, Marques R, Faustino D, Ilheu P, Santos T, Sousa J and Rocha AD. Enhance the Injection Moulding Quality Prediction with Artificial Intelligence to Reach Zero-Defect Manufacturing. Processes. 2023;1162.
[79] Hopmann C, Jeschke S, Meisen T, Thiele T, Tercan H, Liebenberg M, Heinisch J and Theunissen M. Combined learning processes for injection moulding based on simulation and experimental data. AIP Conference Proceedings. 2019;21391.
[80] Ogorodnyk O, Lyngstad OV, Larsen M, Wang K and Martinsen K. Application of machine learning methods for prediction of parts quality in thermoplastics injection moulding. International Workshop of Advanced Manufacturing and Automation. 2021:237–244.
[81] Struchtrup AS, Kvaktun D and Schiffers R. Comparison of feature selection methods for machine learning based injection moulding quality prediction. AIP Conference Proceedings. 2020:2289.
[82] Park HS, Phuong DX and Kumar S. AI based injection moulding process for consistent product quality. Procedia Manufacturing. 2019;28:102–106.
[83] Parka HS, Phuongb DX and Kumarc S. AI based injection moulding process for consistent product quality. Procedia Manufacturing. 2017;28:102–107.
[84] Jung H, Jeon J, Choi D and Park J-Y. Application of machine learning techniques in injection moulding quality prediction: Implications on sustainable manufacturing industry. Sustainability. 2021;13:4120.
[85] Selvaraj SK, Aditya Raj R, Mahadevan R, Chadha U and Paramasivam V. A review on machine learning models in injection moulding machines. Advances in Materials Science and Engineering Volume. 2022:1–28.
[86] Pierleoni P, Palma L, Belli A and Sabbatini L. Using plastic injection moulding machine process parameters for predictive maintenance purposes. In: Proceedings of the International Conference on Intelligent Engineering and Management (ICIEM). June2020; (pp. 115–120). London, UK: IEEE.
[87] Frumosu FD, Ronsch GO and Kulahci M. Mould wear-out prediction in the plastic injection moulding industry: A case study. International Journal of Computer Integrated Manufacturing. 2020;33 (12):1245–1258.
[88] Shiraishi W and Yamamoto K. Injection moulding system for the calculation of optimum operating. US Patent and Trademark OIce, Washington, DC, USA, US Patent No. 10416617, 2019.
[89] Mehta ES and Padhi SN. Quality and defect prediction in plastic injection moulding using machine learning algorithms based gating systems and its mathematical models. International Journal on Recent and Innovation Trends in Computing and Communication. 2022;11(3):216–230.
[90] Ogorodnyk O, Lyngstad OV, Larsen M, Wang K and Martinsen K. Application of machine learning methods for prediction of parts quality in thermoplastics injection moulding. International Workshop of Advanced Manufacturing and Automation. 2021:237–244.
[91] Struchtrup AS, Kvaktun D and Schiffers R. Comparison of feature selection methods for machine learning based injection moulding quality prediction. AIP Conference Proceedings. 2020;22891.
[92] Struchtrup AS, Kvaktun D and Schiffers R. Adaptive quality prediction in injection moulding based on ensemble learning. Procedia CIRP. 2022;99:301–306.
[93] Munir N, Nugent M, Whitaker D and McAfee M. Machine learning for process monitoring and control of hot-melt extrusion: Current state of the art and future directions. Pharmaceutics. 2021;13 (9):1432.
[94] Abeykoon C, McMillan A and Nguyen BK. Energy efficiency in extrusion-related polymer processing: A review of state of the art and potential efficiency improvements. Renewable and Sustainable Energy Reviews. 2021:147.
[95] Gupta A, Guntuku SC, Desu R and Balu A. Optimisation of turning parameters by integrating genetic algorithm with support vector regression and artificial neural networks. The International Journal of Advanced Manufacturing Technology. 2014;77:1–9.
[96] Rasmussen C. E. and Williams C. K. I. Gaussian Processes for Machine Learning. The MIT Press, 2006.

Tamanna Rai and Rishabha Malviya

Chapter 11
Discovering new frontiers: material explorations with AI and polymers

Abstract: Artificial intelligence (AI) has been a key tool in many fields, including the creation and development of polymer materials, since the start of the digital revolution. This chapter thoroughly analyzes the AI approaches used in this setting, including materials informatics. To explore new horizons in material design and innovation, it focuses on the synergistic integration of AI with polymer science. Additionally, the chapter discusses the application of machine learning (ML) to the development and identification of polymer materials. Lastly, the obstacles and potential for ML future development in polymer materials are suggested.

Keywords: Artificial intelligence, machine learning, polymer science, property prediction, deep learning, smart material, polymer synthesis

11.1 Introduction

The advancement of machine learning (ML)-based approaches has resulted in the development of cutting-edge methods in scientific fields such as materials science and polymer engineering. ML has recently advanced materials science by utilizing computer algorithms that improve research, awareness, exploration, and modeling [1]. When combined with human ingenuity, these algorithms can greatly aid in the discovery and innovation of new materials for the next innovations [2]. Appropriate design of experiment (DoE) approaches are crucial for the methodical exploration of the design space and the required attributes when designing new materials [3]. This frequently entails choices about a variety of topics, such as system configurations, operating circumstances, and material qualities. The experiential method is used in classical design approaches, which poses some challenges [4]. Primarily, due to their exclusive provision of point-wise evaluation, the relationships between the dependent and independent variables sometimes lack a definitive mathematical representation and are termed "black-box functions" [5]. Secondly, the design space frequently encompasses numerous independent factors, leading to an extensive array of potential

Tamanna Rai, United College of Engineering and Research, Greater Noida, India; Department of Pharmacy, School of Medical and Allied Sciences, Galgotias University, Greater Noida, India
Rishabha Malviya, Department of Pharmacy, Galgotias University, Greater Noida, India

https://doi.org/10.1515/9783111681870-011

combinations. A detailed search rapidly becomes impractical as dimensionality increases [6].

ML seeks to create and evaluate statistical algorithms that acquire knowledge from data and utilize this knowledge on novel, unobserved data, hence enabling independent task performance without direct programming [7]. This advancement functions as a robust instrument for evaluating current polymer materials and identifying novel ones, signifying a substantial transition from conventional methods that depend largely on individual expertise and frequently necessitate considerable time, occasionally extending over decades, from initial experimentation to market introduction [8].

This chapter distinguishes itself by focusing on ML applications in the formulation and optimization of specific polymer classes, addressing unique issues and customized ML methodologies for each. This work distinguishes itself from broader reviews by focusing on the actual application of ML in expediting the design process, forecasting material attributes, and optimizing formulations to attain desired performance more effectively. Moreover, it examines future trends and obstacles, offering a balanced evaluation that considers both current and prospective limitations, a topic often overlooked in the literature [9]. This chapter claims the importance of ML in boosting polymer material innovation by highlighting the shortcomings of traditional experimentation and time-consuming simulations in polymer design [10].

11.2 Fundamental process for machine learning-enhanced polymer development

A genetic algorithm (GA) was employed to construct numerous polymer networks with specified attributes [11]. Subsequently, ML techniques were applied to calculate the characteristics of polymers and conduct design studies. These models pertain to property prediction or new material discovery and follow the subsequent procedures, as shown in Figure 11.1.

Stage 1: Acquiring molecular structures from literature or accessible databases.

Stage 2: Molecules are fingerprinted into shapes that computers can recognize. To be more specific, this involves turning these three-dimensional chemical shapes into tensors.

Stage 3: Researchers can develop a structure-property prediction algorithm by selecting an appropriate ML framework and entering molecular data.

Stage 4: Selecting a suitable ML model to build novel compounds and using the framework to predict features in Stages 1–3.

Stage 5: Examining these new compounds using domain expertise to confirm that the selected samples adhere to essential chemical principles and can be synthesized. Ultimately, one can acquire the requisite polymer [12–15].

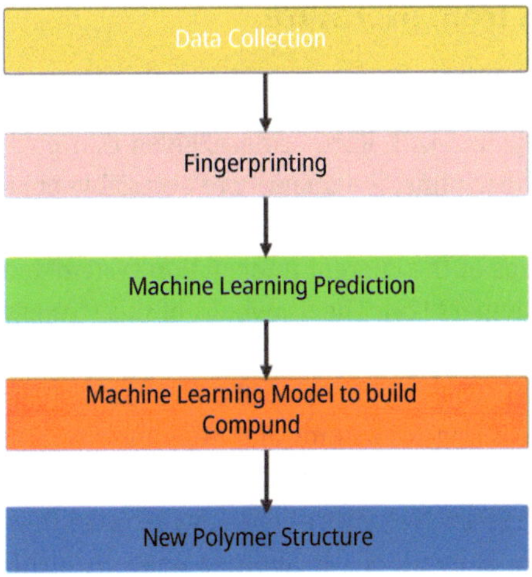

Figure 11.1: Illustrates the basic flowchart for artificial polymer discovery.

11.3 Data collection

Data is crucial when using ML to research polymer membranes. The efficacy of ML models is greatly impacted by their accuracy and the volume of data used to train them [16]. One of the most significant obstacles in constructing strong ML models is the limited availability of data, particularly in terms of maintaining superior quality [17]. Inadequate, low-quality datasets do not accurately reflect the complete diversity of the overall data population, potentially leading to biased forecasts. Training with limited datasets frequently leads to overfitting; thus, the model captures noise instead of the expected basic patterns. As a result, it loses its ability to extrapolate to new, unknown facts [18]. By reducing bias, improving feature representation, and giving models more ability to make predictions across the training data, large, exceptional datasets can help mitigate these disadvantages [19]. The absence of clearly defined criteria for identifying low-quality data complicates the assessment of data trustworthiness. Verifying the data can enhance the dataset's quality by detecting outliers or inaccurate data points [20].

Making sure that the dataset contains at least ten times as many data points as there are input variable types is a widely used strategy to avoid overfitting and improve model dependability. The following subsections outline common data collection methods [21].

11.3.1 The compilation of datasets from literature

The comprehensive global citation database "Web of Science" contains about one million papers related to the term "polymer" [22]. Furthermore, given that polymer research is an expeditiously advancing discipline, the quantity of accessible polymers will perpetually increase [23]. This technique has been accepted by a multiplicity of academics since it alleviates the restrictions of the current information system and allows researchers to engineer particular polymers [24]. The drawback of this strategy is that dataset creation is labor-intensive, and the data points may be fewer than those from existing databases [25]. Recently, researchers developed a mechanical data extraction approach based on language processing algorithms and ML. This could be an excellent technique to avoid the challenges associated with gathering datasets based on books [26]. The fact that almost all research studies only report selected findings, meaning that only partial information is available, is another issue with this data collection process [27]. Consequently, these data do not conform to the Gaussian distribution criteria and may lead to overfitting in ML models.

11.3.2 Compilation of data from theoretical models or scientific computations

Academics frequently utilize scientific computational techniques or conceptual frameworks when they cannot locate the necessary details in current databases or publications [28]. For example, Liu and Cao derived 11 pertinent parameters to predict glass transition temperature using density functional theory (DFT). Using a molecular dynamics (MD) model and ML, Zhu et al. created 300 different kinds of single-chain polymer compounds with high heat conductivity. Using the ADAPT toolkit, researchers identified 100–300 relevant parameters for employing ML to predict the temperature at which glass transitions [29].

11.3.3 Data preprocessing and feature engineering

After being collected, the initial information needs to be cleaned up and set up before it can be used in ML models or for more research. A lot of things go into preprocessing or the preparation of data. Data cleaning includes finding and fixing mistakes, getting rid of duplicates, and getting rid of outliers from a set of data [30]. This is done to make sure that the data is accurate and of higher quality, and that there are no flaws. Adding new numbers to a set of data to fill in missing ones is called data imputation [31]. Data retrieved from the literature frequently contains datasets with missing values for specific variables, necessitating this procedure. Due to its intrinsic chemical and physical properties, raw data is sometimes used by researchers, while others opt to use mean or median values for data calculation [32]. Data normalization is the process of changing the distribution of data so that it is suitable for ML analysis. This means changing the data to fit a normal distribution and normalizing it to have a standard deviation of one and a median of zero [33]. It is essential to translate categorical features into a format that is interpretable by computers, as ML models generally cannot directly analyze categorical data (e.g., types of solutes, polymers, or solvents). Examples of these computer-readable formats include one-hot encoding, label encoding, and binary encoding [34].

11.4 Fingerprinting

During this stage, scientists transform 3D molecules into scalars or tensors, facilitating computer interpretation and subsequent processing. We acknowledge that several fingerprinting methodologies are discussed separately below [35].

11.4.1 Line notation

The Simplified Molecular Input Line Entry System (SMILES) is a commonly utilized format for depicting molecular structures [36]. SMILES strings comprise alphanumeric characters that denote bonds and atoms within a molecule [37]. They contain information about the molecule's atoms, linkages, rings, and branches. SMILES strings are notable for their direct usefulness as descriptors in ML models and their capacity to be transformed into alternative descriptors [38]. As a result, SMILES strings are crucial for molecular expression in a way that ML algorithms can understand which streamlines the application of ML to the creation of polymers.

11.4.2 Morgan fingerprinting

Rogers and Hahn developed Morgan fingerprinting as a re-execution of the extended connectivity fingerprint (ECEP). Each monomer can have its distinct sequential atom numbering obtained using this method. Three essential steps make up this method. First, each atom in the monomer is assigned an integer value. Second, the unique identification number of each atom is updated to match that of its neighbors (see Figure 11.2) [39]. In this method, each repeat generates an identifier that represents progressively bigger spherical components encircling the central atom. Third, to generate a single representative, identical attributes are eliminated [40]. Morgan fingerprinting may be a more effective approach to display 3D structure than linear notation encoding since it can describe each atom and its neighbors. This strategy was also employed

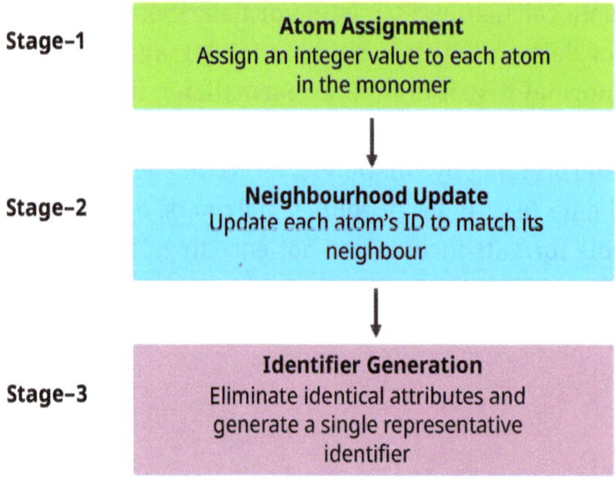

Figure 11.2: Diagrammatic representation of Morgan fingerprinting.

in numerous investigations. To put it into practice, use the open-source cheminformatics software RDKit [41].

11.5 Machine learning algorithm

The ML approach aims to establish a connection between the fingerprinted input and the target property, which is an effective substitute for predicting the target feature [42]. Currently, the model is constructed using both linear and nonlinear regression techniques. The most widely used techniques are kernel-based regression and artificial neural networks (ANNs). Below is a brief description of various techniques [43].

11.5.1 Regression model and nonlinear regression algorithms

The linear regression algorithm was one of the original and most basic approaches, representing the desired characteristic as a linear function using fingerprints. These models fail to adjust for the nonlinear relationship between descriptions and qualities, limiting their predictive effectiveness [44]. The so-called "Kern trick" is often used to solve this problem in complex regressions. Kernel functions, such as polynomial kernels and radial basis functions, translate data into a higher-dimensional feature space. In linear regression, this is used in place of the original fingerprint x [45]. In the original feature space, this process leads to a nonlinear regression. Some kernel-based methods that have been used are Gaussian process regression (GPR), support vector machine (SVM) and kernel ridge regression (KRR) [46]. For example, KRR has been used to predict the features of polymers using extensive DFT calculations

[47]. To expedite the search for high-quality polymers and determine their dielectric constant at various frequencies, particle swarm optimization (PSO) and interval support vector regression were suggested by Yin et al. This study uses the PSO method to enhance the SVM hyper-parameters [48].

11.5.2 Random forests

This approach skillfully handles the complex interaction of attributes by combining a variety of decision trees. The maximum allowable depth for all trees and the total number of trees in the forest were two important factors that were adjusted to enhance their performance [49].

11.5.3 Gradient boosting machines (GBMs)

Methods like LightGBM and XGBoost were applied to improve prediction accuracy. These models are resistant to overfitting and able to handle large datasets, as they repair faults from earlier rounds [50].

11.5.4 Graph convolutional neural network

As previously mentioned, molecular graphs can only use graph convolutional neural network (GCNN) to predict properties. Convolution and pooling are the two ways that GCNN differs from CNN [51]. The GCNN kernel is impacted by its surroundings, whereas the CNN kernel has a standard 2D receptive field and moves at a regular Euclidean distance. Second, CNN polls uniformly in a 2D grid, but GCNN polls no uniformly in its surroundings [52]. Complex network connection in polymer graphs is an example of a strange area that GCNN can handle because it has two unique properties. A study that employed GCNN to identify the locations of interactions between protein kinase inhibitors found that the accuracy was 86% (Wang et al., 2015) [53]. When Tao et al. utilized GCNN to predict a photopolymer's glass transition temperature; their model was accurate up to 79.18% of the time [54]. Altae-Tran et al. employed GCNN in their study to accomplish one-shot acquiring knowledge, and their model performs well in small molecule prediction [55]. When Aldephi et al. used the modified GCNN, they looked at the link after polymerization and came up with a new network called wD-MPNN, which seemed to work better than the old GCNN [56].

11.5.5 The ANN and the deep neural network

Many years of work have gone into making ANN and deep neural networks (CNN) that work like the brain to learn the connections between different traits and descriptions by doing things over and over again. The output layer displays how well the potential polymers function, while the hidden layers demonstrate the connections between the input and output layers [57]. To extract significant features from the input layer and further simplify the features, CNN adds additional standard and pool layers. The convolutional layers are made up of trainable filters that perform distinct convolutions across the entire input. This allows local characteristics to be extracted [58]. Pooling layers further reduce feature dimensions by pooling sub regions into a single result. Maximum pooling is the most prevalent type of pooling. When compared to completely linked layers, conventional and pool layers have significantly less buried neurons. This allows networks to go much deeper [59]. ML models for the appropriate choice of thermal conductivity, glass transmission temperature, and polymer solvent pairings were created using NNs. A binary classification neural network (NN) model that can determine whether a solvent is suitable for a polymer or not was trained using 11,958 pairs of polymers and good solvents and 8,469 pairs of polymers and non-solvents to identify the best polymer-solvent pairs. To forecast how thermally conductive polymers will be, we used a CNN trained on thermal conductivities found by traditional MD models. Wei et al. developed a CNN model using 2D cross-sectional photos of polymer composites, which has been shown to accurately forecast how well materials transport heat [60].

11.5.6 Decision tree

To help in decision-making, decision trees employ a framework similar to a tree to foretell potential outcomes. It has also been widely utilized for polymer discovery because to its easy understanding ability and reliance on graphical depiction [61]. Information gain, gradient boosting, and a variety of other methods have been used to construct decision trees. Li et al. predicted how hydrogels self-assemble using logistic regression, decision trees, and gradient boosting, which may be useful in biomedical settings like cell culture [62]. Bhowmik et al. utilized decision trees and principal component analysis (PCA) to predict the specific heat of polymers [63]. Kumar et al. separated ribonucleoprotein absorption, toxicity, and editing rates into high and low groups using random forest [64].

11.5.7 Variational autoencoders (VAEs)

We looked into the hidden space of polymer structures to come up with different design choices. VAEs used a probabilistic approach to model complex distributions and come up with new polymer options that met certain criteria [65].

11.6 Polymer design algorithms

Upon completion of training, the polymer replacement models can be employed to accelerate the discovery of novel polymers. Here are two distinct approaches. The first approach employs predictions to identify candidates that meet specific property requirements. In contrast, the second approach uses generative and genetic models to directly produce desirable candidates [66].

11.6.1 Enumeration

Artificial intelligence methodologies are frequently employed to expedite the systematic design and advancement of innovative materials. These technologies operate by effectively exploring a designated chemical area [67]. The ML models we've already talked about are used to guess the properties of a big group of possible polymers that are listed according to a physical or chemical scheme. The process of reducing the list according to specific screening criteria comes next [68]. In the end, a ranked list of qualified candidates for the objective application is eventually generated. The first possibilities could be synthetic polymers that have already been made, or they could be hypothetical polymers that have been made by humans or machines, showing the trends in a way that looks like "Ashby plots" of different ML-predicted features for tens of thousands of known or synthesizable polymers [69]. It is possible to choose different combinations of properties based on the needs of specific use [70]. For example, polymers with a high glass transition temperature, a high dielectric constant, and a big band gap can withstand temperature extremes. However, the polymeric electrolytes used in Li-ion batteries must have the proper band gap, electron affinity, and ionization energy. Conversely, gas permeability and selection must be precisely calibrated for polymer membranes [71].

11.6.2 Evolutionary strategy (genetic algorithm GA)

One term for GA's underlying principle is "natural selection." It is simple to treat polymers with GA due to their inherent structure. Polymers consist of several chemical units that are covalently bound to one another. The characteristics of a polymer's chemical components determine the material's final characteristics [72]. To find new potential polymers with the desired characteristics, the GA method employs a sequence of crossover, mutation, and selection procedures [73]. To begin, a large number of polymers are generated at random from a pool of monomers using procedures such as crossover, pruning, mixing, and mutation, which alter the chemical building blocks of the monomers. This process is repeated until a significant number of progeny polymers are obtained [74]. The top candidates for the next generation of polymers are chosen based on a user-defined objective score that takes into account their ideal features. This GA cycle is repeated until a sufficient number of candidates possessing the necessary attributes are identified [75]. GA has been used for more than just finding polymers. It has also been applied to solve problems in materials science, such as uncovering hidden links between material properties and creating functional forms of interatomic potentials. The objective function being evaluated during the selection step is an important part of the GA design scheme. This step has been a big problem for finding new polymers because figuring out properties through tests or calculations is very expensive and takes a long time. However, new, cost-effective, and accurate polymer property models have made this feasible [76]. It is now possible to compute the target function in a fraction of a second. This advancement enables the GA approach to go well beyond preestablished candidate sets and genuinely investigate a very rich chemical polymer space. Additionally, polymers that concurrently meet several property criteria might be targeted by establishing a property-weighted goal function [77].

11.7 Application

11.7.1 Polymer electrolytes design for Li-ion batteries

Rechargeable Li-ion batteries are now commonly employed in a variety of industries, including microelectronics and airplanes. Increasingly, people are interested in developing new, superior solid polymer electrolyte materials because they are required in industry [78]. For Li-ion batteries to perform at their best, the polymer solutions should exhibit high ionic conductivity, high Li-ion transference, a low glass transition temperature, and a wide electrochemical stability window. Experiments to identify the best electrolytes take a lot of time, so data-driven aided polymer design methods are a great way to save time [79]. For instance, to assist with coarse-grained MD simulations, Wang et al. used ML to develop novel polymer solutions with strong ionic conductivity [80].

11.7.2 Rational design of polymer dielectrics

The ability of a polymer to polarize in an electric field is determined by its dielectric constant. For electronics, both extremely high and low dielectric constant polymers are required [81]. In an integrated circuit (IC), low dielectric polymers are used as insulation to keep signal-carrying wires apart and to prevent coupling between metal lines that are close to one another [82]. Conversely, high dielectric polymers are essential for semiconductors and high-density energy storage devices. Many well-known polymers have a low dielectric constant and are frequently used to prevent objects from coming into contact with one another. However, achieving high dielectric constants in plastic materials remains challenging [83]. Several recent initiatives have employed data-driven approaches to develop high dielectric polymers.

11.7.3 Polymer membrane design for gas separation

Polymers are also excellent choices for gas separation due to their enormous surface area [84]. Since the early 1990s, people have been particularly interested in a class of polymers known as polymers of intrinsic microporosity [84]. Researchers must investigate the creation of new polymeric membranes because the existing ones have poor selectivity and degrade over time [85]. It's not easy, though, to come up with polymer membranes that work well and let a lot of gas through while also being selective. Some early research has been conducted to accelerate the search for polymer membranes using data-driven methodologies. For example, gas permeability prediction models have been developed, and the enumeration approach has been utilized to identify polymer membrane candidates for separating CO_2 and CH_4 [86].

11.7.4 Inverse design of thermally conductive polymers

Due to their versatility and ease of manipulation, polymers are increasingly being used as materials for semiconductors, energy storage devices, and several micro/nano fluidics devices. The long-term stability, sustainability, and durability of these devices depend on their ability to effectively dissipate heat. This is because they generate significant heat during operation. The naturally low heat conductivity of polymers needs to be raised so that they can be used successfully in the above-mentioned situations [87]. To comprehend and control the thermal conductivity of polymeric materials, a great deal of research has been conducted. These studies demonstrate that polymer conductivity is intimately related to how they are structured, what chemicals they contain, how they interact with other chemicals, and a variety of other environmental conditions and industrial conditions [88]. These intricate relationships offer thermally conductive polymers a wide range of design options.

11.7.5 Conducting polymer design for electronic applications

Polymers usually don't conduct electricity, but some of them do. These intrinsically conducting polymers are used in electronics such as field-effect transistors, light-emitting diodes, and organic solar cells [89]. Molecular doping is often used to make polymers even more conductive, but it takes longer to find the best pairs of polymers and dopants that work well together. This is because the working conditions, how the dopants and polymers interact, and how the electrons move are all very complicated [90]. Polymer informatics can improve this process by developing models that predict conductivity and utilizing iterative methods to identify the best polymer-dopant pairs. It is supported by the fact that random forest and enhanced regression trees were used to find many good donor/acceptor pairs for organic solar cells [91].

11.8 Future research directions

The promise of artificial intelligence and ML in polymer science is not without its obstacles, which must be overcome before the field can fully benefit from these technologies.

11.8.1 Data quality and quantity

Dataset growth and improvement should be the researchers' priority. This method allows for the collection of a vast amount of high-resolution experimental data, which significantly improves model training and validation.

11.8.2 Algorithm development

It is important to keep coming up with new methods and techniques, especially for dealing with problems that keep coming up, like overfitting, generalization, and processing efficiency. ML techniques that are mixed in hybrid models may also be a good way to boost speed.

11.8.3 Computational resources

Training and using cutting-edge ML models require a lot of work.

11.9 Conclusion

In conclusion, the combination of AI and ML with polymer science has opened up fresh opportunities for material exploration. Researchers can use ML algorithms to quickly screen and optimize polymer qualities, accelerating the development of new materials. AI can predict the behavior of polymers at various stages, including monomer selection, polymerization, and the polymer's final properties. This enables the development of polymers with customized characteristics for specific purposes.

The combination of polymers and AI has far-reaching implications for a variety of areas. For example, the design of polymer membranes for gas separation, polymer electrolytes for lithium-ion batteries, and rational polymer dielectrics. Furthermore, AI can help create polymers for energy storage devices such as super capacitors and batteries. We may expect the development of breakthrough polymer-based goods that will improve many aspects of our lives as long as research in this area continues. Finally, the combination of AI and polymers has the potential to transform material science and other domains.

List of abbreviations

AI	Artificial intelligence
ML	Machine learning
GA	Genetic algorithm
DFT	Density functional theory
MD	Molecular dynamics
SMILES	Simplified Molecular Input Line Entry System
ECEP	Extended connectivity fingerprint
ANN	Artificial neural network
GCNN	Graph convolutional neural network
VAEs	Variational autoencoders

References

[1] Stimpson AJ. A Machine Learning Approach to Modeling and Predicting Training Effectiveness. (Doctoral dissertation, Massachusetts Institute of Technology) Cambridge, MA, USA. 2015.

[2] Yang Z, Sun H, Huang J, Sun Z, Xiong H, Qiao S, Guan Z and Jia X. An efficient destination prediction approach based on future trajectory prediction and transition matrix optimization. IEEE Transactions on Knowledge and Data Engineering. 2018 Nov 28;32(2):203–217.

[3] Chen A, Zhang X and Zhou Z. Machine learning: Accelerating materials development for energy storage and conversion. InfoMat. 2020 May;2(3):553–576.

[4] Butler KT, Davies DW, Cartwright H, Isayev O and Walsh A. Machine learning for molecular and materials science. Nature. 2018 Jul 26;559(7715):547–555.

[5] Jain A, Ong SP, Hautier G, Chen W, Richards WD, Dacek S, Cholia S, Gunter D, Skinner D, Ceder G and Persson KA. Commentary: The materials project: A materials genome approach to accelerating materials innovation. APL Materials. 2013 Jul 1;1(1):12.
[6] Aykol M, Kim S, Hegde VI, Snydacker D, Lu Z, Hao S, Kirklin S, Morgan D and Wolverton C. High-throughput computational design of cathode coatings for Li-ion batteries. Nature Communications. 2016 Dec 14;7(1):13779.
[7] Shrestha A and Mahmood A. Review of deep learning algorithms and architectures. IEEE Access. 2019 Apr 22;7:53040–53065.
[8] Mathew A, Amudha P and Sivakumari S. Deep learning techniques: An overview. Advanced Machine Learning Technologies and Applications: Proceedings of AMLTA 2020. 2021:599–608.
[9] Ge W, De Silva R, Fan Y, Sisson SA and Stenzel MH. Machine learning in polymer research. Advances in Materials. 2025;37:2413695.
[10] Bin Abu Sofian AD, Lim HR, Chew KW and Show PL. Advancing 3D printing through integration of machine learning with algae-based biopolymers. ChemBioEng Reviews. 2024 Apr;11(2):406–425.
[11] Venkatasubramanian V, Chan K and Caruthers JM. Computer-aided molecular design using genetic algorithms. Computers & Chemical Engineering. 1994 Sep 1;18(9):833–844.
[12] Lafuente B, Downs RT, Yang H, Stone N, Armbruster T and Danisi RM. The power of databases: The RRUFF project. Highlights in Mineralogical Crystallography. 2015 Nov 13;1:25.
[13] Chen G, Shen Z, Iyer A, Ghumman UF, Tang S, Bi J, Chen W and Li Y. Machine-learning-assisted de novo design of organic molecules and polymers: Opportunities and challenges. Polymers. 2020 Jan 8;12(1):163.
[14] Elton DC, Boukouvalas Z, Fuge MD and Chung PW. Deep learning for molecular design – A review of the state of the art. Molecular Systems Design & Engineering. 2019;4(4):828–849.
[15] Heid E, Greenman KP, Chung Y, Li SC, Graff DE, Vermeire FH, Wu H, Green WH and McGill CJ. Chemprop: A machine learning package for chemical property prediction. Journal of Chemical Information & Modeling. 2023 Dec 26;64(1):9–17.
[16] Dangayach R, Jeong N, Demirel E, Uzal N, Fung V and Chen Y. Machine learning-aided inverse design and discovery of novel polymeric materials for membrane separation. Environmental Science & Technology. 2024 Dec 16;59(2):993–1012.
[17] Barocas S, Hardt M, Narayanan A. Fairness and Machine Learning: Limitations and Opportunities. 2023 Dec 19; Cambridge, Massachusetts: MIT Press.
[18] Schirmer M, Ijaz UZ, D'Amore R, Hall N, Sloan WT and Quince C. Insight into biases and sequencing errors for amplicon sequencing with the Illumina MiSeq platform. Nucleic Acids Research. 2015 Mar 31;43(6):e37.
[19] Liang W, Tadesse GA, Ho D, Fei-Fei L, Zaharia M, Zhang C and Zou J. Advances, challenges and opportunities in creating data for trustworthy AI. Nature Machine Intelligence. 2022 Aug;4 (8):669–677.
[20] Ahmed SF, Alam MS, Hassan M, Rozbu MR, Ishtiak T, Rafa N, Mofijur M, Shawkat Ali AB and Gandomi AH. Deep learning modelling techniques: Current progress, applications, advantages, and challenges. Artificial Intelligence Review. 2023 Nov;56(11):13521–13617.
[21] Day O and Khoshgoftaar TM. A survey on heterogeneous transfer learning. Journal of Big Data. 2017 Dec;4:1–42.
[22] Mairpady A, Mourad AH and Mozumder MS. Accelerated discovery of the polymer blends for cartilage repair through data-mining tools and machine-learning algorithm. Polymers. 2022 Apr 28;14(9):1802.
[23] Choubey V, Govind M and Arjune V. The emergence of bioplastic as a sustainable polymer in the Indian food packaging industry: A scientometric analysis. Journal of Data Science, Informetrics, and Citation Studies. 2023 Apr 27;2(1):1–2.

[24] Lyu HG, Haider AH, Landman AB and Raut CP. The opportunities and shortcomings of using big data and national databases for sarcoma research. Cancer. 2019 Sep 1;125(17):2926–2934.
[25] Pratt B, Loizos P. Choosing Research Methods: Data Collection for Development Workers. 1992; Oxford, UK: Oxfam.
[26] Chen CP and Zhang CY. Data-intensive applications, challenges, techniques and technologies: A survey on big data. Information Sciences. 2014 Aug 10;275:314–347.
[27] Hong Z, Ward L, Chard K, Blaiszik B and Foster I. Challenges and advances in information extraction from scientific literature: A review. JOM. 2021 Nov;73(11):3383–3400.
[28] National Research Council. Division on Engineering, Physical Sciences, Board on Mathematical Sciences, Their Applications, Committee on Applied, Theoretical Statistics, Committee on the Analysis of Massive Data. Frontiers in Massive Data Analysis. 2013 Oct 3; Washington, DC: National Academies Press.
[29] Zhu MX, Song HG, Yu QC, Chen JM and Zhang HY. Machine-learning-driven discovery of polymers molecular structures with high thermal conductivity. International Journal of Heat and Mass Transfer. 2020 Dec 1;162:120381.
[30] Bach SH, He B, Ratner A and Ré C. Learning the structure of generative models without labeled data. InInternational Conference on Machine Learning. 2017 Jul 17; (pp. 273–282). PMLR.
[31] Polyzotis N, Roy S, Whang SE and Zinkevich M Data management challenges in production machine learning. In Proceedings of the 2017 ACM international conference on management of data 2017 May 9 (pp. 1723–1726).
[32] Donders AR, Van Der Heijden GJ, Stijnen T and Moons KG. A gentle introduction to imputation of missing values. Journal of Clinical Epidemiology. 2006 Oct 1;59(10):1087–1091.
[33] Kohl SM, Klein MS, Hochrein J, Oefner PJ, Spang R and Gronwald W. State-of-the art data normalization methods improve NMR-based metabolomic analysis. Metabolomics. 2012 Jun;8:146–160.
[34] Lookman T, Alexander FJ, Rajan K, editors. Information Science for Materials Discovery and Design. 2016; Switzerland:Springer International Publishing.
[35] Weininger D, Weininger A and Weininger JL. SMILES. 2. Algorithm for generation of unique SMILES notation. Journal of Chemical Information and Computer Sciences. 1989 May 1;29(2):97–101.
[36] Vollmer JJ. Wiswesser line notation: An introduction. Journal of Chemical Education. 1983 Mar;60(3):192.
[37] Gakh AA and Burnett MN. Modular chemical descriptor language (MCDL): Composition, connectivity, and supplementary modules. Journal of Chemical Information and Computer Sciences. 2001 Nov 26;41(6):1494–1499.
[38] Tagade PM, Adiga SP, Pandian S, Park MS, Hariharan KS and Kolake SM. Attribute driven inverse materials design using deep learning Bayesian framework. npj Computational Mater. 2019 Dec 20;5(1):127.
[39] Willett P. Searching techniques for databases of two-and three-dimensional chemical structures. Journal of Medicinal Chemistry. 2005 Jun 30;48(13):4183–4199.
[40] Marchese Robinson RL, Palczewska A, Palczewski J and Kidley N. Comparison of the predictive performance and interpretability of random forest and linear models on benchmark data sets. Journal of Chemical Information & Modeling. 2017 Aug 28;57(8):1773–1792.
[41] Dablander M, Hanser T, Lambiotte R and Morris GM. Sort & Slice: A simple and superior alternative to hash-based folding for extended-connectivity fingerprints. Journal of Cheminformatics. 2024 Dec;16(1):1–20.
[42] Raghu M, Poole B, Kleinberg J, Ganguli S and Sohl-Dickstein J On the expressive power of deep neural networks. Ininternational conference on machine learning. 2017 Jul 17; (pp. 2847–2854). PMLR.
[43] Roushangar K, Ghasempour R and Shahnazi S. Kernel-based modeling. In Handbook of Hydroinformatics 2023 Jan 1. (pp. 267–281). Elsevier

[44] Heinonen M, Shen H, Zamboni N and Rousu J. Metabolite identification and molecular fingerprint prediction through machine learning. Bioinformatics. 2012 Sep 15;28(18):2333–2341.
[45] Schackart KE III and Yoon JY. Machine learning enhances the performance of bioreceptor-free biosensors. Sensors. 2021 Aug 17;21(16):5519.
[46] Pérez-Cruz F, Van Vaerenbergh S, Murillo-Fuentes JJ, Lázaro-Gredilla M and Santamaria I. Gaussian processes for nonlinear signal processing: An overview of recent advances. IEEE Signal Processing Magazine. 2013 Jun 12;30(4):40–50.
[47] Mannodi-Kanakkithodi A, Chandrasekaran A, Kim C, Huan TD, Pilania G, Botu V and Ramprasad R. Scoping the polymer genome: A roadmap for rational polymer dielectrics design and beyond. Materials Today: Proceedings. 2018 Sep 1;21(7):785–796.
[48] Yi Y, Wang LM and Yin FH. Interval support vector regression enables high-throughput machine learning predictions for dielectric constant of polymer dielectrics. Applied Physics Letters. 2021 May 31;118(22):1–5.
[49] Ali MZ, Abdullah A, Zaki AM, Rizk FH, Eid MM and El-Kenway EM. Advances and challenges in feature selection methods: A comprehensive review. Journal of Artificial Intelligence and Metaheuristics. 2024;1:67–77.
[50] Sahin EK. Comparative analysis of gradient boosting algorithms for landslide susceptibility mapping. Geocarto International. 2022 May 3;37(9):2441–2465.
[51] Wang F, Diao X, Chang S and Xu L. Recent progress of deep learning in drug discovery. Current Pharmaceutical Design. 2021 May 1;27(17):2088–2096.
[52] Zamora-Resendiz R and Crivelli S. Structural learning of proteins using graph convolutional neural networks. BioRxiv. 2019 Apr 16;1:610444.
[53] Zhang H, Liu X, Cheng W, Wang T and Chen Y. Prediction of drug-target binding affinity based on deep learning models. Computers in Biology and Medicine. 2024 Apr 8;174:108435.
[54] Tao L, Varshney V and Li Y. Benchmarking machine learning models for polymer informatics: An example of glass transition temperature. Journal of Chemical Information & Modeling. 2021 Oct 18;61(11):5395–5413.
[55] Altae-Tran H, Ramsundar B, Pappu AS and Pande V. Low data drug discovery with one-shot learning. ACS Central Science. 2017 Apr 26;3(4):283–293.
[56] Yan C and Li G. The rise of machine learning in polymer discovery. Advanced Intelligent Systems. 2023 Apr;5(4):2200243.
[57] Osisanwo FY, Akinsola JE, Awodele O, Hinmikaiye JO, Olakanmi O and Akinjobi J. Supervised machine learning algorithms: Classification and comparison. International Journal of Computer Trends and Technology (IJCTT). 2017 Jun;48(3):128–138.
[58] Swapna M, Sharma YK and Prasadh BM. CNN architectures: Alex Net, Le Net, VGG, Google Net, Res Net. International Journal of Recent Technology and Engineering. 2020 Mar;8(6):953–960.
[59] Goodfellow I, Pouget-Abadie J, Mirza M, Xu B, Warde-Farley D, Ozair S, Courville A and Bengio Y. Generative adversarial networks. Communications of the ACM. 2020 Oct 22;63(11):139–144.
[60] Hussein AA, Yaseen ET and Rashid AN. Learnheuristics in routing and scheduling problems: A review. Samarra Journal of Pure and Applied Science. 2023 Mar 30;5(1):60–90.
[61] Hu X, Huang M, Kong N, Han F, Tan R and Huang Q. Enhancing the electrical insulation of highly thermally conductive carbon fiber powders by SiC ceramic coating for efficient thermal interface materials. Composites Part B: Engineering. 2021 Dec 15;227:109398.
[62] Zhou S, Wang S, Wu Q, Azim R and Li W. Predicting potential miRNA-disease associations by combining gradient boosting decision tree with logistic regression. Computational Biology and Chemistry. 2020 Apr 1;85:107200.
[63] Bhowmik S and Howlader J Online payment fraud monitoring and detection: Performance analysis of tree-based ensemble machine learning models. In 2025 17th International Conference on COMmunication Systems and NETworks (COMSNETS) 2025 Jan 6 (pp. 102–107). IEEE.

[64] Kumar R, Le N, Tan Z, Brown ME, Jiang S and Reineke TM Efficient Polymer-Mediated Delivery of Ribonucleoprotein Payloads Through Combinatorial Design & Parallelized Experimentation.
[65] Ge W, De Silva R, Fan Y, Sisson SA and Stenzel MH. Machine learning in polymer research. Advances in Materials. 2025;37:2413695.
[66] Zografos M and Ougiaroglou S Simplifying decision tree classification through the autodtrees web application and service. In International Conference on Intelligent Tutoring Systems 2024 Jun 1 (pp. 162–173). Cham: Springer Nature Switzerland.
[67] Chauhan R and Yafi E. Applicability of classifier to discovery knowledge for future prediction modelling. Journal of Ambient Intelligence and Humanized Computing. 2023Aug;14(8):10347–10362.
[68] Mannodi-Kanakkithodi A, Chandrasekaran A, Kim C, Huan TD, Pilania G, Botu V and Ramprasad R. Scoping the polymer genome: A roadmap for rational polymer dielectrics design and beyond. Mat Today: Proceedings. 2018 Sep 1;21(7):785–796.
[69] Singla S, Mannan S, Zaki M and Krishnan NA. Accelerated design of chalcogenide glasses through interpretable machine learning for composition–property relationships. Journal of Physics: Materials. 2023 Apr 5;6(2):024003.
[70] Swolfs Y, Verpoest I and Gorbatikh L. Recent advances in fibre-hybrid composites: Materials selection, opportunities and applications. International Materials Reviews. 2019 May;64(4):181–215.
[71] Costa CM, Lizundia E and Lanceros-Méndez S. Polymers for advanced lithium-ion batteries: State of the art and future needs on polymers for the different battery components. Progress in Energy and Combustion Science. 2020 Jul 1;79:100846.
[72] Kim C, Batra R, Chen L, Tran H and Ramprasad R. Polymer design using genetic algorithm and machine learning. Computational Materials Science. 2021 Jan 1;186:110067.
[73] Patra TK, Meenakshisundaram V, Hung JH and Simmons DS. Neural-network-biased genetic algorithms for materials design: Evolutionary algorithms that learn. ACS Combinatorial Science. 2017 Feb 13;19(2):96–107.
[74] Hibbert DB. Genetic algorithms in chemistry. Chemometrics and Intelligent Laboratory Systems. 1993 Jan 1;19(3):277–293.
[75] Forrest S, Javornik B, Smith RE and Perelson AS. Using genetic algorithms to explore pattern recognition in the immune system. Evolutionary Computation. 1993 Sep 1;1(3):191–211.
[76] Cox GA, Mortimer-Jones TV, Taylor RP and Johnston RL. Development and optimisation of a novel genetic algorithm for studying model protein folding. Theoretical Chemistry Accounts. 2004 Jul;112:163–178.
[77] Wessner CW and Wolff AW. National Research Council (US) Committee on comparative national innovation policies: Best practice for the 21st century. National support for emerging industries. In Rising to the Challenge: US Innovation Policy for the Global Economy 2012. Washington, DC; National Academies Press (US).
[78] Hu H, Zhang F, Luo S, Chang W, Yue J and Wang CH. Recent advances in rational design of polymer nanocomposite dielectrics for energy storage. Nano Energy. 2020 Aug 1;74:104844.
[79] Dunn B, Kamath H and Tarascon JM. Electrical energy storage for the grid: A battery of choices. Science. 2011 Nov 18;334(6058): 928–935.
[80] Wang Y, Xie T, France-Lanord A, Berkley A, Johnson JA, Shao-Horn Y and Grossman JC. Toward designing highly conductive polymer electrolytes by machine learning assisted coarse-grained molecular dynamics. Chemistry of Materials. 2020 Apr 22;32(10):4144–4151.
[81] Shen KH and Hall LM. Effects of ion size and dielectric constant on ion transport and transference number in polymer electrolytes. Macromolecules. 2020 Nov 3;53(22):10086–10096.
[82] Xu J, Wong M and Wong CP. Super high dielectric constant carbon black-filled polymer composites as integral capacitor dielectrics. In 2004 Proceedings. 54th Electronic Components and Technology Conference (IEEE Cat. No. 04CH37546) 2004 Jun 4; (Vol. 1, pp. 536–541). IEEE.

[83] Obrzut J, Anopchenko A, Kano K and Wang H High frequency loss mechanism in polymers filled with dielectric modifiers. MRS Online Proceedings Library (OPL). 2003 Jan; 783:B3–5.
[84] Astorino C, De Nardo E, Lettieri S, Ferraro G, Pirri CF and Bocchini S. Advancements in gas separation for energy applications: Exploring the potential of polymer membranes with intrinsic microporosity (PIM). Membranes. 2023 Dec 6;13(12):903.
[85] Wang S, Li X, Wu H, Tian Z, Xin Q, He G, Peng D, Chen S, Yin Y, Jiang Z and Guiver MD. Advances in high permeability polymer-based membrane materials for CO 2 separations. Energy and Environmental Science. 2016;9(6):1863–1890.
[86] Rose I, Bezzu CG, Carta M, Comesaña-Gándara B, Lasseuguette E, Ferrari MC, Bernardo P, Clarizia G, Fuoco A, Jansen JC and Hart KE. Polymer ultrapermeability from the inefficient packing of 2D chains. Nature Materials. 2017 Sep 1;16(9):932–937.
[87] Yang X, Wu H, Liu C and Zhang X. A novel analytic model for prediction of the anisotropic thermal conductivity in polymer composites containing aligned 1D nanofillers. International Journal of Thermal Sciences. 2023 Feb 1;184:107980.
[88] Bin Abu Sofian AD, Sun X, Gupta VK, Berenjian A, Xia A, Ma Z and Show PL. Advances, synergy, and perspectives of machine learning and biobased polymers for energy, fuels, and biochemicals for a sustainable future. Energy & Fuels. 2024 Jan 16;38(3):1593–1617.
[89] Ates M, Karazehir T and Sezai Sarac AJ. Conducting polymers and their applications. Current Physical Chemistry. 2012 Aug 1;2(3):224–240.
[90] Liu Y, Gao S, Zhang X, Xin JH and Zhang C. Probing the nature of charge carriers in one-dimensional conjugated polymers: A review of the theoretical models, experimental trends, and thermoelectric applications. Journal of Materials Chemistry C. 2023;11(1):12–47.
[91] Li YH, Wang JY and Pei J. C–H σ-dopants mediated n-doping of conjugated polymers: Mutual designs and multiscale characteristics. Accounts of Materials Research. 2024 Aug 1;5(9):1059–1071.

Harshita Shakya and Sona Gandhi*

Chapter 12
Advancing nanomaterial safety: AI in toxicity and environmental assessments

Abstract: Artificial intelligence (AI) and machine learning (ML) have become crucial in toxicology, particularly for assessing the safety of nanomaterials as they enter clinical and consumer markets. In today's world, it is important to understand the toxicological and environmental effects of these compounds, and we know that AI-powered computational techniques are increasingly effective at organizing and analyzing large datasets, while also supplementing traditional animal research. Nanomaterials should be evaluated thoroughly because they have distinctive nanoscale interactions before they receive certified approvals from regulators to ensure they are utilized safely, particularly in nanomedicine. The use of AI and ML makes it easy to handle the large amount of data on the behavior of nanomaterials and their impacts on human health and the environment by streamlining data processing. The integration of AI with everyday objects like drones and smart devices proves its capacity to support nanotechnology. However, as AI applications growing in great abundance, it is very important to recognize both the advantages and the disadvantages they carry. In this chapter, we have examined the role of AI in anticipating ecological and toxicological dangers, including applications like AI-guided nanobots for air and water purification, also addressing concerns with data quality, regulatory standards, and . For a safe and more sustainable future in nanotechnology, we have summed up this chapter by offering guidelines on how to ethically integrate AI into safety evaluations of nanomaterials.

Keywords: Nanotoxicology, artificial intelligence, machine learning, nanomaterials, ecotoxicity, AI-guided nanobots, sustainable nanotechnology, health implications, computational methods

12.1 Introduction

Nanotechnology refers to the use of nanomaterials composed of extremely tiny particles that range between 1 and 100 nm [1]. These nanomaterials are utilized in modern inven-

***Corresponding author: Sona Gandhi**, School of Engineering and Technology, Manav Rachna International Institute of Research and Studies, Faridabad 121004, Haryana, India,
e-mail: gandhi7hd@gmail.com
Harshita Shakya, Department of Chemistry, Gurugram University, Mayfield Garden, Sector 51, Gurugram 122003, India

https://doi.org/10.1515/9783111681870-012

tions such as better drug delivery systems, better diagnostic imaging, and specific cancer treatments in medicine, as well as improved electronics, cosmetics, and environmental uses like air and water purification [2]. Owing to their distinctive properties at the nanoscale, there is direct interaction with tissues and cells, which raises concerns regarding potential toxicity. Although they offer several advantages, their safety largely depends on factors such as composition, solubility, and interactions with biological systems. Hence, it is of prime importance to conduct scientific investigations to ensure that these materials are used judiciously in both consumer and clinical applications [3].

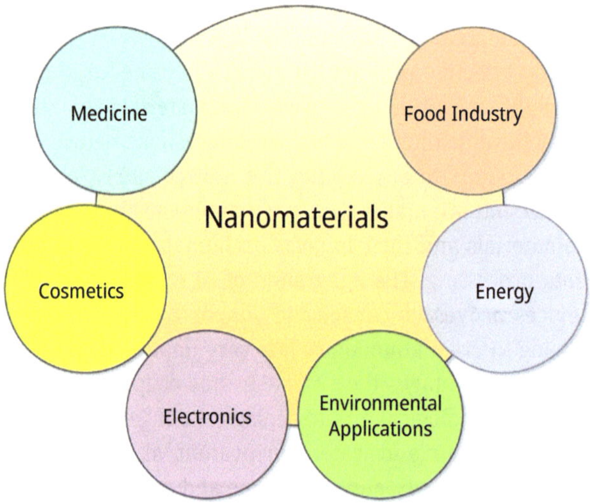

Figure 12.1: Applications of nanomaterials across industries.

There is a need for advanced evaluation techniques considering the peculiar properties, applications (Figure 12.1), potential risks, and safety of nanomaterials. The approach of utilizing animal testing is not effective due to interspecies variation and high costs. Hence, through automation and improvements in the evaluation of safety, machine learning in association with advancements in artificial intelligence (AI) is reforming the field of toxicology [4]. The relationship between a compound's structure, or physicochemical properties and its biological activity, known as quantitative structure–activity relationships (QSAR), is one of the leading developments in predicting toxicity [5]. There has been a significant enhancement in QSAR models as machine learning has reduced the need for manual statistical processes and improved accuracy through algorithms like support vector machines, random forests, and artificial neural networks (ANNs) [6, 7]. Computational tools also utilize chemical structure descriptors (ranging from 2D to complex 4D representations) to analyze structural data and predict toxicological consequences. In an era of research driven by data, the coalition of AI and computational tools has become necessary for evaluating toxicity [8–10].

The ground-breaking role of these technologies in ensuring safe applications of nanomaterials has been discussed in this chapter. There has been a tremendous rise in the use of nanomaterials in electronics, cosmetics, and pharmaceuticals, which poses concerns about their interaction with the environment and biological systems [11]. There is a possibility of hazards upon the interaction of nanomaterials with cells and tissues owing to their peculiar nanoscale properties, so it is essential to incorporate comprehensive AI-driven assessment methodologies. This chapter emphasizes the role of AI in managing large datasets to predict the fate of nanomaterials inside biological and ecological systems. It also analyzes AI led innovations, such as nanobots for the purification of water and air, in addition to critical topics like data quality, regulatory frameworks, and ethical considerations [12–14]. Furthermore, this chapter discusses the role of decision-making frameworks that establish a balance between social, economic, and environmental factors to assist industries in navigating the complexities of nanomaterial safety while working toward sustainability goals. This chapter presents recent advancements and case studies and aims to deliver actionable insights regarding the promotion of sustainable practices in nanomaterial usage while ensuring the safety of humans and the environment [15, 16].

12.2 Nanomaterial toxicity

Nanotoxicology discusses the impact of distinctive properties on biological systems and the environment; their size ranges from 1 to 100 nm, affecting reactivity, dispersion, and cellular interactions [17]. These properties create serious dangers in terms of health and the environment, which necessitate in-depth research, even though they have also led to advancements in the fields of medicine and environmental science.

The effect of artificial nanoparticles on living beings and the environment is examined in the study of nanomaterials toxicity [18]. The study explains that the toxicity of nanoparticles may differ from that of bulk materials because they can pass through biological barriers, such as the blood–brain barrier. In the purview of nanotoxicology, risks are detected, and plans are created by analyzing their properties for safe applications [19]. Even though nanoparticles possess favorable properties in terms of drug delivery and environmental cleaning, their high reactivity and small size can lead to harmful interactions within biological systems, potentially causing cellular damage, inflammation, or oxidative stress [20]. To properly manage their risks and ensure their safe integration into various fields, a comprehensive understanding of these interactions is essential.

12.2.1 Mechanisms of toxicity and health risks

The main reason for nanoparticle (NP) toxicity is oxidative stress, which frequently occurs due to an excess of reactive oxygen species (ROS). For example, the Fenton reaction, which leads to the formation of harmful radicals such as OOH and OH, is catalyzed by iron-based NPs [21–23]. Even if they do not produce ROS on their own, certain inert NPs can induce ROS production by attacking mitochondria [24]. Excessive ROS can damage lipids, proteins, and DNA, potentially resulting in cancer, dementia, cardiovascular disorders, and cell death, even though reasonable ROS levels are crucial for cellular functions [25–28]. For instance, carbon NPs generate oxidative stress in fish brain cells and pulmonary inflammation in rats [29, 30], whereas TiO_2 NPs cause early inflammatory responses in rodents [28]. Because ROS cause unsaturated fatty acids to undergo peroxidation, they also contribute to CNS damage [31, 32].

Oxidative stress from NPs impacts cell signaling in three stages: low stress enhances defense gene transcription via nrf2, moderate stress activates inflammation via NFκB, and high stress triggers apoptosis and necrosis [33]. ROS-induced DNA damage can result in single- and double-strand breaks, leading to severe cellular consequences [34, 35]. Mitochondrial DNA (mtDNA) damage is linked to disorders such as neurogenic muscle weakness and ataxia [36, 37]. To mitigate ROS effects, cerium oxide NPs have been developed, which scavenge free radicals, showing similar protection as *N*-acetyl cysteine in mice [38]. Additionally, certain physicochemical properties of NPs, such as their ability to unfold fibrinogen and activate inflammatory pathways, contribute to toxicity [39].

The potential of nanomaterials to penetrate biological barriers, including the blood–brain barrier, respiratory system, and skin, is another major concern. The possibility of systemic toxicity, which could damage vital organs like the liver, spleen, heart, and brain, is heightened by their ability to reach deeper tissues and organs that are often protected from larger particles or harmful substances. Therefore, it is fundamentally important to understand the intricate interactions that exist between biological systems and nanomaterials, particularly as their use in medicinal applications and other diverse domains continues to grow.

12.2.2 Factors influencing toxicity

The distinctive physical and chemical characteristics of nanomaterials are described below that affect how they act in biological systems, determining their toxicological profile.

12.2.2.1 Size and surface characteristics

The small size of nanomaterials increases their reactivity on the nanoscale and gives them a high ability to penetrate biological systems. Their interactions with cells are significantly affected by their charge and surface functionality. For instance, NPs carrying a positive charge on their surface tend to interact readily with cell membranes, which are negatively charged, and this can cause harmful complications.

12.2.2.2 Shape and morphology

The fate of NPs inside the biological system is greatly affected by their shape, as they can easily pass through cell membranes. Fibrous or rod-shaped NPs can cause greater cellular damage and inflammation compared to spherical NPs. Moreover, the shape of NPs also has a considerable impact on their absorption and release from cells in the body.

12.2.2.3 Chemical composition and reactivity

The reactivity and subsequent interactions within the biological system are determined by the composition of the nanomaterials. For instance, even though silver NPs exhibit antibacterial properties, they can still cause harm to cells by releasing ions. Similarly, carbon-based nanomaterials can generate ROS, which can negatively impact the functioning of normal biological processes. Therefore, it is necessary to understand these consequences when evaluating the true safety of these materials.

Table 12.1 provides an overview of the various types of nanomaterials, including their size, shape, chemical composition, and potential health hazards, in order to better understand how these attributes relate to their toxicological consequences.

12.3 AI-driven approaches in toxicity assessment

By increasing predictive accuracy, decreasing dependency on conventional animal testing, and speeding up chemical evaluations, the use of sophisticated computer tools has completely transformed toxicity assessment. Using labeled data, supervised learning techniques like multiple linear regression, support vector machines, decision trees, and random forests can accurately model pharmacokinetic parameters, dose-response relationships, and bioactivity [40, 41]. These models enable classification and prediction of toxic versus non-toxic chemicals with enhanced precision. To evaluate large and complex datasets, unsupervised learning methods such as principal

Table 12.1: Properties of nanomaterials and their associated toxicological effects.

Nanomaterial type	Size (nm)	Shape	Chemical composition	Toxicological effects
Carbon nanoparticles	1–100	Spherical, nanotube	Carbon	Oxidative stress, pulmonary inflammation, DNA damage, neurotoxicity, and liver damage
Titanium dioxide (TiO_2) nanoparticles	1–100	Spherical, rod	Titanium oxide	ROS generation, inflammatory responses, and potential carcinogenic effects in lung and skin cells
Silver nanoparticles	1–100	Spherical	Silver	Release of toxic silver ions, cytotoxicity, oxidative stress, genotoxicity, and antimicrobial activity
Cerium oxide nanoparticles	1–50	Spherical, nanorods	Cerium oxide	Antioxidant properties (ROS scavenging), reduction of oxidative stress, and potential neuroprotective effects
Iron-based nanoparticles	10–100	Spherical, nanorods	Iron	Fenton reaction catalysis, ROS production (OH · and OOH), oxidative stress, and mitochondrial damage
Silica nanoparticles	5–100	Spherical, rod	Silicon dioxide	Inflammatory responses, oxidative stress, and potential fibrogenic effects in the lung tissue
Gold nanoparticles	1–50	Spherical, rod	Gold	Low toxicity, potential for ROS generation at high concentrations, and minimal genotoxicity
Polymeric nanoparticles	50–200	Spherical, rod	Polymers (e.g., PLA, PLGA)	Biodegradable, low toxicity, but it can cause inflammation and immune system responses at higher doses
Zinc oxide nanoparticles	1–100	Spherical, rod	Zinc oxide	ROS generation, inflammation, cytotoxicity, skin irritation, and genotoxicity at higher concentrations

component analysis and self-organizing maps examine unlabeled data to find hidden patterns, reduce dimensionality, and group chemicals according to structural or functional similarities [42]. Reinforcement learning, while less commonly applied in toxicology, offers potential for optimizing simulations and experimental designs by enabling models to learn decision-making strategies through environmental interactions and feedback [41]. Effective data management plays a crucial role in leveraging these computational techniques, with systematic organization and quality control ensuring reliable analyses of toxicogenomics, high-content imaging, and other diverse datasets. Automation of data cleaning, transformation, and feature extraction streamlines workflows and enables efficient handling of large-scale toxicological data [43, 44]. Rapid chemical screening and the forecasting of important toxicological parameters like bioactivity, dose-response, and toxicokinetic parameters are made possible by AI-driven predictive models, such as deep neural networks and ensemble learning techniques, which have shown accuracy on par with in vivo experiments [45]. Real-world applications, like predicting how well NPs will deliver to tumors and generating QSAR analyses, show off the models' revolutionary potential and highlight how AI-driven methods can effectively prioritize and assess toxicity risks.

The procedure is shown in detail in Figure 12.2 below to better demonstrate the workflow and integration of various AI-driven approaches in toxicity evaluation. It emphasizes how cutting-edge computational methods, such as reinforcement learning, supervised learning, and unsupervised learning, help to increase prediction accuracy and reduce dependency on conventional testing procedures.

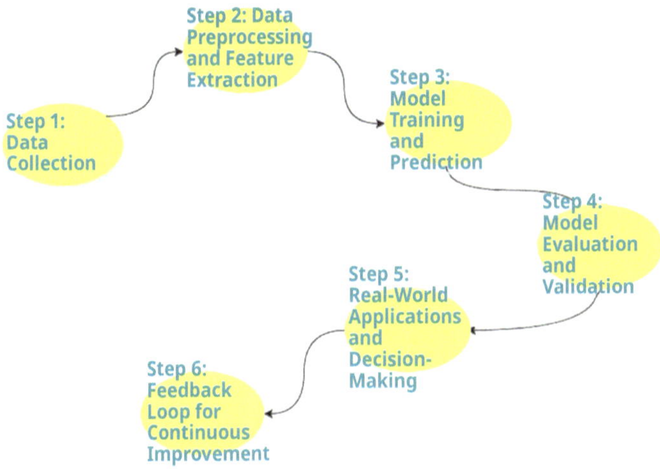

Figure 12.2: AI-driven toxicity assessment process.

12.4 Quantitative structure–activity relationship models (QSAR) in toxicity and environmental assessments

The correlation between a compound's structure and its biological effects, such as ADME (absorption, distribution, metabolism, and excretion) and toxicity, is explained with the help of QSAR models. These models are widely employed in the fields of drug development and toxicology to better understand the potential physiological effects of various compounds and substances. QSAR models are able to identify the effects of a variety of substances more precisely owing to recent advancements in computer technology. Furthermore, with the progression of machine learning, more effective methods have become available, improving these models and increasing the accuracy of predictions. One particularly useful aspect of QSAR models is the 'read-across' strategy, which can predict the effects of novel compounds based on data from comparable known sources. This approach may significantly reduce the need for further laboratory testing and animal testing.

There have been studies that explored the use of machine learning in QSAR modeling. In a study, dataset of about 11,764 compounds (from Tox21 database) was evaluated using deep learning models like DeepTox. It demonstrated extraordinary accuracy across twelve bio-assays related to various toxicological endpoints [46]. Ai et al. [47] predicted aquatic toxicity based on a dataset of 400 chemicals, making use of ensemble support vector machines, whereas Zhang et al. [48] evaluated the carcinogenicity of around 1,003 compounds using ensemble extreme gradient boosting. Such studies show the increasing trend of enhancing the precision and efficiency of QSAR predictions with the use of advanced and sophisticated machine learning techniques.

In a number of recent studies, the application of machine learning in QSAR modeling has been explored, using different methods that provide varying degrees of toxicity prediction power. Some important features, such as dataset size, toxicological endpoints, performance metrics, and limitations and strengths, can help compare machine learning models and their usage in QSAR (as shown in Table 13.2).

Four basic steps are involved in making QSAR models:

1. A training dataset comprising compounds or substances with known physical or biological properties is collected.
2. Molecular descriptors (numerical representations of molecular characteristics) are utilized for encoding purposes.
3. The chemical properties of the compounds are identified based on the descriptors by training the model using various mathematical and machine learning algorithms.
4. In order to ensure robustness and reliability, the performance of the model is evaluated using a validation dataset [49].

The extensive library of PFAS (per- and polyfluorinated alkyl substances), which consists of more than 5,000 compounds, raises serious concerns regarding chemical safety. These PFAS pose significant dangers to human health globally and are continuously contaminating the environment. Since the number of PFAS is enormous, it is not viable to assess and study those individually using conventional experimental techniques. To address this, Cheng and Ng [49] worked out a QSAR model tailor-made for PFAS, which could accurately predict the bio-activity of 3,486 compounds. The model used different machine learning modules such as logistic regression and random forests, working on a dataset of 1,010 compounds and 26 bioassays to categorize the bioactivity of these compounds with great accuracy (AUC > 0.9).

In order to develop computer models for predicting toxicity on the basis of structural aspects, the Tox21 Data Challenge was launched in 2014 by the National Center for Advancing Translational Sciences (NCATS). This program managed to provide a comprehensive dataset of about 11,764 chemicals, along with bio-assays and structural

Table 12.2: Comparison of machine learning models used in QSAR studies for toxicity prediction.

Machine learning model	Application	Dataset size	Toxicological endpoints	Performance metrics	Strengths	Limitations
Support vector machines (SVMs)	Bioactivity and toxicity prediction	Small to moderate	Carcinogenicity and aquatic toxicity	Accuracy, sensitivity, and specificity	Effective with smaller datasets, suitable for classification	Struggles with complex datasets
Random forest	General toxicity, carcinogenicity, and environmental risk	Moderate to large	Various endpoints (e.g., carcinogenicity)	Accuracy, AUC, sensitivity	Handles large datasets well and is robust against overfitting	Less interpretable due to ensemble complexity
Deep learning (e.g., DeepTox)	Predicting toxicity for diverse chemical structures	Large (e.g., Tox21: 11,764 compounds)	12 toxicological endpoints (e.g., liver toxicity, and mutagenicity)	AUC, accuracy, sensitivity, and specificity	High accuracy, suitable for complex datasets	Requires large datasets, computationally intensive
Extreme gradient boosting (XGBoost)	Carcinogenicity prediction and bioactivity classification	Moderate to large	Carcinogenicity and bioactivity	Accuracy, AUC, and *F*1-score	High performance, robust to noise in data	Sensitive to hyperparameter tuning
Ensemble support vector machines	Aquatic toxicity prediction	Moderate (e.g., 400 chemicals)	Aquatic toxicity	Accuracy, sensitivity, and specificity	Combines multiple models and good for generalization	May not capture nonlinear relationships well
Logistic regression	PFAS bioactivity classification	Moderate to large	Bioactivity of PFAS compounds	AUC, accuracy	Simple, fast, and easy to interpret	May underperform with complex relationships

data comprising 12 toxicological endpoints. Mayr et al. [46] created the DeepTox pipeline, which outperformed alternative techniques in toxicity prediction by employing deep learning methodology in a number of subchallenges. The DeepTox's victory demonstrated the potential of deep learning models in the field of toxicity prediction.

On similar lines, machine learning-based models were used to predict acute oral toxicity values (e.g., LD_{50}) for about 11,992 compounds by the U.S. Interagency Coordinating Committee on the Validation of Alternative Methods (ICCVAM) Acute Toxicity Workgroup. This project received contributions from about 35 research groups, and the final models are under evaluation to assess their ability to substitute animal testing in regulatory testing and experiments [50].

Rat studies are long and resource-intensive and have been an important part of chemical safety evaluations. Historically, they have been the backbone of carcinogenicity testing. In order to accelerate the process, QSAR models have been developed to predict the carcinogenicity of compounds. For instance, Li et al. developed the DeepCarc model in 2021, which could predict the carcinogenicity of small compounds/molecules with the help of deep learning. They demonstrated that the DeepCarc model provides an efficient alternative to conventional experimental techniques with exceptional accuracy and sensitivity by training the model with 692 compounds and evaluating it with a dataset of about 171 compounds [51].

The combination of different computational techniques has led to the development of weight-of-evidence models, which rank compounds based on several mechanisms such as QSAR, toxicogenomic, and structural warning models. According to a recent study published, this method significantly improves chemical coverage and performance prediction. They identified around 44 compounds of high concern using this integrated approach to evaluate the carcinogenicity of 1,623 chemicals that come into contact with food, 34 of which had supporting data from public carcinogenicity databases [52].

The developments in machine learning-based QSAR modeling, which provide highly efficient and precise non-animal testing techniques, have significantly changed the perspective of chemical safety. These models possess a great potential to improve toxicity and environmental evaluations for established compounds as well as novel substances like nanomaterials as they continue to develop further in the near future.

12.5 Environmental impacts and risk assessment of nanomaterials

Since the use of nanomaterials is growing, it is essential to understand their environmental impacts. These materials exhibit different behavior at the nanoscale compared to larger materials due to their microscopic size, high reactivity, and surface area. Their time in the environment, movement, clumping behavior, and breakdown can influence their interactions with soil, water, and air. Some NPs may cluster together, which decreases their availability to animals while potentially disrupting sediment ecosystems; on the other hand, some may decompose to release toxic ions that can potentially endanger aquatic ecosystems [53].

Environmental monitoring systems are essential for assessing the quality of the environment, which, in turn, can help identify future ecological problems. These technologies measure multiple elements over a period of time and across particular areas to help determine how materials like NPs impact the environment. As the complexity of these consequences increases, machine learning methods and ANNs are utilized to study the potential risks of nanomaterials by analyzing large environmental datasets [54].

AI models can accurately estimate the toxicity of nanomaterials and their continuing consequences on ecosystems. These models take into account the properties of NPs as well as the surrounding conditions by analyzing complex ecological information [55]. Additionally, the efficiency of risk evaluations has been improved by the integration of experimental and computational data with techniques such as QSARs and probabilistic risk models [56]. Despite its potential, barriers such as a shortage of data, standardized testing, and the complexity of real-world circumstances continue to obstruct progress [57].

The different compositions of nanomaterials and their interactions with pollutants interfere with the prediction of how they affect the environment. The inability of laws and regulations to keep pace with the rapid advancement of nanotechnology has led to monitoring gaps. Although nanomaterials have the ability to bioaccumulate in organisms, particularly in food chains, they may eventually lead to exposure at higher levels of trophic relationships. For example, metal oxide NPs could harm biodiversity by altering microbial populations, nutrient cycles, predator–prey relationships, and aquatic life [58].

Despite these challenges, interdisciplinary collaboration and ongoing research continue to attempt to find solutions and provide effective environmental monitoring and risk assessment of nanomaterials. The following technologies will help preserve ecosystems as well as manage resources effectively [59].

12.6 AI-guided nanobots for environmental purification

AI-guided nanobots, an advanced integration of robotics, artificial intelligence, and nanotechnology, have altered environmental purification by offering targeted and efficient solutions for the remediation of air, water, and soil [60]. These tiny gadgets use machine learning algorithms to navigate, identify pollutants, and carry out precise purification procedures [61]. Air quality in urban and industrial regions is improved through air filtration using nanobots that are outfitted with catalytic NPs, which reduce pollutants such as particulate matter and greenhouse gases [62]. They remove toxins, microplastics, and heavy metals from water and soil treatments, restoring ecosystems at low financial and environmental costs [63]. Large-scale deployment, intelligent technology integration, and efficiency gains are all part of the future, but for sustainable use, challenges such as costs, environmental risks, and regulatory gaps need to be addressed [60].

12.7 AI-guided nanobots for precision drug delivery

AI-guided nanobots enable precise drug delivery by navigating the body to target specific areas, such as tumors, ensuring minimal side effects. The flowchart in Figure 13.3 outlines the stepwise process from diagnosis to the controlled release of therapeutic agents.

Medical micro/nanorobots, integrating robotics, AI, and nanotechnology, are transforming targeted drug delivery by precisely navigating the human body to deliver therapeutic agents to disease sites such as tumors. Magnetically guided nanorobots, for example, have been used to deliver drugs such as fluorouracil to reduce tumor growth in mouse models, with external triggers activating drug release at the target site [64, 65]. Biohybrid nanorobots, combining synthetic structures with motile microorganisms, also show great potential for the precise delivery of genes, proteins, or therapeutic payloads, as demonstrated with Listeria monocytogenes carrying NPs for gene expression monitoring in mice [66].

12.8 Innovative nanomaterials for air, water, and soil pollution control

The persistent problem of environmental pollution, which continues to be a major worldwide concern, has become simpler to solve thanks to recent technological developments. When materials produced by industrial processes are utilized, toxic residues are released into the air, water, soil, and environment, causing pollution. These pollutants, such as particulate matter, heavy metals, pesticides, and organic compounds, can degrade water quality and, in extreme cases, even make tap water flammable [67, 68]. Traditional methods for controlling pollution are often not enough, as these contaminants are complex, including a mix of volatile and low-reactivity compounds. A promising solution lies in the use of nanocatalysts, made from nanomaterials, which can assist in both homogeneous and heterogeneous catalysis [69, 70]. Because of their high surface area-to-volume ratio, adjustable porosity, and potent reactivity, nanomaterials, including metal and metal oxide NPs as well as carbon-based and polymeric

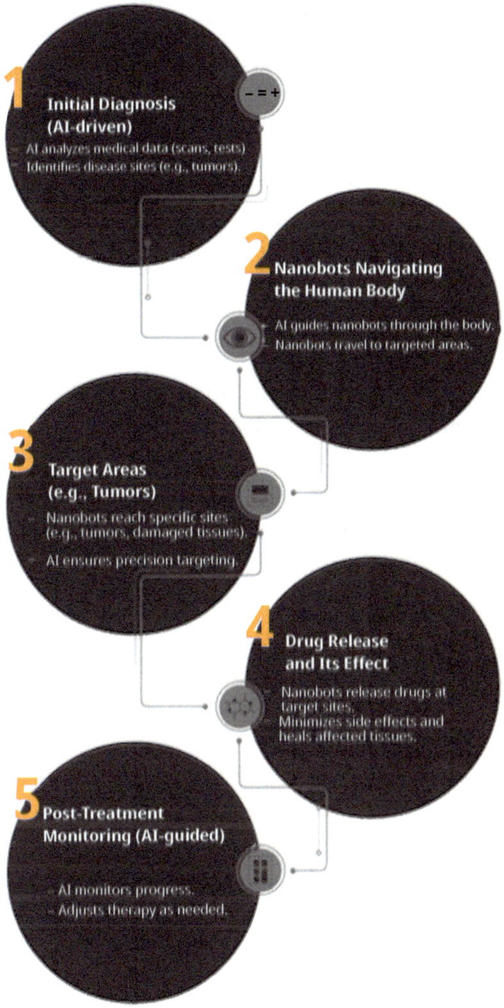

Figure 12.3: Flowchart of AI-guided nanobots for precision drug delivery.

materials, are very effective at eliminating pollutants [71–73]. Specifically, polymeric NPs are highly efficient in removing organic pollutants and heavy metals from water, as well as treating wastewater and groundwater [74–76]. Their affordability, non-toxicity, and biodegradability make them ideal for ecologically responsible repair [77, 78]. Furthermore, when combined with traditional adsorbents like activated carbon and zeolites, their pollutant-removal capabilities are significantly enhanced, demonstrating great potential for future environmental applications [79–81].

12.9 Harnessing AI and nanotechnology for advanced healthcare: transforming trauma and critical illness treatment

Artificial intelligence (AI) and nanotechnology are transforming healthcare by offering precise, efficient, and minimally invasive solutions, especially in managing accident trauma and critical illnesses [82]. Nanobots, guided by AI, perform targeted drug delivery, tissue repair, and chronic disease monitoring. In India, these technologies address challenges such as the urban–rural healthcare disparity, rising costs, and increasing disease burden. AI-driven telemedicine, early detection systems, and resource management tools are improving access and the quality of care. Global advancements, such as AI in medical imaging, federated learning for secure data use, and AI-enabled remote monitoring, provide insights for India's adoption.

Recent investments further underscore this trend. For instance, in October 2024, AI Fund, led by computer scientist Andrew Ng, made its first investment in India by funding Jivi, an AI healthcare firm based in Gurugram. Jivi employs AI to assist in potential diagnoses, treatments, and administrative tasks, reflecting the growing confidence in AI's role in enhancing healthcare services [83, 84]. However, challenges such as data standardization, privacy concerns, regulatory gaps, and skill shortages must be addressed. Collaborative efforts among the government, healthcare providers, IT firms, and academia are essential for India to fully leverage AI and nanotechnology, creating a more inclusive, efficient, and sustainable healthcare system.

12.9.1 AI and nanotechnology: A synergistic approach to healthcare

The integration of AI and nanotechnology offers innovative solutions for managing complex medical conditions. AI's capabilities in data analysis and predictive modeling enable early diagnoses, personalized treatment planning, and remote monitoring [85]. AI algorithms can analyze imaging data, blood tests, and other diagnostic inputs with remarkable accuracy, allowing for quicker identification of internal injuries in trauma patients and optimizing recovery trajectories for patients with critical illnesses. Moreover, AI-powered telemedicine platforms enhance continuous monitoring, making them particularly beneficial for post-accident recovery and the ongoing management of chronic conditions [86].

Nanotechnology offers less intrusive and focused therapies through the use of nanobots. These tiny machines can move through the human body and interact molecularly with cells [87]. Drugs can be delivered directly to tumor areas or damaged tissues, minimizing adverse effects and increasing therapy effectiveness. In trauma situations, they can restore tissues at the cellular level, such as repairing nerve cells or healing damaged bones. Additionally, nanobots equipped with sensors can monitor internal body variables, such as glucose levels or oxygen saturation, and communicate real-time data to healthcare practitioners for automatic therapy modifications, particularly when integrated with AI [88, 89].

12.9.2 Applications of AI and nanobots in trauma and critical illness cases

Nanobots and AI are already showing promise in treating accident trauma and critical illnesses. In trauma cases, nanobots can intervene directly in situations such as internal hemorrhage by delivering clotting agents precisely to the affected area [90, 91]. For brain injuries, AI analyzes neural activity to detect areas needing intervention, while nanobots can deliver therapeutic agents to reduce swelling and promote neuron repair [92]. In spinal cord injuries, nanobots can assist in regenerating nerve cells by delivering growth factors to the injured area, offering hope where traditional treatments fall short [93].

For critical illnesses such as cancer and cardiovascular diseases, AI-driven nanobots have demonstrated exceptional potential [94]. AI models can identify cancer cells, and nanobots are capable of delivering chemotherapy drugs directly to these cells, thereby sparing healthy tissues [95]. Similarly, nanobots are utilized in cardiovascular treatments, such as dissolving blood clots in real time, which can help prevent heart attacks or strokes [96]. In diabetes management, nanobots equipped with glucose sensors continuously monitor blood sugar levels and deliver insulin as needed, with AI ensuring that the delivery is both timely and precise [97, 98].

12.9.3 Challenges and opportunities in integrating AI and nanotechnology in Indian healthcare

Nanotechnology has great potential in healthcare, but its integration faces several challenges, especially in India. The toxicity of nanomaterials, influenced by factors like size, dose, and surface area, can lead to harmful reactions. Additionally, controlling their activity in sensitive environments and ensuring the reliability of experiments remains difficult [99]. High costs, environmental impacts, and impurities add to the barriers [100]. We still don't fully understand how nanomaterials act in controlled laboratory settings (in vitro) and inside living organisms (in vivo). Because of this ambiguity, it is difficult to make precise predictions about how they may impact various biological systems and surroundings. The adoption of nanomaterials is further hindered by the absence of modern instruments for their characterization and standardization. These issues need to be resolved through research, safety procedures, and standardization in order to fully realize the potential of AI and nanotechnology in healthcare, guaranteeing both accessibility and efficacy [101, 102].

12.9.4 The government's role in promoting AI and nanotechnology in healthcare

By fostering research, establishing advantageous legal frameworks, and guaranteeing accessibility, the government plays a critical role in advancing AI and nanotechnology in healthcare. Innovation can be stimulated through research grants, tax breaks, and partnerships with multinational tech companies [103]. AI diagnostics and nanobot treatments can be made available to low-income groups by the government by incorporating them into public health insurance programs like Ayushman Bharat [104]. The healthcare gap between urban and rural areas can be closed by providing financial incentives for rural hospitals to implement AI and nanotechnology. To improve AI efficacy and patient-specific treatments, it is also essential to reinforce the digital health infrastructure and standardize healthcare data across the country [105]. National and international regulations have become necessary due to safety and ethical concerns [106].

The Indian government has been an important contributor to the adoption of AI in the medical field [107]. Another significant effort was the establishment of the National e-Health Authority (NeHA) in 2015, with the objective of developing an integrated healthcare information system to support the use of AI in healthcare operations [108]. NeHA is accountable for providing guidance to various stakeholders on the possible applications of AI in healthcare and ensuring compliance with regulations protecting the privacy and safety of medical data [109]. In 2017, the government established an AI Task Force within the Ministry of Commerce and Industry, in partnership with NITI Aayog, to examine broader uses of AI in other fields, such as healthcare.

Additionally, the Policy Group on AI was formed by the Ministry of Electronics and Information Technology, and its mission is to build an accessible AI adoption strategy that prioritizes skill development, privacy, and security [110]. Another significant move toward developing AI and data science skills throughout the country is the creation of the Center of Excellence (CoE) for Data Science and AI in Karnataka, which is supported by NASSCOM and the state government [111]. Additionally, the government has strengthened intellectual property rights under the National Intellectual Property Rights Policy in acknowledgment of the advantages that AI brings to various areas, including healthcare.

As demonstrated by the US-India Science & Technology Endowment Fund (USISTEF), which provides grants to projects with a major societal impact in order to promote creative healthcare solutions, international cooperation is also essential. Together, these government programs foster an atmosphere that supports the development of AI in healthcare, increasing its accessibility, effectiveness, and innovation throughout India.

12.9.5 Recent real-life use cases

AI and nanotechnology are transforming healthcare globally. In the United States, pancreatic cancer cells are treated with chemotherapy by AI-guided nanobots, which shrink tumors without harming healthy tissue [112]. By encouraging nerve regeneration, they help stroke patients in Japan recover [113]. AI-powered nanobots in Israel are able to identify and break up blood clots in real time, averting strokes and heart attacks [114]. In India, pilot projects are using AI-driven diagnostic tools in cities like Bengaluru and Hyderabad to accurately detect trauma injuries [115]. These applications highlight the potential of AI and nanobots to revolutionize healthcare.

12.10 Nanotechnology in COVID-19 diagnostics and treatment

Nanotechnology has significantly advanced COVID-19 diagnostics and treatment, improving sensitivity and specificity in methods like reverse transcription–polymerase chain reaction (RT-PCR) through signal amplification techniques and the use of magnetic NPs for efficient SARS-CoV-2 RNA extraction [116, 117]. Functionalized NPs streamline the process by binding RNA without elution, reducing time and risks, while paramagnetic iron oxide NPs assist in isolating viral particles for accurate RT-PCR and immunochromatographic tests, particularly aiding non-urban healthcare settings [118, 119]. In therapeutics, nanomedicine enhances antiviral drug delivery using organic NPs to improve bioavailability and reduce cytotoxicity, while metal NPs, such as silver and zinc, exhibit antiviral properties by disrupting viral replication and host attachment [120, 121]. Magnetic nanomedicine enables precise, controlled drug delivery across biological barriers like the blood–brain barrier [122, 123]. However, challenges such as regulatory issues, nanotoxicology concerns, and the need for extensive clinical trials persist, necessitating further research to optimize nanotechnology's potential in combating infectious diseases, including COVID-19 [124, 125].

12.11 Challenges in data quality and regulatory standards

Ensuring data quality is a critical challenge in AI-driven nanomaterial assessments, as the reliability of predictions heavily depends on the accuracy and standardization of the data used [126]. Researchers have emphasized the importance of defining data quality through frameworks such as "fitness for use" [127] and "conformance to requirements" [128]. These principles align with the guidelines from the U.S. National Institute of Statistical Sciences [129], which view data as a product with inherent quality shaped by its generation process. Addressing data quality challenges requires a systematic approach to minimize errors and align datasets with specific research objectives.

Data scarcity exacerbates the challenges of ensuring high-quality information in nanomaterial research. Both international and domestic studies highlight the need to mitigate this issue through innovative strategies. Efforts in China, such as those by the 63rd Research Institute of the PLA General Staff Headquarters [130, 131], have tackled fundamental problems like identifying error sources and enhancing data quality through advanced processes and technologies. The Chinese Academy of Sciences has proposed an assessment system categorizing data quality into external form, content, and utility, emphasizing the need for tailored evaluation criteria [132].

Big data offers transformative potential for advancing nanomaterial research but presents unique challenges in ensuring quality in multisource and multimodal environments. The Total Data Quality Management group at MIT emphasized the importance of aligning quality dimensions with consumer needs [127]. However, integrating diverse datasets in fields such as biology, medicine, and geophysics has revealed the complexities of ensuring consistency and reliability [133]. Developing standardized assessment methods for big data remains a priority to unlock its full potential in nanotechnology.

Navigating regulatory frameworks for nanomaterials requires robust data quality to support evidence-based policies. Early research on data quality criteria, such as those proposed by Alexander and Tate and Katerattanakul and Siau, offers foundational insights into developing evaluation standards for specific applications [134, 135]. In the context of nanomaterials, regulatory frameworks must address the unique characteristics of these materials, balancing innovation with safety and compliance. Harmonized international guidelines are essential to establish consistency across jurisdictions and ensure the safe deployment of nanomaterials.

Ethical concerns play a pivotal role in managing AI-driven nanomaterial assessments. Transparency and accountability are critical to addressing biases, privacy risks, and potential misuse of data. Frameworks such as the semiotic-based model by Shanks and Corbitt and the IQIP model by Knight and Burn emphasize structured approaches to ethical data management [136, 137]. The NISS principles further underscore the need for data quality processes that balance cost, value, and societal impact [129]. Multidisciplinary collaboration is vital to ensure that ethical considerations are integrated into data quality and regulatory practices.

Ensuring compliance and accountability in nanomaterial research is a multifaceted challenge that requires clear metrics and benchmarks. Drawing on methodologies such as those proposed by the Chinese Academy of Sciences (2009) and the IQIP model [137], organizations can implement robust quality management systems. These systems must align with global sustainability goals, fostering public trust, and ensuring that advancements in nanotechnology prioritize societal benefits while mitigating risks.

12.12 Real-world applications and future directions

The increasing use of nanomaterials in various applications necessitates thorough toxicity assessments (Table 13.3). Silver NPs, widely used for their antimicrobial properties, can cause oxidative stress, DNA damage, and disrupt ecosystems at high concentrations [138]. Carbon nanotubes, valued for their strength and conductivity, pose risks of lung inflammation and chronic inflammation when inhaled [139], while polymeric nanomaterials, though useful in drug delivery, can release toxic degradation products. Innovations in nanomaterial design focus on improving biocompatibility and reducing toxicity, such as coating NPs with biopolymers to enhance safety [140, 141]. Collaborative global research and standardized safety guidelines are critical to ensuring consistent testing and minimizing risks [142]. Moving toward safer production methods, including controlling the size, shape, and exposure times of nanomaterials, is vital to mitigate toxicity while maintaining their effectiveness in applications [143].

12.13 Autonomous labs and AI in safety assessment

Although nanotechnology has produced advancements, the environmental and toxicological hazards of nanomaterials remain a source of concern. As demonstrated by the USISTEF, which provides grants to projects with significant societal impact to promote innovative healthcare solutions, international cooperation is essential. Together, such government programs foster an environment that supports the development of AI in healthcare, enhancing its accessibility, effectiveness, and innovation across India.

Table 12.3: Toxicity concerns and innovations in nanomaterial safety.

Nanomaterial	Toxicity concerns	Innovations/strategies for safety	References
Silver nanoparticles	Oxidative stress, genotoxicity, and environmental disruption	Coating with biopolymers to reduce toxicity while maintaining effectiveness	[138]
Carbon nanotubes	Lung inflammation, fibrosis, chronic inflammation, and oxidative stress	Hydrophilic functionalization to reduce aggregation and toxicity risks	[139]
Polymeric nanomaterials	Toxic degradation products, immunogenicity, and cytotoxicity	Surface modification to enhance biocompatibility and reduce toxicity	[140]
General strategies	Balancing functionality with safety	Reducing toxic components, controlling size, shape, and exposure times	[139], [143]
Global research and standards	Lack of standardized testing for nanotoxicity	International collaboration for consistent safety standards	[142]

12.13.1 Potential of autonomous research labs

Autonomous research labs driven by AI and robots could revolutionize the way safety assessments of nanomaterials are carried out. These labs allow high-throughput screening of nanomaterials to evaluate their toxicity and environmental impact by automating experimental procedures. Compared to traditional lab settings, autonomous labs may operate continuously, around the clock, which expedites the research process. For instance, real-time experiment design, execution, and optimization using robotic systems with machine learning algorithms reduce human error and resource waste [144]. In nanotechnology, where assessing a range of material qualities and their biological interactions can be time-consuming, this approach is particularly useful. Additionally, autonomous labs improve reproducibility, a critical component in gaining regulatory approval.

12.13.2 Integration of AI for real-time safety monitoring

Real-time study of nanomaterial interactions in biological and environmental environments is possible with AI-powered safety monitoring systems. Factors such as NP aggregation, release rate, and toxicity levels are continuously monitored during experimental as well as industrial operations by these systems using advanced sensors and internet of things (IoT) technology. Potentially dangerous concentrations of NPs are detected before they become problematic through real-time AI analytics, such as tracking the movement of NPs in air and water [145]. Such proactive approaches can enhance workplace safety and reduce environmental issues. Moreover, AI-based models can rapidly develop a comprehensive toxicity profile by merging information from multiple sources, such as spectroscopic studies and animal testing.

12.13.3 Predictive modeling for safer nanotechnology

One of artificial intelligence's main applications is predictive modeling, which is working toward the production of safer nanomaterials. These models predict the behavior and properties of novel nanomaterials in order to evaluate their potential toxicity even before they are synthesized, using machine learning techniques and available data. This helps researchers identify harmful characteristics prior to the fabrication process and implement modifications to reduce potential dangers and risks. For instance, AI models predicted the cytotoxic effects of NPs with great precision when size, shape, and surface charge are taken into consideration [146]. These methods of prediction are also helpful in the development of environmentally friendly alternatives by evaluating environmental impacts over a period of time, such as bioaccumulation and material disintegration. When combined with green chemistry methods, predictive modeling is crucial for promoting the creation of sustainable nanotechnologies.

12.14 Vision for environmentally friendly nanomaterials

The objective for environmentally friendly nanomaterials is to develop materials that are safe, effective, recyclable, non-toxic, and biodegradable. Researchers are focusing on alternatives that are sustainable by using environmentally friendly techniques, including nontoxic solvents and materials made from renewable sources, rather than conventional chemical techniques that could have negative consequences. For example, biopolymers and bioderived NPs have been created as more secure alternatives to artificial nanomaterials because these materials break down more easily in the natural environment and are therefore less harmful over time [147]. Another important aspect is lowering production energy use; initiatives are being made to create low-energy synthesis techniques that have less of an environmental impact [148].

12.14.1 AI in promoting circular economy in nanotechnology

The circular economy's primary objectives are waste reduction, recycling, and reuse. Artificial intelligence (AI) can play a significant role in putting these concepts into practice in nanotechnology. AI) can improve nanomaterial recovery, reprocessing, and recycling by estimating the entire lifespan of nanomaterials and providing suggestions on how to extract valuable components from waste. For example, AI algorithms can maximize the recycling of metal NPs or carbon nanotubes, increasing the efficiency of the process [149]. In addition, machine learning can help develop new methods for reusing nanomaterials, increasing their versatility in new applications while reducing the need for raw materials and their negative environmental consequences [150].

12.14.2 Building public trust in emerging nanotechnologies

As nanotechnology grows, gaining public trust will be essential to its adoption across sectors. Questions have been raised regarding the morality and safety of using NPs, particularly in areas of concern like consumer products and healthcare. These issues must be openly and responsibly addressed if the technology is to be adopted and utilized safely. Gaining the public's trust requires ensuring transparency in research and development, as well as clearly and concisely communicating the benefits and potential concerns [151]. Ethical concerns including the possibility of abuse or uneven access to nanotechnology must also be considered. In this process, governmental organizations play a vital role because they enforce strict safety regulations and ensure that nanotechnologies have been developed and utilized accordingly. Public involvement like interactive instruction and candid discussion can allay anxieties and increase comprehension [152]. AI and machine learning can support these efforts by providing precise risk assessments and evidence-based data to enhance public communication and legislation.

12.15 Conclusion

This chapter addresses the significant role of AI and machine learning in improving the security and environmental responsibility of NPs. Since NPs are vital for advancements in consumer goods, healthcare, and the environment, AI is now essential for precisely determining toxicity, tracking interactions in real time, and improving material creation. AI maintains a balance between security and technological advances by facilitating well-informed risk assessments. Biodegradable products and AI-guided disposal are two examples of environmentally friendly methods that are important for minimizing environmental impact. Encouraging responsible advancement in nanotechnology requires cooperation, adherence to international standards, and consideration of ethical considerations.

List of abbreviations

AI	Artificial intelligence
ML	Machine learning
NP	Nanoparticle
ROS	Reactive oxygen species
QSAR	Quantitative structure–activity relationship
ADME	Absorption, distribution, metabolism, and excretion
ANN	Artificial neural network
IoT	Internet of things
LD_{50}	Median lethal dose
PFAS	Per- and polyfluorinated alkyl substances
NCATS	National Center for Advancing Translational Sciences
ICCVAM	Interagency coordinating committee on the validation of alternative methods
DeepTox	Deep learning toxicity prediction model
DeepCarc	Deep learning model for carcinogenicity prediction
USISTEF	US-India Science & Technology Endowment Fund
NeHA	National e-Health Authority
NISS	National Institute of Statistical Sciences
CoE	Center of excellence
AUC	Area under the curve

References

[1] Khan I, Saeed K and Khan I. Nanoparticles: Properties, Applications and Toxicities. Arabian Journal of Chemistry. 2019;12(7):908–931. https://doi.org/10.1016/j.arabjc.2017.05.011.

[2] Sriram T, Chakraborty T and Prasanna P. Artificial Intelligence Powered Insights into Nanotoxicology. International Journal *of Advancement in* Life Sciences Research. 2024;7(2):68–80. https://doi.org/10.31632/ijalsr.2024.v07i02.005.

[3] De Jong WH and Borm PJA. Drug Delivery and Nanoparticles: Applications and Hazards. International Journal of Nanomedicine. 2008;3(2):133–149. https://doi.org/10.2147/ijn.s596.

[4] Wu Y and Wang G. Machine Learning Based Toxicity Prediction: From Chemical Structural Description to Transcriptome Analysis. International Journal of Molecular Sciences. 2018;19(8):2358. https://doi.org/10.3390/ijms19082358.

[5] Cherkasov A, Muratov EN, Fourches D, Varnek A, Baskin II, Cronin M, Dearden J, Gramatica P, Martin YC and Todeschini R. QSAR Modeling: Where Have You Been? Where Are You Going To?. Journal of Medicinal Chemistry. 2014;57(12):4977–5010. https://doi.org/10.1021/jm4004285.

[6] Hansch C, Maloney PP, Fujita T and Muir RM. Correlation of Biological Activity of Phenoxyacetic Acids with Hammett Substituent Constants and Partition Coefficients. Nature. 1962;194:178–180. https://doi.org/10.1038/194178b0.

[7] Barratt MD. Prediction of Toxicity from Chemical Structure. Cell Biology and Toxicology. 2000;16 (1):1–13. https://doi.org/10.1023/A:1007688211933.

[8] Cammarata A and Menon GK. Pattern Recognition: Classification of Therapeutic Agents According to Pharmacophores. Journal of Medicinal Chemistry. 1976;19(6):739–748. https://doi.org/10.1021/jm00227a002.

[9] Devillers J and Balaban AT. Topological Indices and Related Descriptors in QSAR and QSPAR. 2000; Boca Raton, FL, USA: CRC Press.

[10] Ataide Martins JP, Ma RDO and Ms ODQ. Web-4D-QSAR: A Web-Based Application to Generate 4D-QSAR Descriptors. Journal of Computational Chemistry . 2018;39(12):917–924. https://doi.org/10.1002/jcc.25155.

[11] Nel A, Xia T, Mädler L and Li N. Toxic Potential of Materials at the Nanolevel. Science. 2006;311 (5761):622–627. https://doi.org/10.1126/science.1114397.

[12] Isensee C, Griese KM and Teuteberg F. Sustainable Artificial Intelligence: A Corporate Culture Perspective. Sustainability Management Forum | NachhaltigkeitsManagementForum. 2021; (vol. 29, pp. 217–230). Berlin/Heidelberg, Germany: Springer .

[13] Yigitcanlar T. Greening the Artificial Intelligence for a Sustainable Planet: An Editorial Commentary. Sustainability. 2021;13:13508. https://doi.org/10.3390/su132413508.

[14] Abid SK, Sulaiman N, Chan SW, Nazir U, Abid M, Han H, Ariza-Montes A and Vega-Muñoz A. Toward an Integrated Disaster Management Approach: How Artificial Intelligence Can Boost Disaster Management. Sustainability. 2021;13:12560. https://doi.org/10.3390/su132112560.

[15] Ahmed QW, Garg S, Rai A, Ramachandran M, Jhanjhi NZ, Masud M and Baz M. AI-Based Resource Allocation Techniques in Wireless Sensor Internet of Things Networks in Energy Efficiency with Data Optimization. Electronics. 2022;11:2071. https://doi.org/10.3390/electronics11132071.

[16] Iqbal M, Ma J, Ahmad N, Hussain K, Usmani MS and Ahmad M. Sustainable Construction through Energy Management Practices in Developing Economies: An Analysis of Barriers in the Construction Sector. Environmental Science and Pollution Research. 2021;28:34793–34823. https://doi.org/10.1007/s11356-021-13950-2.

[17] Nel A, Xia T, Mädler L and Li N. Toxic Potential of Materials at the Nanolevel. Science. 2006;311 (5761):622–627. https://doi.org/10.1126/science.1114397.

[18] Warheit DB, Laurence BR, Reed KL, Roach DH, Reynolds GAM and Webb TR. Comparative Pulmonary Toxicity Assessment of Single-Wall Carbon Nanotubes in Rats. Toxicological Sciences. 2004;77(1):117–125. https://doi.org/10.1093/toxsci/kfg228.

[19] Oberdörster G, Oberdörster E and Oberdörster J. Nanotoxicology: An Emerging Discipline Evolving from Studies of Ultrafine Particles. Environmental Health Perspectives. 2005;113(7):823–839. https://doi.org/10.1289/ehp.7339.

[20] Nel A, Grainger D, Alvarez PJ and Badesha S. Nanotechnology Environmental, Health, and Safety Issues. Nanotechnology Research Directions for Societal Needs in 2020. 1970;(1st ed., pp. 159–220). Dordrecht, Netherlands: Springer. https://doi.org/10.1007/978-94-007-1168-6_5

[21] Nel A, Xia T, Mädler L and Li N. Toxic Potential of Materials at the Nanolevel. Science. 2006;311:622–627. https://doi.org/10.1126/science.1114397.

[22] Møller P, Jacobsen NR, Folkmann JK, Danielsen PH, Mikkelsen L, Hemmingsen JG, Vesterdal LK, Forchhammer L, Wallin H and Loft S. Role of Oxidative Damage in Toxicity of Particulates. Free Radical Research. 2010;44:1–46. https://doi.org/10.3109/10715760903300691.

[23] Unfried K, Albrecht C, Klotz LO, Von Mikecz A, Grether-Beck S and Schins RPF. Cellular Responses to Nanoparticles: Target Structures and Mechanisms. Nanotoxicology. 2007;1:52–71.

[24] Xia T, Kovochich M, Brant J, Hotze M, Sempf J, Oberley T, Sioutas C, Yeh JI, Wiesner MR and Nel AE. Comparison of the Abilities of Ambient and Manufactured Nanoparticles to Induce Cellular Toxicity According to an Oxidative Stress Paradigm. Nano Letters. 2006;6(8):1794–1807. https://doi.org/10.1021/nl061025k.

[25] Halliwell B and Gutteridge JMC. Free Radicals in Biology and Medicine. 2007; New York: Oxford University Press.

[26] Dalle-Donne I, Rossi R, Giustarini D, Colombo R and Milzani A. S-Glutathionylation in Protein Redox Regulation. Free Radical Biology and Medicine. 2007;43(6):883–898. https://doi.org/10.1016/j.freeradbiomed.2007.06.014.

[27] Oberdörster G, Oberdörster E and Oberdörster J. Nanotoxicology: An Emerging Discipline Evolving from Studies of Ultrafine Particles. Environmental Health Perspectives . 2005;113(7):823–839. https://doi.org/10.1289/ehp.7339.
[28] Sayes CM, Gobin AM, Ausman KD, Mendez J, West JL and Colvin VL. Nano-C-60 Cytotoxicity Is Due to Lipid Peroxidation. Biomaterials. 2005;26(36):7587–7595. https://doi.org/10.1016/j.biomaterials.2005.05.027.
[29] Bermudez E, Mangum JB, Wong BA, Asgharian B, Hext PM, Warheit DB, Everitt JI and Moss OR. Pulmonary Responses of Mice, Rats, and Hamsters to Subchronic Inhalation of Ultrafine Titanium Dioxide Particles. Toxicological Sciences. 2004;77(2):347–357. https://doi.org/10.1093/toxsci/kfh019.
[30] Oberdörster E. Manufactured Nanomaterials (Fullerenes, C60) Induce Oxidative Stress in the Brain of Juvenile Largemouth Bass. Environmental Health Perspectives. 2004;112(10):1058–1062. https://doi.org/10.1289/ehp.7021.
[31] Warheit DB, Laurence BR, Reed KL, Roach DH, Reynolds GAM and Webb TR. Comparative Pulmonary Toxicity Assessment of Single-Wall Carbon Nanotubes in Rats. Toxicological Sciences. 2004;77(1):117–125. https://doi.org/10.1093/toxsci/kfg228.
[32] Adibhatla RM and Hatcher JF. Lipid Oxidation and Peroxidation in CNS Health and Disease: From Molecular Mechanisms to Therapeutic Opportunities. Antioxidants & Redox Signaling. 2010;12(1):125–169. https://doi.org/10.1089/ars.2009.2668.
[33] Griendling KK and FitzGerald GA. Oxidative Stress and Cardiovascular Injury: Part I: Basic Mechanisms and In Vivo Monitoring of ROS. Circulation. 2003;108(16):1912–1916. https://doi.org/10.1161/01.CIR.0000093660.86242.BB.
[34] Rallo R, France B, Liu R, Nair S, George S, Damoiseaux R, Giralt F, Nel A, Bradley K and Cohen Y. Self-Organizing Map Analysis of Toxicity-Related Cell Signaling Pathways for Metal and Metal Oxide Nanoparticles. Environmental Science & Technology. 2011;45(4):1695–1703. https://doi.org/10.1021/es103606x.
[35] Li N, Xia T and Nel AE. The Role of Oxidative Stress in Ambient Particulate Matter-Induced Lung Diseases and Its Implications in the Toxicity of Engineered Nanoparticles. Free Radical Biology and Medicine. 2008;44(9):1689–1699. https://doi.org/10.1016/j.freeradbiomed.2008.01.028.
[36] Petersen EJ and Nelson BC. Mechanisms and Measurements of Nanomaterial-Induced Oxidative Damage to DNA. Analytical and Bioanalytical Chemistry. 2010;398(2):613–650. https://doi.org/10.1007/s00216-010-3881-7.
[37] Esposito LA, Melov S, Panov A, Cottrell BA and Wallace DC. Mitochondrial Disease in Mouse Results in Increased Oxidative Stress. Proceedings of the National Academy of Sciences of the United States of America. 1999;96(9):4820–4825. https://doi.org/10.1073/pnas.96.9.4820.
[38] Kirkinezos IG and Moraes CT. Reactive Oxygen Species and Mitochondrial Diseases. Seminars in Cell and Developmental Biology . 2001;12(6):449–457. https://doi.org/10.1006/scdb.2001.0282.
[39] Deng ZJ, Liang MT, Monteiro M, Toth I and Minchin RF. Nanoparticle-Induced Unfolding of Fibrinogen Promotes Mac-1 Receptor Activation and Inflammation. Nature Nanotechnology. 2011;6(1):39–44. https://doi.org/10.1038/nnano.2010.250.
[40] Baskin II Machine Learning Methods in Computational Toxicology. In Computational Toxicology, Methods in Molecular Biology, Nicolotti O, Ed., 2018; (Vol. 1800, pp 119–139). New York, NY: Humana Press. https://doi.org/10.1007/978-1-4939-7899-1_5.
[41] Russell S and Norvig P. Artificial Intelligence: A Modern Approach, (4th ed.). 2020; (pp. 1–1069). London, United Kingdom: Pearson Education, Inc..
[42] Lin Z, Chou WC, Cheng YH, He C, Monteiro-Riviere NA and Riviere JE. Title of the Article. International Journal of Nanomedicine. 2022;17:1365–1379. https://doi.org/10.2147/IJN.S344208.
[43] Cheng Y-H, He C, Riviere JE, Monteiro-Riviere NA and Lin Z. Meta-Analysis of Nanoparticle Delivery to Tumors Using a Physiologically Based Pharmacokinetic Modeling and Simulation Approach. ACS Nano. 2020;14(3):3075–3095. https://doi.org/10.1021/acsnano.9b08142.

[44] Kamiya Y, Handa K, Miura T, Yanagi M, Shigeta K, Hina S, Shimizu M, Kitajima M, Shono F, Funatsu K and Yamazaki H. In Silico Prediction of Input Parameters for Simplified Physiologically Based Pharmacokinetic Models for Estimating Plasma, Liver, and Kidney Exposures in Rats after Oral Doses of 246 Disparate Chemicals. Chemical Research in Toxicology. 2021;34(2):507–513. https://doi.org/10.1021/acs.chemrestox.0c00336.

[45] Pradeep P, Patlewicz G, Pearce R, Wambaugh J, Wetmore B and Judson R. Using Chemical Structure Information to Develop Predictive Models for In Vitro Toxicokinetic Parameters to Inform High-Throughput Risk-Assessment. Computational Toxicology. 2020;16:100136. https://doi.org/10.1016/j.comtox.2020.100136.

[46] Mayr A, Klambauer G, Unterthiner T and Hochreiter S. DeepTox: Toxicity Prediction Using Deep Learning. Frontiers of Environmental Science. 2016;3:80.

[47] Zhang L, Ai H, Chen W, Yin Z, Hu H, Zhu J, Zhao J, Zhao Q and Liu H. CarcinoPred-EL: Novel Models for Predicting the Carcinogenicity of Chemicals Using Molecular Fingerprints and Ensemble Learning Methods. Scientific Reports. 2017;7(1):2118. https://doi.org/10.1038/s41598-017-02365-0.

[48] Ai H, Wu X, Zhang L, Qi M, Zhao Y, Zhao Q, Zhao J and Liu H. QSAR Modelling Study of the Bioconcentration Factor and Toxicity of Organic Compounds to Aquatic Organisms Using Machine Learning and Ensemble Methods. Ecotoxicology and Environmental Safety. 2019;179:71–78. https://doi.org/10.1016/j.ecoenv.2019.04.035.

[49] Cheng W and Ng CA. Using Machine Learning to Classify Bioactivity for 3486 Per- and Polyfluoroalkyl Substances (PFASs) from the OECD List. Environmental Science & Technology. 2019;53(23):13970–13980. https://doi.org/10.1021/acs.est.9b04833.

[50] Mansouri K, Karmaus AL, Fitzpatrick J. et al.. CATMoS: Collaborative Acute Toxicity Modeling Suite. Environmental Health Perspectives. 2021;129(4):47013. https://doi.org/10.1289/EHP8495.

[51] Li T, Tong W, Roberts R. et al.. DeepCarc: Deep Learning-Powered Carcinogenicity Prediction Using Model-Level Representation. Frontiers in Artificial Intelligence. 2021;4:757780. https://doi.org/10.3389/frai.2021.757780.

[52] Wang Y-W, Huang L, Jiang S-W. et al.. CapsCarcino: A Novel Sparse Data Deep Learning Tool for Predicting Carcinogens. Food and Chemical Toxicology. 2020;135:110921. https://doi.org/10.1016/j.fct.2019.110921.

[53] Zhang X, Shu K, Rajkumar S and Sivakumar V. Research on Deep Integration of Application of Artificial Intelligence in Environmental Monitoring System and Real Economy. Environmental Impact Assessment Review. 2021;86:106499. https://doi.org/10.1016/j.eiar.2020.106499.

[54] Zupan J and Gasteiger J. Neural Networks: A New Method for Solving Chemical Problems or Just a Passing Phase?. Analytica Chimica Acta. 1991;248(1):1–30. https://doi.org/10.1016/0003-2670(91)80215-B.

[55] Hopfield JJ. Neural Networks and Physical Systems with Emergent Collective Computational Abilities. Proceedings of the National Academy of Sciences of the United States of America. 1982;79(8):2554–2558. https://doi.org/10.1073/pnas.79.8.2554.

[56] Kaviani S and Sohn I. Application of Complex Systems Topologies in Artificial Neural Networks Optimization: An Overview. Expert Systems with Applications. 2021;180:115073. https://doi.org/10.1016/j.eswa.2021.115073.

[57] Cabaneros SM, Calautit JK and Hughes BR. A Review of Artificial Neural Network Models for Ambient Air Pollution Prediction. Environmental Modelling & Software. 2019;119:285–304. https://doi.org/10.1016/j.envsoft.2019.06.007.

[58] Lakshmi D. et al.. Artificial Intelligence (AI) Applications in Adsorption of Heavy Metals Using Modified Biochar. Science of the Total Environment. 2021;801:149623. https://doi.org/10.1016/j.scitotenv.2021.149623.

[59] Zhang L. et al.. A Review of Machine Learning in Building Load Prediction. Applied Energy. 2021;285:116452. https://doi.org/10.1016/j.apenergy.2020.116452.

[60] Mitchell S, Qin R, Zheng N and Pérez-Ramírez J. Nanoscale Engineering of Catalytic Materials for Sustainable Technologies. Nature Nanotechnology. 2021;16(2):129–139. https://doi.org/10.1038/s41565-020-00799-8.
[61] Roco MC, Hersam MC, Mirkin CA and Tuominen M. Synthesis, Processing, and Manufacturing of Components, Devices, and Systems. Nanotechnology Research Directions for Societal Needs in 2020. 2011:109–158. https://doi.org/10.1007/978-94-007-1168-6_4..
[62] Shivalkar S, Gautam PK, Chaudhary S, Samanta SK and Sahoo AK. Recent Development of Autonomously Driven Micro/Nanobots for Efficient Treatment of Polluted Water. Journal of Environmental Management. 2021;281 Article 111750. https://doi.org/10.1016/j.jenvman.2020.111750.
[63] Singh AV, Ansari MH, Laux P and Luch A. Micro-Nanorobots: Important Considerations When Developing Novel Drug Delivery Platforms. Expert Opinion on Drug Delivery. 2019;16(11):1259–1275. https://doi.org/10.1080/17425247.2019.1676228.
[64] Wang J. Nanomachines: Fundamentals and Applications. 2013; Weinheim; Wiley-VCH. https://doi.org/10.1002/9783527651450.
[65] Wang W, Duan W, Ahmed S, Mallouk TE and Sen A. Small Power: Autonomous Nano- and Micromotors Propelled by Self-Generated Gradients. Nano Today. 2013;8:531–554. https://doi.org/10.1016/j.nantod.2013.08.009.
[66] Akin D, Sturgis J, Ragheb K, Sherman D, Burkholder K, Robinson JP. et al.. Bacteria-Mediated Delivery of Nanoparticles and Cargo into Cells. Nature Nanotechnology. 2007;2:441–449. https://doi.org/10.1038/nnano.2007.149.
[67] Vaseashta A, Vaclavikova M, Vaseashta S, Gallios G, Roy P and Pummakarnchana O. Nanostructures in Environmental Pollution Detection, Monitoring, and Remediation. Science and Technology of Advanced Materials. 2007;8(1–2):47. https://doi.org/10.1016/j.stam.2006.11.003.
[68] Kadi MW and Mohamed RM. Enhanced Photocatalytic Activity of ZrO2-SiO2 Nanoparticles by Platinum Doping. International Journal of Photoenergy. 2013;https://doi.org/10.1155/2013/812097.
[69] Rao PN. Nanocatalysis: Applications in the Chemical Industry. 2010.
[70] Jena B, Ningthoujam R, Pattanayak S, Dash S, Panda MK, Jit BP, Das M and Singh YD. Nanotechnology and Its Potential Application in Postharvest Technology. Bio-Nano Interface. 2022:93–107.
[71] Baig N, Kammakakam I and Falath W. Nanomaterials: A Review of Synthesis Methods, Properties, Recent Progress, and Challenges. Materials Advances. 2021;2(6):1821–1871. https://doi.org/10.1039/D0MA00807A.
[72] Adewuyi A and Lau WJ. Nanomaterial Development and Its Applications for Emerging Pollutant Removal in Water. Handbook of Nanotechnology Applications.2021; (pp. 67–97). Elsevier. https://doi.org/10.1016/B978-0-12-821506-7.00003-X.
[73] Luo J, Fu K, Yu D, Hristovski KD, Westerhoff P and Crittenden JC. Review of Advances in Engineering Nanomaterial Adsorbents for Metal Removal and Recovery from Water: Synthesis and Microstructure Impacts. ACS ES&T Engineering. 2021;1(4):623–661. https://doi.org/10.1021/acsestengg.0c00174.
[74] Yogalakshmi KN, Das A, Rani G, Jaswal V and Randhawa JS. Nano-Bioremediation: A New Age Technology for the Treatment of Dyes in Textile Effluents. Bioremediation of Industrial Waste for Environmental Safety. 2020; (pp. 313–347). Singapore: Springer. https://doi.org/10.1007/978-981-13-1891-7_15
[75] Jain K, Patel AS, Pardhi VP and Flora SJS. Nanotechnology in Wastewater Management: A New Paradigm towards Wastewater Treatment. Molecules. 2021;26(6):1797. https://doi.org/10.3390/molecules26061797.
[76] Bharagava RN, Saxena G and Mulla SI. Introduction to Industrial Wastes Containing Organic and Inorganic Pollutants and Bioremediation Approaches for Environmental Management.

Bioremediation of Industrial Waste for Environmental Safety. 2020; (pp. 1–18). Singapore: Springer. https://doi.org/10.1007/978-981-13-1891-7_1
[77] Rizwan M, Singh M, Mitra CK and Morve RK. Ecofriendly Application of Nanomaterials: Nanobioremediation. Journal of Nanoparticle Research. 2014;2014:431787. https://doi.org/10.1155/2014/431787.
[78] Pandey G and Jain P. Assessing the nanotechnology on the grounds of costs, benefits, and risks. Beni-Suef University Journal of Basic and Applied Sciences. 2020;9(1):1–10. https://doi.org/10.1186/s43088-020-00085-5.
[79] El-Sayed ME. Nanoadsorbents for water and wastewater remediation. Science of the Total Environment. 2020 Article 139903; https://doi.org/10.1016/j.scitotenv.2020.139903.
[80] Kobya M, Demirbas E, Senturk E and Ince M. Adsorption of Heavy Metal Ions from Aqueous Solutions by Activated Carbon Prepared from Apricot Stone. Bioresource Technology. 2005;96 (9):1518–1521. https://doi.org/10.1016/j.biortech.2004.12.005.
[81] Khin MM, Nair AS, Babu VJ, Rajendiran M and Ramakrishna S. A Review on Nanomaterials for Environmental Remediation. Energy & Environmental Science. 2012;5(8):8075–8109. https://doi.org/10.1039/C2EE21818F.
[82] Malik S, Muhammad K and Waheed Y. Emerging Applications of Nanotechnology in Healthcare and Medicine. Molecules. 2023;28(18):6624. https://doi.org/10.3390/molecules28186624.
[83] Pan D. Nanomedicine and Nanobiotechnology in India. WIREs Nanomedicine and Nanobiotechnology. 2024 *First published: 25 January 2024*; https://doi.org/10.1002/wnan.1939.
[84] Liu Q, Zou J, Chen Z, He W and Wu W. Current Research Trends of Nanomedicines. Acta Pharmaceutica Sinica B. 2023;13(11):4391–4416. https://doi.org/10.1016/j.apsb.2023.05.018.
[85] Al Kuwaiti A, Nazer K, Al-Reedy A, Al-Shehri S, Al-Muhanna A, Subbarayalu AV, Al Muhanna D and Al-Muhanna FA. A Review of the Role of Artificial Intelligence in Healthcare. Journal of Personalized Medicine. 2023;13(6):951. https://doi.org/10.3390/jpm13060951.
[86] Anjum S, Ishaque S, Fatima H, Farooq W, Hano C, Abbasi BH and Anjum I. Emerging Applications of Nanotechnology in Healthcare Systems: Grand Challenges and Perspectives. Pharmaceuticals (Basel). 2021;14(8):707. https://doi.org/10.3390/ph14080707.
[87] Abbasi N. Artificial Intelligence in Remote Monitoring and Telemedicine. Journal of Artificial Intelligence General Science. 2024;1(1):202. https://doi.org/10.60087/jaigs.v1i1.202.
[88] John L, Talati D, Remi G and Samuel AH. Telemedicine and AI in remote patient monitoring. Dec. 2023.
[89] Dubey A and Tiwari A. Artificial Intelligence and Remote Patient Monitoring in US Healthcare Market: A Literature Review. Journal of Market Access & Health Policy . 2023;11(1):2205618. https://doi.org/10.1080/20016689.2023.2205618.
[90] Rajendran S, Sundararajan P, Awasthi A and Rajendran S. Nanorobotics in Medicine: A Systematic Review of Advances, Challenges, and Future Prospects with a Focus on Cell Therapy, Invasive Surgery, and Drug Delivery. Precision Nanomedicine. 2024:7(1. https://doi.org/10.33218/001c.92224.
[91] Bony BA and Kievit FM. A Role for Nanoparticles in Treating Traumatic Brain Injury. Pharmaceutics. 2019;11(9):473. https://doi.org/10.3390/pharmaceutics11090473.
[92] Baker CR, Pease M, Sexton DP, Abumoussa A and Chambless LB. Artificial Intelligence Innovations in Neurosurgical Oncology: A Narrative Review. Journal of Neuro-Oncology. 2024;169(3):489–496. https://doi.org/10.1007/s11060-024-04757-5.
[93] Tyler JY, Xu X-M and Cheng J-X. Nanomedicine for Treating Spinal Cord Injury. Review Nanoscale. 2013;5(19):8821–8836. https://doi.org/10.1039/c3nr00957b.
[94] Aggarwal M and Kumar S. The Use of Nanorobotics in the Treatment Therapy of Cancer and Its Future Aspects: A Review. Cureus. 2022;14(9):e29366. https://doi.org/10.7759/cureus.29366.

[95] Bhinder B, Gilvary C, Madhukar NS and Elemento O. Artificial Intelligence in Cancer Research and Precision Medicine. Cancer Discovery. 2021;11(4):900–915. https://doi.org/10.1158/2159-8290.CD-21-0090.
[96] Hu Q, Fang Z, Ge J and Li H. Nanotechnology for Cardiovascular Diseases. Innovation (Camb.). 2022;3(2):100214. https://doi.org/10.1016/j.xinn.2022.100214.
[97] Klonoff DC. The Artificial Pancreas: How Sweet Engineering Will Solve Bitter Problems. Journal of Diabetes Science and Technology. 2007;1(1):72–81. https://doi.org/10.1177/193229680700100112.
[98] Qin W, Wu Y, Liu J, Yuan X and Gao J. A Comprehensive Review of the Application of Nanoparticles in Diabetic Wound Healing: Therapeutic Potential and Future Perspectives. International Journal of Nanomedicine. 2022;17:6007–6029. https://doi.org/10.2147/IJN.S386585.
[99] Kanwar JR, Sun X, Punj V, Sriramoju B, Mohan RR, Zhou S-F, Chauhan A and Kanwar RK. Nanoparticles in the Treatment and Diagnosis of Neurological Disorders: Untamed Dragon with Fire Power to Heal. Nanomedicine: Nanotechnology, Biology and Medicine. 2012;8(4):399–414. https://doi.org/10.1016/j.nano.2011.08.006.
[100] Jamkhande PG, Ghule NW, Bamer AH and Kalaskar MG. Metal Nanoparticles Synthesis: An Overview on Methods of Preparation, Advantages and Disadvantages, and Applications. Journal of Drug Delivery Science and Technology. 2019;53:101174. https://doi.org/10.1016/j.jddst.2019.101174.
[101] Patel S, Nanda R and Sahoo S. Nanotechnology in Healthcare: Applications and Challenges. Medicinal Chemistry. 2015;5(12):528–533. https://doi.org/10.4172/2161-0444.1000312.
[102] Klein PM and Wagner E. Bioreducible Polycations as Shuttles for Therapeutic Nucleic Acid and Protein Transfection. Antioxidants & Redox Signaling. 2014;21(5):804–817. https://doi.org/10.1089/ars.2013.5714.
[103] Marwaha JS, Landman AB, Brat GA, Dunn T and Gordon WJ. Deploying Digital Health Tools within Large, Complex Health Systems: Key Considerations for Adoption and Implementation. npj Digital Medicine. 2022;5:13. https://doi.org/10.1038/s41746-022-00557-1.
[104] Liao F, Adelaine S, Afshar M and Patterson BW. Governance of Clinical AI Applications to Facilitate Safe and Equitable Deployment in a Large Health System: Key Elements and Early Successes. Frontiers in Digital Health. 2022;4:931439. https://doi.org/10.3389/fdgth.2022.931439.
[105] Forcier MB, Gallois H, Mullan S and Joly Y. Integrating Artificial Intelligence into Health Care through Data Access: Can the GDPR Act as a Beacon for Policymakers?. Journal of Law and the Biosciences. 2019;6:317–335. https://doi.org/10.1093/jlb/lsz013.
[106] Schaake M The European Commission's Artificial Intelligence Act. *Standford University Human-Centered Artificial Intelligence (BHAI)*, 2021; Canada: Standford. https://hai.stanford.edu/sites/default/files/2021-06/HAI_Issue-Brief_The-European-Commissions-Artificial-Intelligence-Act.pdf (accessed May 5, 2023).
[107] Chatterjee S and Dohan MS. Artificial Intelligence for Healthcare in India: Policy Initiatives, Challenges, and Recommendations. International Journal of Healthcare Information Systems Informatics. 2021;16(4):1–11. https://doi.org/10.4018/IJHISI.20211001.oa17.
[108] Chatterjee S. Security and Privacy Issues in E-Commerce: A Proposed Guidelines to Mitigate the Risk. IEEE International Advanced Computing Conference (IACC). 2015:393–396. https://doi.org/10.1109/IADCC.2015.7154737.
[109] Chatterjee S. AI Strategy of India: Policy Framework, Adoption Challenges and Actions for Government. Transforming Government: People, Process and Policy. 2020;14(5):757–775. https://doi.org/10.1108/TG-05-2019-0031.
[110] PTI. Centre Forms Policy Group to Study Artificial Intelligence: Nasscom. *The Economic Times*. https://economictimes.indiatimes.com/tech/internet/centre-forms-policy-groupto-study-artificial-intelligence-nasscom/articleshow/60499213.cms (accessed Dec 19, 2024).

[111] NASSCOM. Government of Karnataka and NASSCOM Partner to Launch Centre of Excellence for Data Science and Artificial Intelligence. *NASSCOM*. http://www.nasscom.in/sites/default/files/media_pdf/government-of-karnataka-and-nasscom (accessed Dec 19, 2024).
[112] Chehelgerdi M, Chehelgerdi M, Allela OQB, Pecho RDC, Jayasankar N, Rao DP, Thamaraikani T, Vasanthan M, Viktor P, Lakshmaiya N, Saadh MJ, Amajd A, Abo-Zaid MA, Castillo-Acobo RY, Ismail AH, Amin AH and Akhavan-Sigari R. Progressing Nanotechnology to Improve Targeted Cancer Treatment: Overcoming Hurdles in Its Clinical Implementation. Molecular Cancer. 2023;22:169. https://doi.org/10.1186/s12943-023-01865-0.
[113] Senadheera I, Hettiarachchi P, Haslam B, Nawaratne R, Sheehan J, Lockwood KJ, Alahakoon D and Carey LM. AI Applications in Adult Stroke Recovery and Rehabilitation: A Scoping Review Using AI. Sensors. 2024;24(20):6585. https://doi.org/10.3390/s24206585.
[114] Zargarzadeh A, Javanshir E, Ghaffari A, Mosharkesh E and Anari B. Artificial Intelligence in Cardiovascular Medicine: An Updated Review of the Literature. Journal of Cardiovascular and Thoracic Research. 2023;15(4):204–209. https://doi.org/10.34172/jcvtr.2023.33031.
[115] Levy BE, Castle JT, Virodov A, Wilt WS, Bumgardner C, Brim T, McAtee E, Schellenberg M, Inaba K and Warriner ZD. Artificial Intelligence Evaluation of Focused Assessment with Sonography in Trauma. The Journal of Trauma and Acute Care Surgery. 2023;95(5):706–712. https://doi.org/10.1097/TA.0000000000004021.
[116] Xu J, Zhao S, Teng T, Abdalla AE, Zhu W, Xie L. et al.. Systematic Comparison of Two Animal-to-Human Transmitted Human Coronaviruses: SARS-CoV-2 and SARS-CoV. Viruses. 2020;12(2):244. https://doi.org/10.3390/v12020244.
[117] Xu L, Li D, Ramadan S, Li Y and Klein N. Facile Biosensors for Rapid Detection of COVID-19. Biosensors and Bioelectronics. 2020;170:112673. https://doi.org/10.1016/j.bios.2020.112673.
[118] Mujawar MA, Gohel H, Bhardwaj SK, Srinivasan S, Hickman N and Kaushik A. Nano-Enabled Biosensing Systems for Intelligent Healthcare: Towards COVID-19 Management. Materials Today Chemistry. 2020;17:100306. https://doi.org/10.1016/j.mtchem.2020.100306.
[119] Shan B, Broza YY, Li W, Wang Y, Wu S, Liu Z. et al.. Multiplexed Nanomaterial-Based Sensor Array for Detection of COVID-19 in Exhaled Breath. ACS Nano. 2020;14(9):12125–12132. https://doi.org/10.1021/acsnano.0c05657.
[120] Teixeira MC, Sanchez-Lopez E, Espina M, Calpena AC, Silva AM and Veiga FJ. Advances in Antibiotic Nanotherapy: Overcoming Antimicrobial Resistance. In Emerging Nanotechnologies in Immunology, Shegokar R and Souto ED(Eds.). 2018; (pp. 233–259). Boston, MA: Elsevier. https://doi.org/10.1016/B978-0-12-813833-1.00013-8.
[121] Goscianska J, Freund R and Wuttke S. Nanoscience Versus Viruses: The SARS-CoV-2 Case. Advanced Functional Materials. 2022;32(14):2107826. https://doi.org/10.1002/adfm.202107826.
[122] Kaushik A. Manipulative Magnetic Nanomedicine: The Future of COVID-19 Pandemic/Endemic Therapy. Expert Opinion on Drug Delivery. 2021;18(5):531–534. https://doi.org/10.1080/17425247.2021.1860938.
[123] Mohammadi K, Sani MA, Azizi-Lalabadi M and McClements DJ. Nano-Enabled Personalized Nutrition: Developing Multicomponent-Bioactive Colloidal Delivery Systems. Advances in Colloid and Interface Science. 2022;282:102211. https://doi.org/10.1016/j.cis.2020.102211.
[124] Germain M, Caputo F, Metcalfe S, Tosi G, Spring K, Åslund AK. et al.. Delivering the Power of Nanomedicine to Patients Today. Journal of Controlled Release. 2020;326:164–171. https://doi.org/10.1016/j.jconrel.2020.07.007.
[125] Hashemi B, Akram FA, Amirazad H, Dadashpour M, Sheervalilou M, Nasrabadi D. et al.. Emerging Importance of Nanotechnology-Based Approaches to Control the COVID-19 Pandemic; Focus on Nanomedicine Iterance in Diagnosis and Treatment of COVID-19 Patients. Journal of Drug Delivery Science and Technology. 2021;67:102967. https://doi.org/10.1016/j.jddst.2021.102967.

[126] Cai L and Zhu Y. The Challenges of Data Quality and Data Quality Assessment in the Big Data Era. Journal of Data Science. 2015;14 Article 2. https://doi.org/10.5334/dsj-2015-002.

[127] Wang RY and Strong DM. Beyond Accuracy: What Data Quality Means to Data Consumers. Journal of Management Information Systems. 1996;12(4):5–33.

[128] Crosby PB. Quality Is Free: The Art of Making Quality Certain. 1988; New York: McGraw-Hill.

[129] Alan FK, Sanil AP, Sacks J. et al.. Workshop Report: Affiliates Workshop on Data Quality. 2001; North Carolina: NISS.

[130] Cao JJ, Diao XC, Wang T. et al.. Research on Some Basic Problems in Data Quality Control. Microcomputer Infection. 2010;09:12–14.

[131] Zong W and Wu F. The Challenge of Data Quality in the Big Data Age. Journal of Xi'an Jiaotong University (Social Sciences). 2013;33(5):38–43.

[132] Chinese Academy of Sciences. Data Application Environment Construction and Service. Data Quality Evaluation Method and Index System. 2009. Retrieved October 30, 2013, from http://www.csdb.cn/upload/101205/1012052021536150.pdf.

[133] Song M and Qin Z. Reviews of Foreign Studies on Data Quality Management. Journal of Infection. 2007;2:7–9.

[134] Alexander JE and Tate MA. Web Wisdom: How to Evaluate and Create Information on the Web. 1999; Mahwah, NJ: Erlbaum.

[135] Katerattanakul P and Siau K. Measuring Information Quality of Web Sites: Development of an Instrument. In Proceedings of the 20th International Conference on Information Systems Security. 1999; (pp. 279–285). North Carolina: ACM.

[136] Shanks G and Corbitt B. Understanding Data Quality: Social and Cultural Aspects. Proceedings of the 10th Australasian Conference on Information Systems Security. 1999; (pp. 785–797). Wellington: MCB University Press Ltd.: .

[137] Knight S and Burn J. Developing a Framework for Assessing Information Quality on the World Wide Web. Journal of Information Science. 2005;18:159–171.

[138] Costa LG, Pellacani C and Guizzetti M. Chapter 14 – In Vitro and Alternative Approaches to Developmental Neurotoxicity. In Reproductive and Developmental Toxicology (2nd ed.). Gupta RC (Ed.). 2017; (pp. 241–253). Cambridge, MA, USA: Academic Press.

[139] Teleanu D, Chircov C, Grumezescu A, Volceanov A and Teleanu R. Impact of Nanoparticles on Brain Health: An Up-to-Date Overview. Journal of Clinical Medicine. 2018;7:490. https://doi.org/10.3390/jcm7040490.

[140] Maurizi L, Papa A-L, Boudon J, Sudhakaran S, Pruvot B, Vandroux D, Chluba J, Lizard G and Millot N. Toxicological Risk Assessment of Emerging Nanomaterials: Cytotoxicity, Cellular Uptake, Effects on Biogenesis and Cell Organelle Activity, Acute Toxicity, and Biodistribution of Oxide Nanoparticle. In Unraveling the Safety Profile of Nanoscale Particles and Materials, Gomes AC and Sárria MP, (Eds.), (pp. **2018**). London, UK: IntechOpen.

[141] Banik BL, Fattahi P and Brown JL. Polymeric Nanoparticles: The Future of Nanomedicine. Wiley Interdisciplinary Reviews: Nanomedicine *and* Nanobiotechnology. 2016;8:271–299. https://doi.org/10.1002/wnan.1347.

[142] Crucho CIC and Barros MT. Polymeric Nanoparticles: A Study on the Preparation Variables and Characterization Methods. Mater. Sci. Eng. C. 2017;80:771–784. https://doi.org/10.1016/j.msec.2017.06.025.

[143] Singh N, Joshi A, Toor AP and Verma G. Chapter 27 – Drug Delivery: Advancements and Challenges. In Nanostructures for Drug Delivery, (Andronescu E and Grumezescu AM). (Eds.), 2017; (pp. 865–886). Amsterdam, The Netherlands: Elsevier.

[144] Burger B, Maffettone PM, Gusev VV and Aitchison CM. A Mobile Robotic Chemist. Nature. 2020;583 (7815):237–241. https://doi.org/10.1038/s41586-020-2442-2.

[145] Akintola AA. AI-Driven Monitoring Systems for Bioremediation: Real-Time Data Analysis and Predictive Modelling. World J. Adv. Res. Rev. 2024;24(01):788–803. https://doi.org/10.30574/wjarr.2024.24.1.3099.

[146] Ji Z, Guo W, Wood EL, Liu J, Sakkiah S, Xu X, Patterson TA and Hong H. Machine Learning Models for Predicting Cytotoxicity of Nanomaterials. Chem. Res. Toxicol. 2022;35(2):125–139. https://doi.org/10.1021/acs.chemrestox.1c00310.

[147] Zhang Y, Poon K, Masonsong GSP, Ramaswamy Y and Singh G. Sustainable Nanomaterials for Biomedical Applications. Pharmaceutics. 2023;15(3):922. https://doi.org/10.3390/pharmaceutics15030922.

[148] Samir A, Ashour FH, Hakim AAA. et al.. Recent Advances in Biodegradable Polymers for Sustainable Applications. Npj Mater. Degrad. 2022;6:68. https://doi.org/10.1038/s41529-022-00277-7.

[149] Reis WF, Barreto CG and Capelari MGM. Circular Economy and Solid Waste Management: Connections from a Bibliometric Analysis. Sustainability. 2023;15(22):15715. https://doi.org/10.3390/su152215715.

[150] Negrete-Cardoso M, Rosano-Ortega G, Álvarez-Aros EL, Tavera-Cortés ME, Vega-Lebrún CA and Sánchez-Ruíz FJ. Circular Economy Strategy and Waste Management: A Bibliometric Analysis in Its Contribution to Sustainable Development, Toward a Post-COVID-19 Era. Environ. Sci. Pollut. Res. Int. 2022;29(41):61729–61746. https://doi.org/10.1007/s11356-022-18703-3.

[151] Remya RR, Julius A, Suman TY, Mohanavel V, Karthick A, Pazhanimuthu C, Samrot AV and Muhibbullah M. Role of Nanoparticles in Biodegradation and Their Importance in Environmental and Biomedical Applications. J. Nanomaterials. 2022 Article 6090846; https://doi.org/10.1155/2022/6090846.

[152] Prediger P, De Figueiredo Neves T, Camparotto NG and Rodrigues EA Green Nanomaterials for Environmental Remediation. In Handbook of Green and Sustainable Nanotechnology. Shanker U, Hussain CM and Rani M, Eds., 2023;1031–1064. Cham: Springer. https://doi.org/10.1007/978-3-031-16101-8_23.

Fatima Haider, Uswa Mansoor, Maham Naeem, and Gul Shahnaz*

Chapter 13
Innovative approaches in drug delivery systems (DDS) leveraging AI and advanced material characterization

Abstract: The convergence of artificial intelligence (AI) and advanced materials science is reshaping the landscape of drug delivery systems (DDSs). This chapter delves into the application of AI-powered computational techniques to optimize the design and performance of smart polymer-based nanocarriers. By leveraging machine learning algorithms, including supervised, unsupervised, and deep learning paradigms, researchers can predict drug–polymer interactions, optimize release profiles, and enhance targeting specificity.

Molecular dynamics simulations and perturbation theory are employed to refine the architecture of nanocarriers, particularly biodegradable polymers like polycaprolactone, for applications in oncology. The chapter explores the potential of various smart polymer configurations, such as pH-responsive and thermosensitive materials, to achieve tailored drug release. Case studies illustrate the successful integration of AI with polymer science to develop bioinspired DDS that maximize therapeutic efficacy while minimizing adverse effects.

While the potential of AI-driven DDS is immense, challenges such as data variability and regulatory hurdles persist. Addressing these limitations requires collaborative efforts among AI experts, material scientists, and clinicians. By embracing AI-powered innovations, we can pave the way for personalized medicine and revolutionize the treatment of complex diseases.

Keywords: Drug delivery systems, smart polymers, predictive modeling, virtual screening

***Corresponding author: Gul Shahnaz**, Saulat Institute of Pharmaceutical Sciences and Drug Research, Quaid-i-Azam University, Islamabad 45320, Pakistan, e-mail: gshahnaz@qau.edu.pk
Fatima Haider, Saulat Institute of Pharmaceutical Sciences and Drug Research, Quaid-i-Azam University, Islamabad 45320, Pakistan, e-mail: fatima.haider3107@gmail.com
Uswa Mansoor, Saulat Institute of Pharmaceutical Sciences and Drug Research, Quaid-i-Azam University, Islamabad 45320, Pakistan, e-mail: uswamansoorwork@gmail.com
Maham Naeem, Saulat Institute of Pharmaceutical Sciences and Drug Research, Quaid-i-Azam University, Islamabad 45320, Pakistan, e-mail: mahamnaeem.1013@gmail.com

https://doi.org/10.1515/9783111681870-013

13.1 Introduction

Imagine a world where medications know exactly when and where to act in your body, delivering treatment with unmatched precision and minimizing side effects. This is no longer the realm of science fiction but a reality made possible by modern drug delivery systems (DDSs). Innovative DDS improve the safety, effectiveness, and targeting of medications, which are essential for developing therapeutic interventions. To meet a variety of medical needs, these systems are designed to deliver active pharmaceutical ingredients (APIs) at precise rates, timings, and locations within the body. To transform medication delivery and patient care, modern DDS combine cutting-edge materials like smart polymers and nanomaterials with cutting-edge technology like artificial intelligence (AI).

Smart polymers and AI-driven modeling combine to provide customized drug release, targeting, and bioavailability solutions that have the potential to revolutionize disease management [1, 2].

13.1.1 Drug delivery systems

How do medications navigate the complexities of the human body to deliver treatment precisely where and when it's needed? The answer lies in advanced DDS, a cornerstone of modern pharmaceutical innovation. A DDS is a pharmaceutical formulation or device designed to administer an active pharmaceutical ingredient (API) into the human body with enhanced efficacy and safety, sometimes with a controlled rate, timing, and location of the therapeutic substance's release. The administration of the therapeutic product, the release of its active ingredients, and their subsequent transportation across biological membranes to the target site are all included in the drug delivery process [1].

A DDS acts as the critical interface between the patient and the drug; it may be a formulation intended for therapeutic use or a device facilitating drug delivery [1]. A distinction between a drug formulation and a medical device is crucial, as it determines the regulatory framework under which they fall, controlled by drug control agencies. The categorization of a DDS as a drug or device is determined on a case-by-case basis, reflecting the broad spectrum between these two categories [1].

DDS have evolved from traditional techniques, such as oral and transdermal formulations that can provide therapeutic effects for small compounds for up to 24 h, to sophisticated technologies that offer improved efficacy, precision, automation, and performance [2].

Currently, oral dosage forms and injections are the most widely used drug formulations. Convenience, accurate dosing, portability, non-invasiveness, and established administration schedules are just a few of the many benefits that oral dosage forms offer, though their ability to deliver larger therapeutic molecules is constrained. Parenteral injections, despite their ability to deliver macromolecules, are invasive and can be misused. These drawbacks highlight how traditional oral and parenteral formulations are unable to satisfy sophisticated therapeutic needs such as on-demand activation, broad applicability for macromolecules, and customized administration. Although not all pharmaceutical compounds require these capabilities, many do, driving significant research on novel DDS [3].

Advances in drug delivery research have aimed at enhancing oral and injectable systems while exploring alternative administration routes too, which include transdermal, ocular, pulmonary, and nasal pathways. A DDS must meet the essential criteria of safety, therapeutic efficacy, ease of administration, and manufacturing feasibility [3].

By increasing drug targeting, optimizing controlled release, and boosting the bioavailability of pharmaceutical substances, developments in DDSs have completely transformed therapeutic interventions. Higher therapeutic efficacy and fewer systemic side effects result from stimuli-responsive DDS, such as those activated by pH or temperature changes, which guarantee drug release at the right location and time [4].

By accelerating systemic drug delivery to the exact target site, modern DDS are designed with state-of-the-art technology to optimize therapeutic efficacy and minimize off-target accumulation in the body. As a result, they significantly influence how illness is managed and treated. Recent DDS offer greater advantages than conventional drug delivery techniques due to their enhanced performance, automation, precision, and efficacy. They are composed of multipurpose, biocompatible, biodegradable nanomaterials or miniature devices with high viscoelasticity and a long half-life in circulation [5].

13.1.2 Importance of smart polymers in drug delivery

Smart polymers represent a groundbreaking advancement in DDSs, seamlessly integrating environmental responsiveness with precision to revolutionize therapeutic efficacy and patient care. Smart polymers have emerged as pivotal components in DDSs, offering controlled and responsive release mechanisms that enhance therapeutic efficacy and patient compliance. Small changes in environmental conditions, such as pH, temperature, light, or the presence of particular biomolecules, can cause these polymers to experience substantial and reversible changes in their chemical or physical characteristics.

This adaptability makes them ideal candidates for developing advanced DDS tailored to specific physiological conditions. For example, regulated medication release is made possible by thermosensitive polymers, which can modify their solubility with only slight temperature changes. Similarly, pH-responsive polymers can alter their structure in reaction to changes in pH, enabling the delivery of drugs to particular body locations with precision. Smart polymers' adaptability to different stimuli enables the development of complex DDSs that can release therapeutic chemicals under controlled conditions, enhancing treatment results [6].

13.1.3 Role of AI in advancing drug delivery technologies

AI speeds up the creation of new medicines, optimizes therapeutic results, and makes it easier to construct complex DDSs by utilizing cutting-edge algorithms and machine learning (ML) models.

Designing and refining nanoparticles and other medication delivery systems has been made possible in large part by AI algorithms. By designing these systems to target particular tissues or cells, medicinal drugs can be delivered precisely where they are required, increasing efficacy and reducing unwanted effects. AI-driven models, for example, have been used to forecast how DDSs will behave in biological settings, allowing for the development of safer and more effective treatment alternatives [7]. Pharmaceutical manufacturing has been completely transformed by the incorporation of AI into 3D-printed dosage forms. This collaboration enhances treatment results by enabling the development of customized drugs based on the requirements of each patient.

AI helps to optimize these dosage forms' design and manufacturing processes, guaranteeing exact control of overdoses and drug release characteristics [8].

AI makes it possible to monitor and regulate drug release from delivery devices in real-time. AI systems can make well-informed decisions on the timing and dosage of medications by evaluating patient-specific data, including bloodstream drug concentrations and physiological characteristics. By maintaining ideal therapeutic levels, this dynamic modulation improves patient safety and treatment effectiveness [9]. AI has been used to create sophisticated medicine delivery systems in the field of chronic diseases.

To forecast the effectiveness of medication delivery systems for long-term eye disorders like glaucoma, for instance, researchers have developed ML models. By optimizing drug release characteristics, these AI-powered systems can enhance treatment results and patient adherence [10].

Drug delivery nanosystems have been designed, characterized, and manufactured using AI techniques. Researchers can forecast the ideal characteristics of nanoparticles for targeted medication delivery by applying ML algorithms, thereby improving treatment precision, and reducing systemic adverse effects [11].

13.2 Smart polymers in drug delivery

Smart polymers have become essential components of DDS, providing regulated and responsive release mechanisms that improve patient compliance and therapeutic efficacy. Slight changes in pH, temperature, light, or the presence of specific biomolecules can cause these polymers, often referred to as smart polymers, to undergo substantial and reversible changes in their physical or chemical characteristics. They are perfect candidates for creating sophisticated DDS suited to particular physiological situations because of their versatility [6].

13.2.1 Types of smart polymers in drug delivery

1. **pH-responsive polymers**: These polymers are designed to react to changes in pH in various body regions. The pH levels of the gastrointestinal tract, for example, vary from neutral to slightly alkaline in the intestines to acidic in the stomach. To ensure targeted drug release, pH-responsive polymers can be engineered to be stable in the stomach's acidic environment while breaking down or swelling in the intestines' higher pH. When it comes to safeguarding acid-labile medications or directing drug release to the intestines, this feature is especially helpful [12].
2. **Temperature-sensitive polymers:** These polymers exhibit a phase transition at specific temperatures, known as the lower critical solution temperature (LCST). The polymer is soluble below the LCST and precipitates above it. This property can be used to make injectable gels that provide a depot for long-term medication release by solidifying at body temperature. These systems minimize systemic side effects and are beneficial for localized therapy [6].
3. **Light-responsive polymers:** When exposed to particular light wavelengths, these polymers alter structurally. Because light exposure may be accurately controlled in applications like dermatological therapies or ophthalmic treatments, this characteristic makes it possible to control medication release both spatially and temporally [6].
4. **Magnetic and ultrasound-responsive polymers:** Externally triggered medication release can be achieved by incorporating magnetic nanoparticles or by creating polymers that respond to ultrasound. This technique improves the accuracy of treatment plans by offering noninvasive control over medication administration [13].

Research has focused on developing smart polymers for specific medical applications. Chemotherapeutic chemicals are released directly at the cancer site using smart polymers that react to the acidic environment of tumors. This focused strategy improves the treatment's efficacy and lowers systemic toxicity [14]. To replicate the body's natural insulin control and offer a more physiological approach to diabetes care, glucose-responsive polymers are being developed that release insulin in reaction to rising blood glucose levels [6]. As carriers for gene delivery, smart polymers shield genetic material from deterioration and make it easier for it to enter target cells. This application has potential for regenerative medicine and the treatment of genetic diseases [15].

By reacting to local environmental signals, smart polymers can be designed to release medications at specific bodily locations, such as tumor tissues or inflammatory areas. By increasing the medication concentration at the intended location and limiting exposure to healthy tissues, this focused strategy improves therapeutic efficacy and reduces side effects [15]. The pace and timing of drug release can be precisely controlled thanks to smart polymers' capacity to react to stimuli. Maintaining ideal medication concentrations within the therapeutic window, enhancing treatment results, and lowering dosage frequency all depend on this regulation [16].

Many medications, particularly biologics such as proteins and peptides, are prone to breaking down before they reach their intended location. These agents can be encapsulated by smart polymers, which shield them from early deterioration and guarantee their stability until release [17]. Patient adherence to treatment plans is enhanced by smart polymer-based DDS, which decreases the frequency of medication administration and permits noninvasive delivery methods. Injectable gels with continuous release, for instance, can replace daily injections, improving patient compliance and comfort [18].

Notwithstanding the encouraging benefits, there are still several obstacles to overcome in the creation and use of smart polymers in medication delivery. For clinical use, it is essential that smart polymers are biocompatible and nontoxic. To evaluate these materials' long-term impact on the body, thorough research is needed [15]. Large-scale production may be hampered by the intricacy of creating smart polymers. For them to be widely used, manufacturing procedures must be developed that are both affordable and repeatable [17]. It can take a long time and a lot of resources to navigate the regulatory environment for novel compounds, which calls for extensive safety and effectiveness documentation [17].

13.2.2 Case study

This study aims to advance the field of materials science with specific applications in biomedicine, focusing on the development of innovative DDS. The research sought to engineer novel "smart" materials, particularly thermoresponsive polymers like poly (*N*-isopropylacrylamide) (PNIPAM), which exhibit LCSTs near physiological levels (~32 °C). Such materials are of significant interest for applications in tissue engineering, regenerative medicine, and controlled drug delivery due to their ability to undergo reversible phase transitions in response to thermal stimuli.

The study introduced an innovative approach utilizing the magnetocaloric effect (MCE) of FeRh alloys to noninvasively modulate the properties of PNIPAM. Unlike conventional heating methods that rely on intrusive components or direct thermal inputs, this technique leverages external magnetic fields to trigger phase transitions. FeRh alloys were selected for their exceptional MCE performance near physiological temperatures and reduced cytotoxicity compared to traditional materials like gadolinium (Gd). The overarching objective was to address the challenges of developing biocompatible, magnetically triggered drug-release systems that are clinically viable.

A novel PNIPAM/FeRh composite material was engineered to explore magnetically controlled drug delivery mechanisms. The composite was fabricated using PNIPAM polymer, $Fe_{49}Rh_{51}$ alloy, and the chemotherapeutic agent doxorubicin (DOX), with ethanol as the solvent. The FeRh alloy substrate was prepared via arc melting high-purity Fe and Rh powders, followed by annealing at 1,000 °C and rapid quenching. PNIPAM was deposited onto the FeRh substrate through solvent-casting, forming a polymer layer approximately 20 µm thick. To enhance drug-loading efficiency and adhesion, laser-engraved microwells were created on the FeRh surface. These wells were subsequently filled with DOX and overlaid with a PNIPAM coating, serving as reservoirs for drug storage and enabling controlled release.

The composite demonstrated temperature-responsive behavior at the LCST of PNIPAM (32 °C), driven by MCEs induced by a 3 T magnetic field – parameters compatible with magnetic resonance imaging (MRI) systems. Advanced optical imaging and spectroscopic analyses, including UV-VIS and Raman spectroscopy, validated the controlled release of DOX. The biocompatibility of the PNIPAM/FeRh scaffold was evaluated using mouse embryonic fibroblasts, which exhibited effective adhesion, spreading, and proliferation over 72 h. Cells were observed to localize both on the polymer's surface and within the laser-engraved wells, underscoring the material's cytocompatibility.

These findings underscore the potential of the PNIPAM/FeRh composite for developing magnetically activated DDSs and advanced biomedical implants [19].

13.3 AI-driven predictive modeling

Predictive modeling in materials science, is the use of computational techniques, particularly ML. AI is employed to predict the properties of materials based on their structural and compositional data. It is critical for rational design strategies to improve the understanding of the properties of materials and biomacromolecules. To increase efficacy and minimize adverse effects, by simulating and predicting the behavior of drug delivery devices, we can utilize sophisticated computational approaches. Characteristics like droplet size and drug loading efficiency in self-microemulsifying DDSs (SMEDDSs) have been optimized using ML methods such as linear regression. By analyzing intricate relationships between formulation elements, these models can produce drug delivery methods that are more successful [20].

Customized therapy regimens that adapt to each patient's specific physiological circumstances are created through integration with real-time data and medical imaging. When combined with in vitro characterization, in silico modeling provides a reliable framework for predicting DDS formulation performance in vivo, thereby reducing the need for animal testing. By enhancing understanding of how formulations behave in human physiology, this approach improves treatment outcomes [21].

13.3.1 Machine learning techniques in predictive modeling

In ML, the focus is on making an accurate prediction, or the "what," rather than the "how." For instance, in image recognition for cancer diagnosis, the relationship between the individual features (pixels) and the result is relatively irrelevant if the prediction is accurate. Algorithms are mathematical procedures that describe the relationships between variables [22]. They are of two types:

i) Easily interpretable, that is, Auditable Algorithms
ii) Not easily interpretable, that is, black boxes.

A comparison between Auditable Algorithms and Black Boxes is given in the table 13.1.

Table 13.1: Comparison between auditable algorithms and black boxes [23].

Auditable algorithms	Black boxes
Simple models, better for interpretation.	Complex models, better for complex data.
Includes multiple regression and decision trees.	Includes neural networks and certain support vector machines.
Can be used on consumer PCs, so they are computationally cheap.	Computationally expensive, as they require an increase in processor time to build models.
Interpretation is easier due to the linear relationship between predictors and outcomes.	Nonlinear relationships make interpretation difficult.
More like statistical techniques.	Few commonalities exist with statistical techniques.

Most ML techniques fall into one of two categories: supervised and unsupervised.

13.3.1.1 Supervised learning

In supervised ML, a model is trained for a variety of input ranges, also known as features, and an association is made on the basis of the outcome. It establishes a relationship between independent and dependent variables. For example, relating the weight and smoking status of a person to greater chances of getting diabetes. After training the algorithm, the model can predict the outcome in new data as well.

There are different types of predictions made by supervised learning (SL) models, including:

i) **Continuous**, offering a range, for example, a number from 0 to 100. Also called a regression algorithm [24], as it relates to the prediction of continuous values. Regression algorithm in ML predicts a dose for chemotherapy that is tolerable or can be used to predict an individual's life expectancy.
ii) **Discrete** classifies the output in a yes-or-no format, for example, positive or negative results; cancer is either benign or malignant, based on which the diagnosis is predicted. It is also referred to as a classification algorithm.

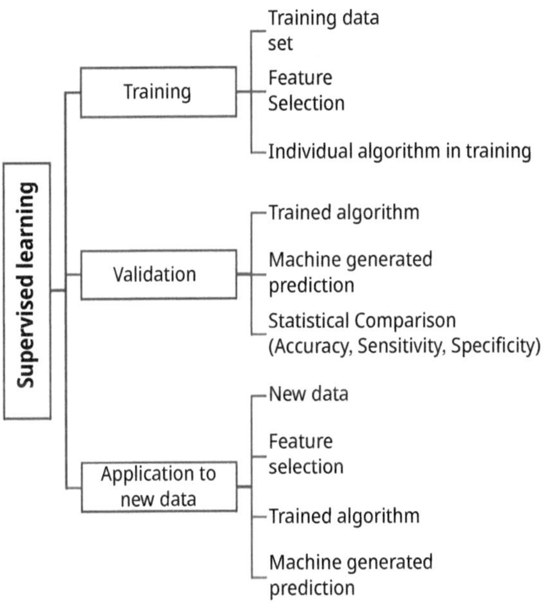

Figure 13.1: Flowchart of supervised learning algorithm training and output prediction.

SL has some limitations, such as the input data used to train the algorithm might not fit well with new data. However, this risk of overfitting can be eradicated by employing various techniques. For example, splitting a dataset into training and testing datasets. This randomization allows the algorithm to associate the characteristics with outcomes, and as a result, the algorithm also generalizes well to new datasets [23].

A flow chart showing steps involved in output prediction by supervised learning is shown in figure 13.1.

13.3.1.2 Unsupervised learning

Unsupervised learning, as the name implies, does not require supervision; that is, user input data is not necessary. There is no predetermined result; instead, patterns are created by algorithms. As a result, this method is exploratory and aids in identifying unknown patterns or clusters in the dataset.

Because they are often used to decrease the number of features or dimensions in the analysis, unsupervised learning techniques are also known as **dimension reduction techniques**. The algorithm receives raw data and clusters it into a collection of related data points. The algorithm excludes data if it is insufficient for grouping. The workflow for unsupervised learning is represented in figure 13.2. Principal component analysis (PCA), t-distributed stochastic neighbor embedding (t-SNE), and latent Dirichlet analysis are the procedures used in these methods [25].

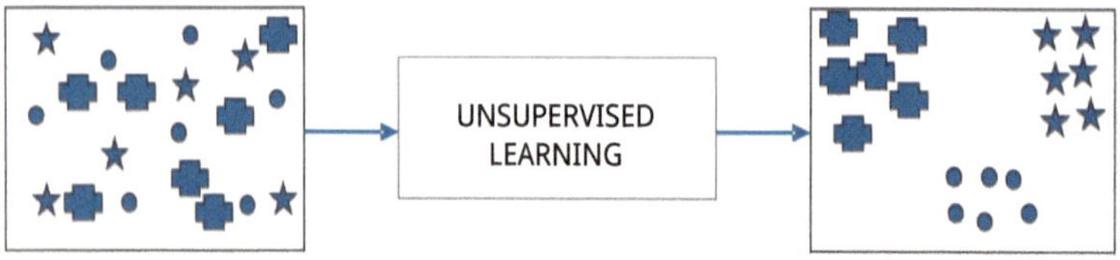

Figure 13.2: Workflow of unsupervised learning.

Supervised and unsupervised learning techniques are further classified into three fundamental ML techniques.

i. **Linear regression**: An algorithm for SL. A linear relationship is established by fitting the data along a line between the features (an independent variable) and the target (a dependent variable) [26].
ii. **K means clustering:** An algorithm for unsupervised learning. The main objective is to find groups in the data that are represented by the variable K. K applied to unlabeled data.
iii. **Decision trees**: Hierarchical structure algorithm for nonparametric SL that may be applied to both regression and classification problems.

It includes key components such as the **root node** (representing the entire dataset), **decision nodes** (where splits occur), **branches** (connecting nodes), and **leaf nodes** (final outcomes).

The tree-building process involves splitting the dataset into subsets and reducing, to enhance performance and prevent overfitting. Each path from the root to a leaf represents a unique decision rule. Decision trees are appreciated for their simplicity, effectiveness, and interpretability in decision-making [27]. A simple structure of decision tree is represented in figure 13.3.

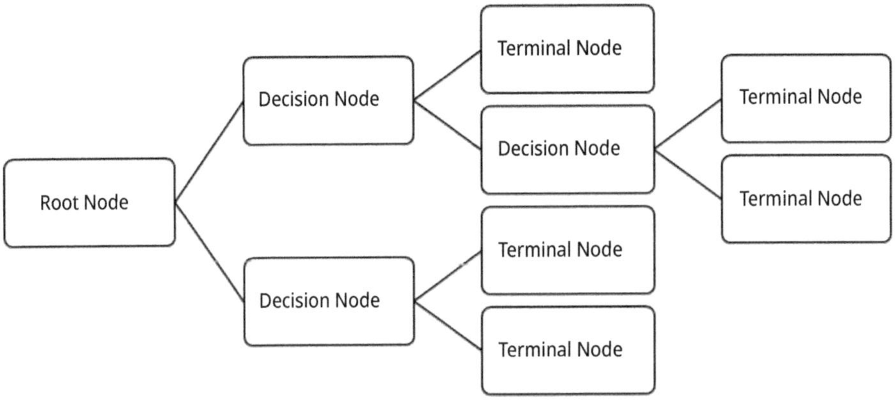

Figure 13.3: Basic structure of decision trees.

13.3.2 Supervised algorithms for multiple inputs dataset

Other supervised ML techniques for multi-input datasets include:

13.3.2.1 Logistic regression

These are close to multivariate logistic regression. It involves the utilization of regularization functions that reduce the magnitudes of features. A study demonstrated the use of logistic regression in identifying prostate cancer. The technique employed was, desorption electrospray ionization mass spectrometric imaging of small metabolites and lipids.

13.3.2.2 Support vector machine

Provide accurate predictions for scenarios when there is a nonlinear relationship between features and the outcomes. It employs a kernel trick model that separates two groups by hyperplane or a decision boundary [28].

Artificial neural network (ANNs): ANNs have a complex architecture because these are somewhat modeled on the neuronal structure of the mammalian cortex. These networks make it possible to combine the study of artificial intelligence and biology, and they offer a rich environment for simulating human memory and learning [29]. Recurrent and convolutional networks are examples of specialty neural networks that have been added to ANNs and have produced remarkable results on a variety of tasks. ANNs are prone to over-fitting due to their highly parameterized designs. They can perform better if regularization techniques are used [23]. Figure 13.4 illustrates the integration of Graph Neural Networks and Multitask Learning as 'Fast' and 'Accurate' respectively.

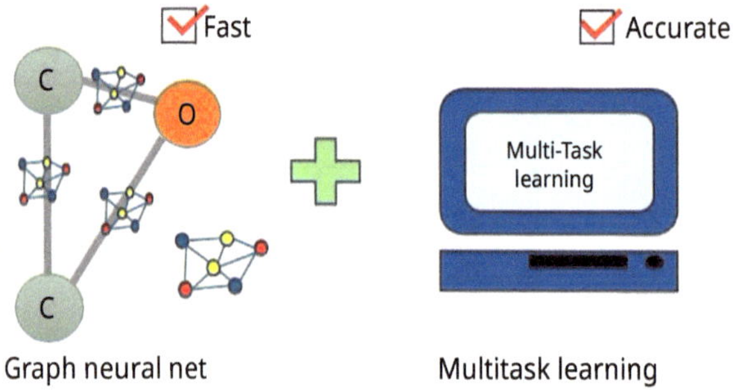

Figure 13.4: Graph neural net and multitask learning.

13.3.3 Neural networks and deep learning

Neural networks are built with several neurons that reflect the information extracted from each feature in the dataset. As information goes between 'neurons' or nodes, it is multiplied by the neuron's weight and changed by an activation function.

13.3.3.1 Deep neural networks (DNNs)

Deep neural networks (DNNs) have many hidden layers. The term "deep" in DNNs refers to the depth of the network, which is determined by the number of hidden layers. Usually, it contains three hidden layers. Deep learning utilizes DNNs and produces impressive results when used for complex tasks, such as image recognition and computer-assisted diagnosis of melanoma. Deeper networks require additional weights and biases to be optimized during training because of the growing number of connections between nodes. The successful training of DNNs has been made possible by improvements in hardware and optimization methods, even though this may present difficulties in terms of computing complexity and training time [31].

13.3.3.2 Convolutional neural network (CNN)

Convolutional neural network (CNN) is a deep learning approach that has been widely used in computer vision tasks, including picture categorization and object recognition.

Despite its impressive achievements, deep learning still has limitations when compared to human intelligence [29]. Complex techniques are required to improve model performance and interpretability in the face of issues like overfitting and vanishing gradients.

An analysis on various ML algorithms for ML model development is provided in table 13.2. Lower MSE and higher R^2 indicate better model performance [32].

Table 13.2: Properties of machine learning algorithms [32].

Algorithm	Size of dataset	Mean squared error (MSE)	R^2	Avg prediction time
ANNs	1,000	1.89	0.92	0.015
SVM	800	2.15	0.88	0.010
Decision trees	600	2.50	0.82	0.003

13.3.4 Deep learning in polymer behavior prediction and drug delivery

Drug-phospholipid complexing is a formulation approach used to enhance the bioavailability of active pharmaceutical ingredients (APIs). However, predicting the candidacy of lipidic drug candidates in in-vitro experiments is a time-consuming and not very cost-effective technique. A study predicted the production of drug-phospholipid complexes using ML algorithms. However, the decrease in test performance caused by the small size of the training data was not adequately addressed.

In 2023, Yoo et al. [33] presented a novel deep learning-based prediction algorithm to address this issue. In order to improve prediction performance, it used PCA and variational autoencoder (VAE) approaches. And to accurately depict the intricate link between medications and lipid molecules, the model makes use of a multilayer one-dimensional CNN with a skip connection. With a high sensitivity of 97.5%, the suggested model was able to identify interesting candidate medications, underscoring the need of creating effective prediction models [33].

The efficacy of ML in handling intricate predictive tasks pertaining to polymer behavior and drug delivery in pharmaceutical formulations is demonstrated by another study, which shows that different neural network architectures can precisely predict drug release profiles from polymer-based long-acting injectables [34].

13.4 Virtual screening techniques

Virtual screening techniques (VSTs) involve the use of computational tools and approaches to screen a large compound library to identify and predict the biological activity and affinity of a drug molecule for a specific target [24].

The history of computer-aided drug design dates back to the 1950s, when computational approaches were first used to generate NMR simulations, marking the initial steps in the field's evolution. Multiple modeling tools were used to solve complex regression equations; however, the intricate spatial structure of molecules limited the scope of structural activity relationships. In the 1970s, graphical programs representing HOMO, LUMO, and bipolar moments of molecules were developed. These advancements overcame the barriers associated with in silico prediction of molecular properties with their theoretical results. Similarly, powerful genetic and biochemical screening tools for the prediction of desired active compounds from various extracts have been developed in recent decades [24, 35].

AI-driven virtual screening expedites the discovery of novel two-dimensional materials with tailored physical and chemical properties, offering significant potential for energy conversion and storage applications. ML-assisted virtual screening facilitates the identification of promising low-potential candidate molecules for organic redox-active polymers in battery applications. The incorporation of information technology has revolutionized drug discovery, empowering researchers to efficiently screen extensive compound libraries and make precise predictions about their biological functions [36].

13.4.1 AI algorithms in virtual screening

ML, deep learning, and other computer techniques are used by AI-based virtual displays to evaluate pharmacological parameters and biological data. Artificial intelligence is defined as a superset of ML, which includes supervised, unsupervised, and reinforcement learning (RL). Furthermore, deep learning, a subset of ML, is widely used in drug research and development. The algorithms used in drug design and discovery include ANNs, deep neural networks, support vector machines, classification and regression, generative adversarial networks (GANs), symbolic learning, and meta-learning.

A few of the drug discovery processes that have used these algorithms are structure and ligand-based virtual screening, toxicity prediction, drug monitoring and release, pharmacophore modeling, and quantitative structure–activity relationship, etc. Furthermore, newly created modeling algorithms were crucially supported by innovative data mining, curation, and management strategies [37].

13.4.1.1 Genetic algorithms

Genetic algorithms, a nature-mimicking optimization method, are stochastic global optimization approaches based on evolutionary biology. This concept is analogous to the genotype-phenotype relationship in evolutionary biology, which relates to the chemical issue of comprehending the structure-property relationship. Together with theoretical techniques like force-fields and high-throughput first-principles methods, GA has been widely applied in chemistry and materials science. Consequently, the approach enables precise and automated structure determination.

Plane–cut–splice (PCS) crossover, rattle mutation (ratt.), and permutation mutation (perm.) are some of the operators implemented in GAMaterial for atomic clusters and interfaces [38].

13.4.1.2 GA algorithm in polymer design

The GA algorithm can help with polymer design by quickly and consistently producing novel polymer candidates for specified applications. It can be used to generate polymer candidates depending on user-defined properties like glass transition temperature (T_g) and bandgap [39]. The GA algorithm uses crossover and mutation operations to change the sequence of chemical building blocks in polymers and generate new progeny. The fitness of the offspring is assessed using ML models, which yield rapid property estimations. Only the best-performing kids are kept as parent polymers, and the procedure is repeated until sufficient polymer candidates with the requisite qualities are discovered [36].

13.4.1.3 Genetic algorithm software and polymer development

To clarify the architectures of atomic clusters, doped clusters or materials, and atomic clusters on surfaces, Lourenço et al. offer a new genetic algorithm program called GA-Material, which is written in Python3.x and conducts global searches.

The following is the workflow for GAMaterial's implementation of the genetic algorithm method for automatic polymer discovery and development:

1. **Initial structure generation:** DFT (density functional theory) or DFTB(density-functional tight binding) methods are used to build a set of random polymer structures, which are then locally optimized.
2. **Progenitor structure selection**: After the population has been established, progenitor polymer structures are chosen based on a proportion of the lowest-energy configurations.
3. **Crossover and mutation:** The selected progenitors are permitted to undergo crossover and/or mutation to produce offspring polymer structures. To assure dissimilarity between offspring structures and current structures, a filter operator can be utilized based on parameters such as relative cumulative difference between polymer isomers and maximum interatomic distance difference.
4. **Random selection and ML integration**: Next, a subset of structures is randomly chosen from the newly formed progeny. Alternatively, a ML regression model can be applied to the offspring structures to forecast total energies, allowing for the

selection of the most promising structures to generate the next population following local optimization.

5. **Local optimization and iteration:** The chosen structures are then locally optimized. An elitist option can be used to ensure that the next population always has the lowest-energy local minimum structure [38]. A simple workflow for GA Material in polymer discovery is shown in figure 13.5.

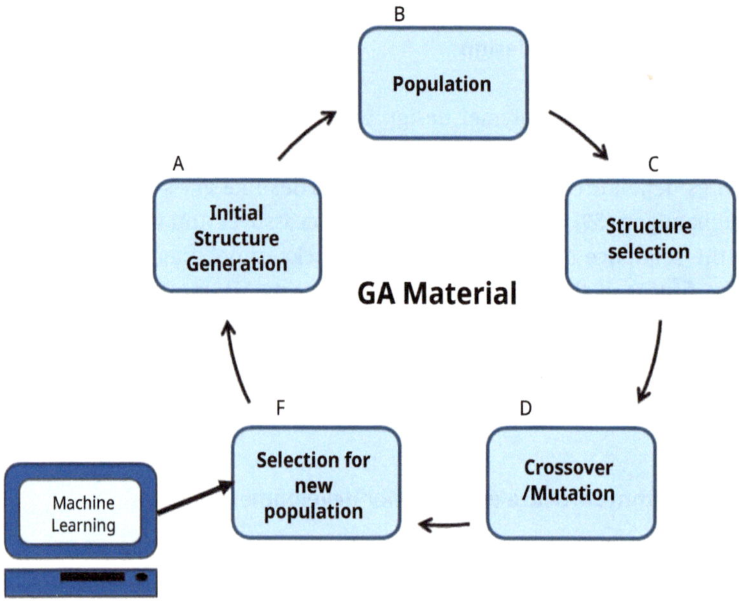

Figure 13.5: Workflow for GAMaterial in polymer discovery.

13.4.2 Reinforcement learning in polymerization control

RL is a subfield of ML in which machines gradually learn control behaviors through self-exploration of their surroundings. Deep RL (DRL) is a versatile solution to control problems. The fundamental advantage of the RL algorithm is that it learns to optimize control policies by exploring the environment, regardless of the system's linearity or multivariability.

Polymerization reactions are difficult to control because the product quality and various polymeric architectures are dictated by molecular mass distribution (MMD). DRL approaches, such as deep deterministic policy gradient (DDPG), can handle continuous control activities, such as altering monomer and initiator flow rates in real time. The controller learns from historical data and adapts its tactics in response to performance feedback, allowing it to efficiently deal with polymerization process fluctuations [40].

13.4.3 Smart polymer screening

Smart polymer screening is utilized to identify and optimize polymer formulations rapidly by predicting their properties using advanced computational techniques, including RL. Similar to its role in polymerization control, RL can adaptively refine screening processes based on real-time feedback regarding polymer characteristics.

Repeating units of polymers act as chemo-structural entities that help in the screening of polymers from libraries of candidate polymers. ML exploits this feature, that is, the polymer repeat unit, for extraction and screening. Few approaches utilized for this purpose are neural networks, multitask learning,, and other deep learning techniques.

Neural network is one of the approaches utilized in ML for natural language processing. In this technique, material is represented either in the form of a simplified molecular input line entry system (SMILES) or as graphs.

The main concept of multitask learning involves training a model to learn various target properties simultaneously. Hence, the risk of overfitting one target property is ruled out and overall performance is improved [30].

13.4.4 Case study: screening of polymeric-nanoparticle system for treatment of cancer

Inadequate loading and premature drug release are the key limitations of drug delivery by nanoparticles for tumors. Both properties are essential for cancer treatment. To combat cytotoxicity, nanoparticles are exploited for targeted drug delivery; however, reduced retention time poses toxicity threats.

A study conducted by Animasawun and his colleagues in 2020 employed reverse engineering techniques. They investigated drug–polymer interactions between etoposide, etoposide phosphate, and teniposide, as well as nine polymers: PGA, PGA–PEG, PGA40%C8, PGA40%C18, PGA85%C18, PGA30%CH, PGA50%TRY, PGA50%Phe, and PGA50%SA. Polymers employed for nanoparticle preparation were PGA50%TRY for teniposide and PGA85%C18 for etoposide and etoposide phosphate. The drug loading obtained from the nanoprecipitation technique was 13% for teniposide, 5% for etoposide, and less than 5% for etoposide phosphate.

The study demonstrated that the selection of polymers for each drug produced nanoparticles with improved properties. For screening, three different approaches were employed, including high-throughput drug release experiments, FTIR spectroscopy, and contact angle measurements. Contact angle measurements and FTIR explained the results of drug release using solid dispersion films. Table 13.3 provides the percentage of drug released in 24h for different drug-polymer films and drug-loaded nanoparticles. The major factor that affected the drug release behavior of polymers was the polymer–drug interaction. This interaction resulted in peak shifts due to side-chain rearrangement.

Table 13.3: Percentage of drug released in 24 h (drug–polymer films and drug-loaded nanoparticles) [41].

System	% drug released in 24 h
Teniposide–PGA50%TRY film	29
Teniposide–PGA50%TRY nanoparticles	35
Etoposide–PGA85%C18 film	64
Etoposide–PGA85%C18 nanoparticles	22
Etoposide phosphate–PGA85%C18 film	23.8
Etoposide phosphate–PGA85%C18 nanoparticles	16

13.4.4.1 Results

The three polymer–drug combinations employed in this study demonstrated that the formulation with the best drug loading had the poorest drug retention, whereas the formulation with the lowest drug loading had the best drug retention. This study showed that drug–polymer interactions should be considered when creating polymer nanoparticle formulations. In conclusion, screening for polymer interactions can help in choosing the right polymer-drug matches to increase drug loading and lower drug release rates [41].

13.4.5 Case study: computational technique for enhancing transdermal delivery of peptides

Skin-penetrating peptides (SPPs) are becoming increasingly popular as a non-invasive method of transdermal medication delivery. However, it is time-consuming to identify SPP sequences through experimental screening of peptide libraries. According to recent research, SPPs need to exhibit an affinity for both the medication of interest and skin keratin to function as effective enhancers.

To find SPPs that could improve transdermal CsA delivery, Menegatti and colleagues in 2016 developed a computational procedure. They created and screened virtual libraries of disulfide-cyclic peptides against keratin and cyclosporine A (CsA). The chosen sequences were evaluated experimentally and were found to bind both CsA and keratin and enhance transdermal permeation of CsA. This was determined by mass spectrometry and affinity chromatography. Four heptameric sequences stood out as top candidates. An octameric peptide produced noticeably higher delivery of CsA than heptameric SPPs. Incubation with skin keratinocytes further confirmed the chosen sequences' safety profile. This approach proved to be efficient for delivering the required ingredients for medicinal and cosmeceutical purposes [39].

13.5 Advanced characterization of nanomaterials

Nanotechnology represents a seminal technological advancement of the twenty-first century. Nanoparticles find applications across diverse domains such as medicine, materials science, agriculture, and environmental engineering, facilitating the achievement of substantial outcomes within these disciplines. Notwithstanding the considerable progress in this technological sphere, a substantial disparity in the understanding of the surfaces of these nanoparticles persists. As a result, defining these particles is essential for shedding light on their traits, functions, and impacts. The most critical attributes encompass dimensions, morphology, elemental composition, and surface charge density, among others.

Various strategies are applied in pursuit of this aim, highlighting the importance of spectroscopic, microscopic, and zetapotential assessments of the nanoparticles. The primary aim is to formulate an appropriate dosage form that maximizes the potential benefits derived from these particles [42].

Nanomaterials exhibit interactions with biological macromolecules. The degree of such interactions is contingent upon the physicochemical characteristics of these nanoparticles. These interactions exert a considerable influence on the efficacy of targeted delivery, bioimaging, and diagnostic applications, all contributing to favorable outcomes. Nonetheless, the surface charge of nanoparticles can induce conformational alterations or perturbations in membranes or proteins, which are generally deemed unfavorable. As the discipline of nanomaterial synthesis transitions from a predominantly empirical approach to a more predictable framework, it becomes feasible to achieve anisotropic growth across various facets when the capping ligand selectively engages with a specific facet of the nanocrystal, thereby facilitating shape control during nanoparticle synthesis.

This advancement is facilitated by the capacity to characterize the capping ligands and monitor the surface composition of the nanoparticles, which is essential for elucidating the roles of ligands in the modulation of nanoparticle growth [43]. As a result, enhanced characterization of nanomaterials helps establish structure–property relationships and also aids in the generation of design recommendations, boosting the benefit–risk ratio [44].

13.5.1 AI-enhanced techniques for characterization

By leveraging advanced algorithms, one can enhance the accuracy and efficiency of techniques for the characterization of nanomaterials. Various methodologies are employed for increasing the comprehension of nanomaterial applications, which are described further.

13.5.1.1 Image processing

AI-enhanced techniques for nanomaterial characterization include generative models, graph neural networks, transfer learning, adversarial training, attention mechanisms, and federated learning.

Generative models and graph neural networks are used for image processing techniques such as **scanning electron microscopy (SEM), transmission electron microscopy (TEM), X-ray diffraction, and atomic force microscopy** [45].

Deep learning algorithms are based on neural networks that provide real-time identification and characterization of 2D materials. These extract complex graphical features from imaging data to predict physical properties effectively. Recently, denoising frameworks that need no supervision have been developed. These algorithms enhance the clarity of nanoparticle surface dynamics. And the time for resolution through electron microscopy for atomic resolution observations of nanomaterials, is decreased down to 10 ms. This allows better surface characterization [46].

13.5.1.2 Optimization of nanomaterial synthesis

High-throughput screening (HTS) is one of the traditional methods for the screening of candidate drugs. Newer AI optimization methods are shown to outperform these traditional techniques by enhancing chemical material discovery, efficiency, and precision. They are utilized in nanocrystal synthesis, producing better-quality products with limited resource utilization and less carbon emissions [47].

13.5.1.3 Toxicological assessment

The identification of critical parameters for assessing the safety of nanomaterials is facilitated by ML techniques. Decision tree algorithms are employed to analyze and compare datasets, along with the integration of text and data mining in nanotoxicology. Some frameworks, such as ISA-TAB Nano, are utilized for systemic data collection. It facilitates insights into structure–toxicity relationships and predictive modeling [48].

13.5.2 Application of AI in spectroscopy, microscopy, and rheology of smart polymers and nanomaterials

AI algorithms facilitate the design and simulation of smart polymers, such as LCST polymers with diverse stimulus responses, optimizing their properties and biodegradability for biomedical applications [49]. ML methods can predict the behavior of shape memory polymers (SMPs). The approach to describe SMPs combines video data analysis and predictive modeling. As a result, we can achieve greater accuracy, specificity, and sensitivity in predicting their behavior [50].

13.5.2.1 AI-assisted spectroscopy

Artificial intelligence, particularly neural networks, improves the accuracy and efficiency of spectral interpretation. Studies have shown that ANNs may efficiently measure polymer-based cocrystal formulations using Raman and FTIR spectroscopy. By reaching low root mean square error (RMSE) values, these indicate strong predictive accuracy. In 2018, Barmpalexis et al. investigated the presence of carbamazepine–nicotinamide (CBZ/NIC) and ibuprofen–nicotinamide (IBU/NIC) cocrystals in Soluplus® formulations. Raman and attenuated total reflectance Fourier-transform infrared spectroscopy (ATR-FTIR) were employed for this purpose.

The spectrum peaks were demonstrated using feed-forward, back-propagation ANNs, and partial least squares regression analysis. They concluded that incorporating AI-based model techniques into FTIR and Raman studies increased the root mean squared error of prediction (RMSEp) [51].

AI is used in massively parallel spectroscopy (MPS), which allows the observation of bio conjugated nanoparticles. MPS analyzes complicated biological matrices without increasing background noise, making it useful in real-world applications with a detection limit of 1.6 fmol/L.. ML models have been developed to predict the mechanical performance of polymer nanocomposites reinforced with silver-coated gold nanoparticles (AgC NPs) using tensile test data, indicating ideal compositions for enhanced characteristics [52].

13.5.2.2 AI in microscopy of smart polymers and nanomaterials

It is challenging to characterize the surface dynamics of nanoparticles at atomic precision since it requires high temporal and spatial resolution. Traditional ultrafast TEM has a nanoscale spatial resolution limit, despite its ability to achieve picosecond temporal resolution. AI-driven unsupervised denoising frameworks allow for clearer visualization of metal nanoparticle surfaces in TEM. This improves spatiotemporal characterization, achieving time resolutions down to 10 ms while minimizing noise and electron damage. This advancement aids in understanding atomic-level structural dynamics in materials and regulating diffusion, reactivity, and catalysis functionalities [46].

Researchers can effectively visualize nanoscale morphologies by converting SEM pictures into feature representations using a deep learning-based approach [53]. In order to improve nanoparticle categorization and segmentation, GANs produce synthetic SEM images that may be used. These are employed to train CNNs, expanding the amount of training data available for deep learning models [54]. AI pipelines enhance the identification and classification of DNA molecules under electron microscopy (EM). These specifically employ gold nanoparticle-tagged peptides for visualization, highlighting the potential for AI in biological applications of microscopy [55].

13.5.2.3 AI in rheology for material behavior prediction

The storage and loss moduli of 3D-printed polyacrylamide hydrogels have been successfully predicted using deep learning models, such as multilayer perceptrons, which achieve an R^2 value of 0.89. This makes it possible to accurately map material behavior from a small number of input factors [56]. The dynamic behavior of magnetorheological elastomers in tension-compression modes is predicted using ANNs. With R^2 values above 0.97, this demonstrates how well ANNs can represent intricate nonlinear interactions between input parameters and material stress [75].

13.5.3 Integration of AI for data analysis in polymer nanocomposites

The organic matrix materials, known as polymer nanocomposites, contain nanoparticles with at least one dimension smaller than 100 nm [57]. To forecast the characteristics of nanocomposite materials, researchers have developed various numerical techniques, such as multiscale simulations and continuum mechanical approaches [58].

An open-source, data-driven, web-based platform called NanoMine has been created in the field of nanocomposite technology. Composition, processing techniques, microstructural pictures, and measurable properties are among the experimental and computational data on nanocomposites that are archived [59].

Wang et al. performed an experiment demonstrating the integration of deep learning to predict the mechanical properties of nanocomposites. The results demonstrated that the suggested deep learning model increases the accuracy of predicting the rubbery modulus by **34.2%**, the glass modulus by **45.2%**, and the tan δ peak by **19.7%** [60].

13.5.4 AI tools providing deeper insights

Drug development is being greatly improved by AI tools, which offer deeper insights through sophisticated methods such as PCA and clustering.

These are ML techniques that are employed to classify polymer morphologies, aiding in the design of multicomponent systems for applications like drug delivery. By reducing dimensionality in large datasets, the PCA technique makes it possible to visualize and comprehend molecular interactions and therapeutic efficacy more clearly. AI identifies biomarkers, therapeutic targets, and integrates multi-omics data [61].

Similar biological data can be grouped by AI algorithms using the clustering technique, which helps identify targets related to diseases and possible treatment candidates.

13.5.6 Case study: AI-aided polymeric structure optimization for enhanced gene delivery

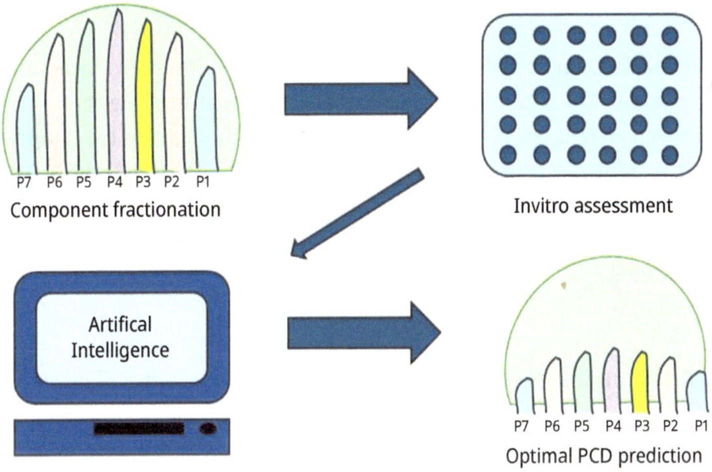

Figure 13.6: Schematic illustration of PCD prediction by XGBoost.

Development of gene therapy vectors, especially polymeric vectors, has always been a challenge due to a lack of understanding of their internal structure. A study by Li et al. in 2023 investigated the impact of polymeric structure on gene transfection efficiency. They utilized a well-known polymeric vector, poly(β-amino ester) (PAE), for experimentation. AI was employed to understand the relationship between polymer component distribution (PCD) and gene delivery. The study examined the characteristics of individual polymeric components and their combinations to understand how they exert synergistic effects on gene delivery.

13.5.6.1 Results

The effect of PCD on transfection performance was investigated, and AI was incorporated. PCD prediction by XG Boost is illustrated in figure 13.6. AI adjusted the PCDs, targeting high transfection efficiency, and further optimized the composition of polymers. A library of polymers (9×10^5) with different PCDs was computer-generated. Next, a trained AI XGBoost model was employed to identify the optimal polymers and evaluate the transfection efficiency. In vitro experiments validated the predictions made by AI. In vitro analysis showed that the transfection efficiency of the PAE-B1-based polyplex was roughly 1.5 times that of Lipo3000 and 2 times that of jetPEI. This analysis served as the basis for the development of several highly effective polymer vectors [62].

13.6 Challenges and limitations

Despite having several advantages over the typical ways of DDS design, AI-integrated drug designing faces some shortcomings and practical challenges limiting its further usage, and using modern AI technology for better research. In this part of the chapter, we will have a deep insight into the current limitations in AI integration with DDS, and we will consider how AI-aided predictive modeling and VSTs are currently facing real-world setbacks. Furthermore, the ethical parameters and considerations accounted for AI applications will also be discussed.

13.6.1 Data quality and accuracy

The fundamentals of AI-integrated designing are the availability and access to accurate and reliable datasets with up-to-date, valid information accessible to researchers, as well as the availability of reliable resources that are detailed for best results.

The basic consideration to be focused on in developing DDS designs using AI and the future innovation of DDS is the lack of availability of current literature with modern narratives and the variation in current available datasets, which lack transparency in the available literature. Literature is available in the form of datasets, statistical sets, or raw unstructured information, which is being integrated by modern AI tools. These data arrays make all the necessary information and knowledge needed for the future advancement and successful implementation of any new or older research and practical implementation of ideas being executed for the most reliable results and in this case (AI-based integration), the bases of integration are remodeling the datasets and data collections [5].

13.6.2 Data scarcity and limited availability

As mentioned previously, the basis of truthful AI-based integration is high-quality, valid, and comprehensive datasets, and these narratives are essential for training AI in order to get elevations in DDS and ML models. However, access to such data can be limited or restricted to preexisting narrow figures, particularly in rare disease conditions or new therapeutic areas that are still underdeveloped or require special considerations [63].

Accurate predictions in developing delivery systems or remodeling existing designs using AI models majorly demand a significant amount of data so the best possible results can be obtained. In many situations or disease states, there may be limited data available for drug under use or population in which the condition prevails, this leads us to less accurate predictions with unclear progression or biased results which lack credibility [64].

For instance, there are many rare diseases that may have limited data available, or they might have regional data with limited variations considered valid or accurate for targeting a larger population or people with different genetics, which can be a significant challenge in making the development of AI models less refined and resourceful. In addition to that, the data used to model and train AI systems may not be able to represent the population of interest or, even so, will not be able to consider special parameters that the population acquires, which can lead to biased discriminatory results lacking validity to be used for drug design development [65].

In addition to that, some types of datasets, such as longitudinal data or real-world evidence, may not be easily available. All these parameters lead to the limited use of AI models. By keeping in mind all the mentioned shortcomings, it becomes crucial to develop careful data considerations that aim quality and evident representation of the data that is to be used to develop AI models [64].

13.6.3 Data privacy and security

The ground for AI-integrated studies aims at using patient data, which is extremely sensitive and private to every patient and institution dealing with the condition; therefore, it is crucial to provide strict privacy protection policies and prevent data breaches at all costs. This involves complete compliance with data protection regulations (such as the General Data Protection Regulation). Due to strict privacy considerations, data collection and subsequent validation as per the rules and policies of privacy protection regulations become a challenge for AI model design and its further use in developmental studies [63].

The use of data is aimed at providing decision-making models that demand a huge dataset to make accurate and better-informed decisions; hence, data privacy and security are major parameters in this. Therefore, it means that data collected from clients or obtained by requesting patients' need to be shared and validated with the agencies and authorities running the drug development programs and legal measures of AI modeling. All these considerations make it necessary to safeguard and protect the client's or patient's information in the whole process. This results in developing public trust in AI-driven drug development, which leads to transparency of the system and information about the AI system will be utilized, its current and possible benefits, and its limitations [66].

In order to ensure high morality and sensibility in AI integration design, it is necessary to address these challenges. When handling sensitive patient or special cases, their privacy and integrity must be kept intact so that the data can be effectively used for AI-driven decision-making. This gives patients confidence that their data and information are kept secure and are only used for the betterment of existing complications, for their own good and social benefit. So better encryption and safety give evident trust in regulating the datasets and literature this involves the creation of policies for the usage of data and regulations related to consent. Moreover, patients should be given full access to their data and provided insight into how the data is being used [67].

13.6.4 Transparency and regulatory obstacles

The use of AI models leads us to understand the use of complex algorithms that serve as the backbone of AI integration within DDS. These algorithms are known as "black boxes" The reason to call them black boxes is that it is quite complicated to develop an understanding of how the model arrives at its predictions and anticipates possibilities. Because of the lack of transparency, this becomes an obstacle in acquiring public trust and a challenge in gaining regulatory approval from related authorities to use AI-based drug development tools. This is mainly due to the difficulty in demonstrating how the model provides accurate and reliable predictions with authentic results.

Further issues with the lack of transparency lead to limited trust in the model's predictions and analyzing tools. This becomes more evident if the model's prediction is not satisfactory and conflicts with the theoretical research conducted by clinicians and drug development researchers, with the prediction failing to accurately pinpoint the outcome [64].

AI models demand training, and for this validation, AI algorithms require enormous, high-quality, well-defined datasets. Such data is not easily available, or even if it is, the quantity is scarce, making it challenging to obtain. These are the considerations one must think about, particularly when acquiring data from medical professionals, academics, and data scientists to guarantee authentic and trustworthy data. Additionally, from the perspective of patient privacy, data-sharing agreements should be developed with full attention to detail [67].

13.6.5 Drug delivery hurdles and model drift

Over time, AI predictive models face a variety of challenges in the development of DDS. Among these challenges, one to consider is the requirement for accurately designed predictive models that offer regulated drug discharge. This is because, over the course of time, the data used to develop the predictive model becomes outdated, mainly due to changes in clinical practices or treatment protocols. Hence, this phenomenon is called model drift. For AI integration to work accurately, the predictive model should be kept updated, and regular retraining of the model becomes necessary. This can become resource-draining and a challenge for specialists to implement.

In using AI, we overcome this issue by utilizing nanocarriers and primarily developing nano systems that are aided and stimulated via artificial intelligence. For example, the penetration of drugs in the ocular tissue of the eyes is generally limited, the reason being the physiological and anatomical barriers within the eye leading to reduced penetration of any foreign matter. So the AI predictive models keep this physiology in consideration and use nanotechnology like nano micelles, nanoparticles, and liposomes, to give accurate delivery keeping in mind the ocular barriers [65].

To identify and develop newer therapeutic targets and agents, as well as to enhance delivery systems to revolutionize currently available therapeutics, a detailed and thoughtful collaboration between AI researchers and pharmaceutical experts is extremely necessary. The problem arises in collaboration because of differences in learning attitudes and methodology in available knowledge, information, educational culture, and internal communication. This can be overcome by using user-friendly consumer interfaces that will promote clinicians in using AI collaboration [67]

13.6.6 Ethical considerations in AI applications in biomedical fields

AI integrations demand data, so the use of data, its privacy, biases equity, and accessibility are always questioned, which demands AI to utilize given datasets morally and sensibly, specifically when dealing with patient-sensitive data [67].

13.6.7 Data privacy and security

Data privacy is an essential consideration in ethical parameters, as biomedical AI systems demand a huge dataset and a vast quantity of sensitive health data; hence, this causes a potential risk of data breaches. Patient health data are essential for AI to make wise decisions; therefore, the protection of this data is crucial for patient privacy. There are certain regulations and laws explained by the general data protection regulation (GDPR) that are mainly implemented in the European Union, and the

Health Insurance Portability and Accountability Act (HIPAA) implemented in the United States [67].

13.6.8 Fairness and bias

Training data are firstly needed by AI systems to generate AI algorithms; hence, if the training datasets do not represent a special population or underserved communities, developing a biased pickup of data resulting in discrimination against a particular socioeconomic status. To prevent this, regular audits must be run and measured to ensure mitigation should be implemented [67].

13.6.9 Explainability and transparency

AI systems operate on algorithms that are not completely understood; hence, they are termed as black boxes. They lack interpretability, leading to a lack of explainability regarding why and to whom the result of an AI prediction could be implemented. This also leads to patient concerns. Due to these parameters, it becomes difficult to distinguish and choose between AI predictive models and the more understandable traditional models of drug designing [68].

13.6.10 Accountability and liability

Like any model, AI is also a software-run model of predictions; hence, it can also provide false information or incomplete results. It is necessary to develop a legal regulation that targets who to hold accountable if the AI model fails to give desired results or gives false data. This is a dual-use concerned matter where both the AI developer as well as the person at the consumer end should be accountable for errors [68].

13.7 Limitations of predictive modeling and virtual screening

13.7.1 Data quality and reliability

When we are using AI predictive models, the basic focus is on high-quality data. In case the training data is biased, the model will not give accurate results, especially if newer material or a rare disease state is being studied. This leads to researchers and material scientists making inaccurate predictions and limiting their trust in computing infrastructure. To overcome this, we need to opt for advanced optimization technology and modify robust models [69].

13.7.2 Experimental validation and model accuracy

AI models and virtual screening need to be validated by experimental analysis, although this requires extensive testing and costly handling that can be time-consuming. Yet, this offers great predictability regarding how efficiently a model works under given conditions or real-world challenges. Additionally, when provided with newer datasets or a complex real-world setting, the accuracy of an AI model might vary from its training dataset. Hence, careful consideration related to data modeling must be undertaken. For example, when dealing with in silico models, virtual screening has to be validated with experimental analysis that leads to slow discovery [69].

13.7.3 Limited applicability and need for assistance

When designing or discovering a newer material the researchers should consider that a predictive model might work only effectively within the dataset used as their training data; hence, it becomes necessary to adjust their research methodology following these tools [69].

13.7.4 Performance misestimation and biological complications

In AI predictive models, the performance of a system is often judged using benchmarks (meaning how existing datasets influence the model's performance). In AI-integrated ML or VS, we first need to introduce a training set of data, and then a comparative study is conducted to estimate how well the model works. In the case of biological testing, things get a little complicated, as the dynamic system of biological system makes it difficult for ML and predictive models to give a very accurate result. However, modern AI predictive models use special considerations in this regard for excellent results [70].

13.8 Future directions

The rapid progression in ML and the convergence of AI in drug discovery and development have proven to be fruitful in all fields of medicine, whether it's the advancement of novel drugs, AI-driven polymer designs, multi-use polymers,, or the potential use of personalized medication. All these advancements are now possible with the aid of AI integration and ML [71].

13.8.1 Emerging trends in AI and smart polymer research

13.8.1.1 Drug discovery and virtual testing

AI predictive modeling is being highly effective in target identification and development of DDS, either by HTS or virtual screening [65]. The accurate predictability and larger dataset availability make it possible for researchers to use AI-integrated virtual clinical trials, which result in cutting costs and reducing time-consuming data generation; hence, the success rates and the correct way to conduct the study design are predicted [71].

13.8.1.2 AI-driven drug delivery systems

AI integration in the development of DDS that respond to parameters studied and embedded into ML models helps design drugs with special triggers for release [72]. Moreover, AI-driven nanotechnology is revolutionizing drug carrier systems, offering accurate and effective drug delivery with limited off-target effects and precise dosing. Smart pills are also a new concept introduced by AI [71].

13.8.1.3 Autonomous experimentation a new progress

A new concept in the development of DDS is the active learning by ML agents, in which datasets are generated specially for materials that are expensive to produce or their measure parameters are tedious. One of the most common approaches among them is Bayesian optimization. All these utilize active learning algorithms, hence requiring no intervention from a researcher [72].

Their major advantage is that they help in discovering desired materials and polymers using HTS or ML, which require no time to rest. Among the new progression is the use of reversible addition–fragmentation chain transfer (RAFT); it helps in finding the most desired polymer characteristic properties, such as 19 F magnetic resonance signal (MRI) signal for contrast agents [72].

13.8.1.4 Potential for personalized medicine through AI-enhanced drug delivery systems

The modern concept of personalized medication has revolutionized the traditional medication and healthcare system. This approach proves to be effective as it aims at using genetics in order to understand medical needs and to intervene them accordingly. This approach helps enhance patient outcomes and provides a smooth path for evolving precise healthcare and treatment [73].

AI and ML prove to be an excellent helping hand in predicting pharmacokinetic and pharmacodynamic profiles of special genetics and target populations, aiding in the development of personalized therapeutics. By using predictive algorithms and analyzing patient datasets, AI predicts accurate treatment strategies with responses obtained via trial designs. This helps healthcare providers use accurate measures to diagnose and then treat conditions as per individual patient characteristics and biomarkers as helping tools. AI and ML have a huge potential in evolving the pharmaceutical industry with high accuracy drug designing and healthcare delivery systems [73].

13.8.1.5 Enhancing drug efficacy with AI

Drug efficacy is a major reason why AI is remodeling and evolving personalized medicine. With up-to-date datasets and advanced algorithms, AI is able to analyze and optimize possible drug responses, helping to develop better treatment strategies that improve patient outcomes. This becomes more evident in chronic conditions, helping in development of treatment as per individual patient needs. By drawing comparisons, considering genetic factors, and evaluating drug effectiveness, a predictive model can estimate the right drug in the right dosage form, for the right period of time, in the right amount, and for the right patient. AI enhancement in personalized medication is very fruitful in chemotherapy as it prevents trial-and-error-based issues. Chronic conditions like diabetes and hypertension can also be better treated using AI modeling [73].

13.8.1.6 Collaborative approaches between AI researchers and material scientists

A cooperation and collaboration between AI and material scientists help in improving discovery and development, providing breakthrough advancements in both fields.

13.8.1.7 Material discovery and experimental investigations

New material discovery faces a major challenge of experimental analysis that aims at predicting properties. However, the use of AI integration helps fasten up the process, providing effective and efficient experimentation [74].

13.8.1.8 Material optimization

AI-driven models can predict material characteristics more accurately and also optimize them using physical modeling at atomic or macromolecular levels. This can be done using past data or the generation of new datasets that enable faster predictions and help organize and obtain data and literature from the tremendous datasets and databases available through AI models [74].

13.9 Conclusion

Over the past few years, DDS have gone through significant advancements. Smart polymer drug delivery designs proved to be of great effectivity making them excellent drug carriers and considered for further innovations. AI proved its part in the pharmaceutical field by advancing DDS and integrating smart polymers through AI-driven predictive modeling, ML, deep learning, and VSTs. The advanced characterization of nanomaterials, enhanced by AI that involves image processing, toxicology assessment, and rheological studies, have given complete insight for advanced drug delivery and polymer designing.

Despite having many advantages, AI integrations also face some challenges that mainly revolve around the availability of quality, precise, and reliable data that is unbiased following all the ethical standards.

The future innovations in drug discovery, efficacy, and personalized medication are all integrated by modern AI models. AI integration will lead material scientists to develop optimal polymer designs, paving brighter pathways for efficient DDS.

Abbreviations

AgC NPs	Silver-coated gold nanoparticles
AI	Artificial intelligence
ANNs	Artificial neural networks
API(s)	Active pharmaceutical ingredient(s)
ATR-FTIR	Attenuated total reflectance Fourier-transform infrared spectroscopy

CBZ/NIC	Carbamazepine–nicotinamide
C8	Octyl chain
C18	Octadecyl chain
CH	Cholesterol
CNN	Convolutional neural network
CsA	Cyclosporine A
DDS	Drug delivery systems
DFT	Density functional theory
DFTB	Density-functional tight binding
DNNs	Deep neural networks
DOX	Doxorubicin
DRL	Deep reinforcement learning
DDPG	Deep deterministic policy gradient
FeRh	Iron–rhodium alloy
FTIR	Fourier-transform infrared spectroscopy
GA	Genetic algorithms
GANs	Generative adversarial networks
GDPR	General data protection regulation
HIPAA	Health Insurance Portability and Accountability Act
HOMO	Highest occupied molecular orbital
HTS	High-throughput screening
IBU/NIC	Ibuprofen–nicotinamide
ISA	Investigation–study–assay
ISA-TAB	ISA tab-delimited file format
LCST/LCSTs	Lower critical solution temperature(s)
LUMO	Lowest unoccupied molecular orbital
MCE	Magnetocaloric effect
ML	Machine learning
MMD	Molecular mass distribution
MRI	Magnetic resonance imaging
MPS	Massively parallel spectroscopy
MSE	Mean squared error
NMR	Nuclear magnetic resonance
PAE	Poly (β-amino ester)
PCA	Principal component analysis
PCS	Plane-cut-splice
PGA	Polyglutamic acid
PEG	Polyethylene glycol
Phe	Phenylalanine
PCD	Polymer component distribution
PNIPAM	Poly(*N*-isopropylacrylamide)
RAFT	Reversible addition–fragmentation chain transfer
RA	Reinforcement algorithm
RMSE	Root mean square error
SA	Stearic acid
SEM	Scanning electron microscopy
SL	Supervised learning
SMEDDS	Self-microemulsifying drug delivery systems
SMILES	Simplified molecular input line entry system

SMPs	Shape memory polymers
SPPs	Skin-penetrating peptides
t-SNE	t-Distributed stochastic neighbor embedding
TEM	Transmission electron microscopy
TRY	Tryptophan
VAE	Variational autoencoder
VST	Virtual screening techniques
VS	Virtual screening

References

[1] Jain KK An overview of drug delivery systems. 2019. Vol. 1, New York, NY: Humana.

[2] Haesun Park AO. Kinam park evolution of drug delivery systems: From 1950 to 2020 and beyond. Journal of Controlled Release, Elsevier. 2022; 342:342.

[3] Anselmo AC and M S. An overview of clinical and commercial impact of drug delivery systems. Journal of Controlled Release, Elsevier. 2014; 190:190.

[4] Pattni BS, C VV and Torchilin VP. New developments in liposomal drug delivery. 2015; 115:10938–10966.

[5] Ezike TC, Okpala US, Onoja UL, Nwike CP, Ezeako EC, Okpara OJ, et al. Advances in drug delivery systems, challenges and future directions. Heliyon. 2023; 9(6):e17488.

[6] Balcerak-Woźniak A, D-z M and Kabatc-Borcz J. A comprehensive review of stimuli-responsive smart polymer materials. Recent Advances and Future Perspectives. 2024; 17:17.

[7] Jena GK, P CN and Chand S. Artificial intelligence and machine learning implemented drug delivery systems: A paradigm shift in the pharmaceutical industry. Journal of Bio-X Research. 2024; 7:7.

[8] Dolores R, Serrano O, Luciano FC, Anaya BJ, Ongoren B, Kara A, Molin G, Ramirez BI, Sánchez-Guirales SA, Simon JA, Tomietto G, Rapti C, Ruiz HK, Rawat S, Kumar D and Lalatsa A. Artificial Intelligence (AI) Applications in Drug Discovery and Drug Delivery: Revolutionizing Personalized Medicine. 2014; 16:1328.

[9] Vidhya KS, S A, Naveen Kumar M and Rangareddy H. Artificial intelligence's impact on drug discovery and development from bench to bedside. 2023; 15:1–10.

[10] Hsueh HT, C RT, Rai U, Liyanage W, Kim YC, Appell MB, Pejavar J, Leo KT, Davison C, Kolodziejski P, Mozzer A, Kwon H, Sista M, Anders NM, Hemingway A, Kiran Rompicharla SV, Edwards M, Pitha I, Hanes J, Cummings MP and Ensign LM. Machine learning-driven multifunctional peptide engineering for sustained ocular drug delivery. 2023; 14:2509.

[11] Das KP and J C. Nanoparticles and convergence of artificial intelligence for targeted drug delivery for cancer therapy: Current progress and challenges. 2022; 6:1067144.

[12] Jagtar Singh PN. pH-responsive polymers for drug delivery: Trends and opportunities. Journal of Polymer Science. 2023; 61:61.

[13] Wei P, C EJ and D J. Ultrasound-responsive polymer-based drug delivery systems. 2021; 1:11.

[14] Sarangi MK, R MEB and Parcha V. Smart polymers for colon targeted drug delivery systems: A review. International Journal of Polymeric Materials and Polymeric Biomaterials. 2020; 70.

[15] Christy Hunter A and M SM. Smart polymers in drug delivery: A biological perspective. Polymer Chemistry. 2017; 8:41–51.

[16] Moghanjoughi AA, K D and Zarrabi A. A concise review on smart polymers for controlled drug release. 2016; 6:333–340.

[17] James HP Smart polymers for the controlled delivery of drugs – A concise overview. 2014; 4:120–127.

[18] Huang H-J, T Y-L, Lin S-H and Hsu S-H. Smart polymers for cell therapy and precision medicine. Journal of Biomedical Science. 2019; 26:73.
[19] Amirov AA, P ES, Yusupov DM, Savintseva IV, Murliev EK, Rabadanov KS, Popov AL, Chirkova AM and Aliev AM. Thermoresponsive PNIPAM/FeRh smart composite activated by magnetic field for doxorubicin release. 2024; 3:410–418.
[20] Aumklad P, Suriyaamporn P, Panomsuk S, Pamornpathomkul B and Opanasopit P Artificial intelligence-aided rational design and prediction model for progesterone-loaded selfmicroemulsifying drug delivery system formulations. Science, Engineering and Health Studies. 2024; 1(18):24050002
[21] Kambayashi A In silico modeling approaches coupled with in vitro characterization in predicting in vivo performance of drug delivery system formulations. Molecular Pharmaceutics. 2023; 20(9): 4344–4353.
[22] Ribeiro MT, Singh S and Guestrin C. editors. " Why should i trust you?" Explaining the predictions of any classifier. Proceedings of the 22nd ACM SIGKDD International Conference on Knowledge Discovery and Data Mining. 2016; 1:97–101.
[23] Sidey-Gibbons JA and Sidey-Gibbons CJ. Machine learning in medicine: A practical introduction. BMC Medical Research Methodology. 2019; 19: 1–18.
[24] Gimeno A, Ojeda-Montes MJ, Tomás-Hernández S, Cereto-Massagué A, Beltrán-Debón R, Mulero M, et al. The light and dark sides of virtual screening: What is there to know?. International Journal of Molecular Sciences. 2019; 20(6):1375.
[25] Van der Maaten L and Hinton G Visualizing data using t-SNE. Journal of Machine Learning Research. 2008; 9(11): 2579–2605.
[26] Hemachandran K, Khanra S, Rodriguez RV and Jaramillo J Machine Learning for Business Analytics: Real-time Data Analysis for Decision-making. 2022; CRC Press.
[27] Barros RC, Carvalho A and Freitas AA. (Eds.). On the automatic design of decision-tree induction algorithms. In: Congresso da Sociedade Brasileira de Computação, XXXIV; Concurso de Teses E Dissertações, XXVII Sociedade Brasileira de Computação (SBC). 2014; Springer.
[28] Banerjee S, Zare RN, Tibshirani RJ, Kunder CA, Nolley R, Fan R, et al. Diagnosis of prostate cancer by desorption electrospray ionization mass spectrometric imaging of small metabolites and lipids. Proceedings of the National Academy of Sciences. 2017; 114(13): 3334–3339.
[29] McClelland JL and Botvinick MM Deep Learning: Implications for Human Learning and Memory. The Oxford Handbook of Human Memory: Foundations and Applications.2020;1:215–43
[30] Gurnani R, Kuenneth C, Toland A and Ramprasad R Polymer informatics at scale with multitask graph neural networks. Chemistry of Materials. 2023; 35(4): 1560–1567.
[31] Esteva A, Kuprel B, Novoa RA, Ko J, Swetter SM, Blau HM, et al. Dermatologist-level classification of skin cancer with deep neural networks. Nature. 2017; 542(7639): 115–118.
[32] Bhuvaneswari S, Abirami A, Kurup D, Sherin RH and Harishma S. (Eds.). Predictive modelling of polymer properties from polymer genomic data using machine learning approaches. In: 2023 3rd International Conference on Innovative Mechanisms for Industry Applications (ICIMIA). 2023; IEEE.
[33] Yoo S, Lee H and Kim J Deep learning for identifying promising drug candidates in drug–phospholipid complexes. Molecules. 2023; 28(12): 4821.
[34] Bannigan P, Häse F, Aldeghi M, Bao Z, Aspuru-Guzik A and Allen C Machine learning predictions of drug release from polymeric long acting injectables. 2021.
[35] Long PQ and Quan PM Virtual screening stategies in drug discovery–A brief overview. Vietnam Journal of Science and Technology. 2021; 59(4): 415–440.
[36] Kim C, Batra R, Chen L, Tran H and Ramprasad R Polymer design using genetic algorithm and machine learning. Computational Materials Science. 2021; 186: 110067.

[37] Gupta R, Srivastava D, Sahu M, Tiwari S, Ambasta RK and Kumar P Artificial intelligence to deep learning: Machine intelligence approach for drug discovery. **Molecular Diversity**. 2021; 25(3): 1315–1360.

[38] Lourenço MP, Hostaš J, Herrera LB, Calaminici P, Köster AM, Tchagang A, et al. GAMaterial – A genetic-algorithm software for material design and discovery. Journal of Computational Chemistry. 2023; 44(7): 814–823.

[39] Menegatti S, Zakrewsky M, Kumar S, De Oliveira JS, Muraski JA and Mitragotri S De novo design of skin-penetrating peptides for enhanced transdermal delivery of peptide drugs. Advanced Healthcare Materials. 2016; 5(5): 602–609.

[40] Ma Y, Zhu W, Benton MG and Romagnoli J Continuous control of a polymerization system with deep reinforcement learning. Journal of Process Control. 2019; 75: 40–47.

[41] Animasawun RK, Taresco V, Swainson SM, Suksiriworapong J, Walker DA and Garnett MC Screening and matching polymers with drugs to improve drug incorporation and retention in nanoparticles. Molecular Pharmaceutics. 2020; 17(6): 2083–2098.

[42] Napte NA and Saha S Advanced instrumentation for characterization of nanoparticles: A discourse. In: Synthesizing and Characterizing Plant-Mediated Biocompatible Metal Nanoparticles. 2025; (pp. 287–306). IGI Global.

[43] Rossi LM, Fiorio JL, Garcia MA and Ferraz CP The role and fate of capping ligands in colloidally prepared metal nanoparticle catalysts. Dalton Transactions. 2018; 47(17): 5889–5915.

[44] Park SJ Protein–nanoparticle interaction: Corona formation and conformational changes in proteins on nanoparticles. International Journal of Nanomedicine. 2020; 1(15): 5783–5802.

[45] Krishnamoorthy U and Balasubramani S Intelligent nanomaterial image characterizations–a comprehensive review on ai techniques that power the present and drive the future of nanoscience. Advanced Theory and Simulations. 2024; 7(12): 2400479.

[46] Crozier PA, Leibovich M, Haluai P, Tan M, Thomas AM, Vincent J, et al. Atomic resolution observations of nanoparticle surface dynamics and instabilities enabled by artificial intelligence. arXiv Preprint arXiv:240717669. 2024;2407(6737); 949–954. 17669.

[47] Wang Z and Zhu X Enhancing nanocrystal synthesis: A comparative study of online artificial intelligence optimization and offline high-throughput experimentation in chemical material discovery. ACS Applied Nano Materials. 2024; 7(6): 6499–6505.

[48] Sriram T, Chakraborty T and Prasanna PM Artificial intelligence powered insights into nanotoxicology. International Journal of Advancement in Life Sciences Research. 2024; 7(2): 68–80.

[49] He Z, Chen W, Zhou Y, Weng H and Shen X The importance of ai algorithm combined with tunable LCST smart polymers in biomedical applications. Frontiers in Computing and Intelligent Systems. 2023; 6(3): 92–95.

[50] Dutta R, Renshaw D, Chen C and Liang D Machine learning based approach for shape memory polymer behavioural characterization. Array. 2020; 7: 100036.

[51] Barmpalexis P, Karagianni A, Nikolakakis I and Kachrimanis K Artificial neural networks (ANNs) and partial least squares (PLS) regression in the quantitative analysis of cocrystal formulations by Raman and ATR-FTIR spectroscopy. Journal of Pharmaceutical and Biomedical Analysis. 2018; 158: 214–224.

[52] Alshoraihy A and Ibrahim A. (Eds.). Innovative vision: Exploring artificial intelligence and machine learning for advancing polymer nanocomposites. In: 2024 XXVII International Conference on Soft Computing and Measurements (SCM). 2024; IEEE.

[53] Rajagopal A, Nirmala V, Andrew J and Vedamanickam AM. (Eds.). AI visualization in nanoscale microscopy. In International Conference on Big Data, Machine Learning, and Applications. 2021; Springer.

[54] Bals J and Epple M Artificial scanning electron microscopy images created by generative adversarial networks from simulated particle assemblies. Advanced Intelligent Systems. 2023; 5(7): 2300004.

[55] Sundharbaabu PR, Chang J, Kim Y, Shim Y, Lee B, Noh C, et al. Artificial intelligence-enhanced analysis of genomic DNA visualized with nanoparticle-tagged peptides under electron microscopy. Small. 2024; 21(12): 2405065.
[56] Mohammad S, Akand R, Cook KM, Nilufar S and Chowdhury F Leveraging deep learning and generative ai for predicting rheological properties and material compositions of 3D printed polyacrylamide hydrogels. Gels. 2024; 10(10): 660.
[57] Schadler L, Brinson L and Sawyer W Polymer nanocomposites: A small part of the story. Jom. 2007; 59: 53–60.
[58] Huang Y, Zhao H, Wang Y, Ratcliff T, Breneman C, Brinson LC, et al. Predicting the breakdown strength and lifetime of nanocomposites using a multi-scale modeling approach. Journal of Applied Physics. 2017; 122(6): 065101.
[59] Zhao H, Li X, Zhang Y, Schadler LS, Chen W and Brinson LC Perspective: NanoMine: A material genome approach for polymer nanocomposites analysis and design. APL Materials. 2016; 4(5): 053204.
[60] Wang Y, Zhang M, Lin A, Iyer A, Prasad AS, Li X, et al. Mining structure–property relationships in polymer nanocomposites using data driven finite element analysis and multi-task convolutional neural networks. Molecular Systems Design & Engineering. 2020; 5(5): 962–975.
[61] Inguva P, Mason LR, Pan I, Hengardi M and Matar OK Numerical simulation, clustering, and prediction of multicomponent polymer precipitation. Data-Centric Engineering. 2020; 1: e13.
[62] Li Y, He Z, Wang X, Li Z, Johnson M, Foley R, et al. Artificial Intelligence (AI)-aided structure optimization for enhanced gene delivery: The effect of the Polymer Component Distribution (PCD). ACS Applied Materials & Interfaces. 2023; 15(30): 36667–36675.
[63] Jena GK, Patra CN, Jammula S, Rana R and Chand S Artificial intelligence and machine learning implemented drug delivery systems: A paradigm shift in the pharmaceutical industry. Journal of Bio-X Research. 2024; 7: 0016.
[64] Vora LK, Gholap AD, Jetha K, Thakur RRS, Solanki HK and Chavda VP Artificial intelligence in pharmaceutical technology and drug delivery design. Pharmaceutics. 2023; 15(7): 1916.
[65] Deori C, Hujuri L, Sarma G, Sonowal T, Head. Artificial Intelligence (AI): It's Role in Drug Discovery and Novel Drug Delivery System. International Journal of Science and Research (IJSR). 2024;13:13.
[66] Mohammed S. AI-Driven Drug Discovery: Innovations and Challenges. IJARCCE. 2024;1:13.
[67] Ekpan F-DM, Ori MO, Samuel H and Egwuatu O The synergy of AI and drug delivery: A revolution in healthcare. International Journal of Advanced Biological and Biomedical Research. 2023; 12: 44–66.
[68] Golriz Khatami S, Mubeen S, Bharadhwaj VS, Kodamullil AT, Hofmann-Apitius M and Domingo-Fernández D Using predictive machine learning models for drug response simulation by calibrating patient-specific pathway signatures. Npj Systems Biology and Applications. 2021; 7(1): 40.
[69] Visan AI and Negut I. Integrating artificial intelligence for drug discovery in the context of revolutionizing drug delivery. Life. 2024; 14(2): 233.
[70] Ghislat G, Hernandez-Hernandez S, Piyawajanusorn C, Ballester PJ. Data-centric challenges with the application and adoption of artificial intelligence for drug discovery. Expert Opin Drug Discov. 2024; 19(11):1297–307
[71] Martin TB and Audus DJ Emerging trends in machine learning: A polymer perspective. ACS Polymers Au. 2023; 3(3): 239–258.
[72] Bhatt MS, Maurya MA, Maurya MV. EMERGING TRENDS IN PHARMACEUTICAL DEVELOPMENT. MOLECULES TO MEDICINES: A Comprehensive Look at Pharmaceutical Development. 2024; 1:111.
[73] Nwankwo E, Ajegbile M, Olaboye J, Emeihe and Maha C AI in personalized medicine: Enhancing drug efficacy and reducing adverse effects. International Medical Science Research Journal. 2024; 4: 806–833.

[74] Abolhasani M, Brown KA and Guest E Role of AI in experimental materials science. MRS Bulletin. 2023; 48(2): 134–141.

[75] Vatandoost, Hossein, Seyed Masoud Sajjadi Alehashem, Mahmood Norouzi, Hamid Taghavifar, and Yi Qing Ni. 2019. "A Supervised Artificial Neural Network-Assisted Modeling of Magnetorheological Elastomers in Tension–Compression Mode." *IEEE Transactions on Magnetics* 55(12):1–8. doi: 10.1109/TMAG.2019.2942804.

Komal Gupta*, Niranjan Kaushik, Amit Singh, Vikram Sharma, and Tanya Biswas

Chapter 14
Robotics innovations in nanotechnology: redefining polymer science with AI

Abstract: Artificial intelligence (AI), nanotechnology, and robotics are integrating to transform polymer science and open up previously unattainable opportunities for improved functionalization at the nanoscale and material innovation. This chapter examines how AI-driven robotics is revolutionizing the generation of specialized nanomaterials, with an emphasis on the synthesis, properties, and uses of polymers. Researchers can speed up discovery processes, anticipate molecular behavior, and optimize polymer design by combining robotic equipment and machine learning algorithms. This minimizes time-to-market for novel materials. AI also makes it possible for robotic platforms to regulate atoms with great accuracy, which makes it possible to synthesize polymers with improved mechanical, electrical, and thermal properties.

The chapter elaborates the challenges and future prospects of this comprehensive subject while exhibiting modern advancements in AI-assisted robotics such as autonomous material testing and self-assembling polymer systems. In addition to advancing the field of nanotechnology, the ability to change a variety of industries, constituting electronics, healthcare, and environmental sustainability is shown by such collaborative approach.

Keywords: Nanotechnology, robotics, algorithms, sustainability, healthcare

***Corresponding author: Komal Gupta**, Department of Pharmacy, School of Medical and Allied Sciences, Galgotias University, Greater Noida, Uttar Pradesh, India; Galgotias College of Pharmacy, Greater Noida, Uttar Pradesh, India
Niranjan Kaushik, Department of Pharmacy, School of Medical and Allied Sciences, Galgotias University, India
Amit Singh, School of Pharmacy, Monad University, Hapur, Uttar Pradesh, India
Vikram Sharma, Galgotias College of Pharmacy, Greater Noida, India
Tanya Biswas, Galgotias College of Pharmacy, Greater Noida, India; Delhi Institute of Pharmaceutical Sciences and Research

https://doi.org/10.1515/9783111681870-014

14.1 Introduction

Integrating nanotechnology and artificial intelligence (AI) has become an emerging field with revolutionary promise in the rapidly evolving environment of contemporary innovation and technology [1]. Together, these two fields – each revolutionary in its own right – have the potential to expand human potential and make possible innovations that were previously only found in fictional realms. Nanotechnology and AI together mark a significant advancement in exploration of technological and scientific discovery. This is crucial to identify the history of both nanotechnology and AI and comprehend how each has influenced the modern world separately in order to fully grasp the significance of this convergence.

Originating in the middle of the twentieth century, AI is frequently considered the apex of science and machine learning (ML) [2]. The goal of early AI research was to imitate human cognitive processes through expert systems and symbolic reasoning. However, a lack of computing power and irrational expectations caused the industry to experience multiple "AI winters." Three main elements have contributed to the rebirth of AI in recent decades: computational power, big data, and algorithmic innovation [3, 4]. AI algorithms now have a wealth of training data thanks to the surge in data generation brought about by the digital age. This resulted in the research of technical methods, especially deep learning, which fueled advances in reinforcement learning (RL), processing of natural language, and picture identification [5, 6]. The learning and implementation of models of AI were greatly hastened by hardware advancements such as graphics processing units and AI chips, which made formerly difficult jobs possible [7]. Recurrent neural networks and convolutional neural networks (CNNs), two more modern AI algorithms created by researchers, are particularly good at picture and speech recognition. AI is now pervasive in many aspects of our daily life, from driverless cars to predictive systems of healthcare. It keeps developing quickly and expanding its powers into new areas.

On the other hand, the multidisciplinary discipline of nanotechnology arose from the discovery that matter had special characteristics at the nanoscale, which is often described as structures with dimensions less than 100 nanometers [8]. The foundation for nanotechnology was established by Richard Feynman's seminal 1959 talk, "There's plenty of space at Bottom," which visualized the modification of atoms and molecules [9]. Since then, nanotechnology has advanced significantly, making it possible to design and produce nanomaterials and nanodevices with remarkable qualities [10, 11]. Nanomaterials with specific properties, such as carbon nanotubes (CNTs), quantum dots, and graphene, have been created by scientists and engineers, creating new opportunities in materials science, electronics, and energy storage. Scientists are producing nanoscale machines and robots for use in everything, from ecological restoration to medication delivery [12, 13].

Because AI and nanotechnology have complementary strengths, their convergence is a logical development. The accuracy of nanotechnology at the atomic and mo-

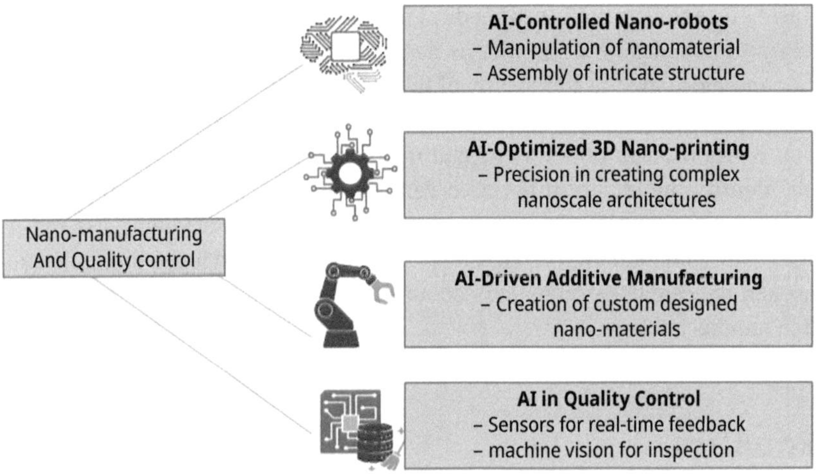

Figure 14.1: Advancement of nanosensors for theranostic purposes.

lecular level complements AI's data processing skills and capacity to manage intricate algorithms [14]. This convergence presents special difficulties that demand careful investigation, but it also offers huge potential in a number of areas, for example, materials research, healthcare, and environment monitoring. Interaction of both AI and nanotechnology offers novel details to some important worldwide problems as both domains develop further, bringing in a new era of creativity (Figure 14.1). Scientists and society at large must comprehend this synergy and its consequences.

This chapter attempts to address these issues head-on through offering a thorough examination of the state of AI in nanotechnology today, pointing out important knowledge gaps and providing perspectives into the collaborative, ethical, and regulatory aspects of the developing area. Through resolving numerous problems, the chapter aims to widen the door for more ethical and informed introduction of nanotechnology and AI into the rapidly developing technical framework. This chapter's rationale is the understanding that both AI and nanotechnology have already significantly impacted society on their own. AI has impacted various fields such as manufacturing, transportation, and finance, thanks to the potential to carry enormous data and complex functions with high precision [15, 16].

However, nanotechnology has rendered it possible to engineer and modify materials at the nanoscale levels, offering a world where matter's properties behave differently than they do at the macroscopic level [17]. However, the potential for previously unknown breakthroughs that might transform entire industries and resolve crucial global issues lies at the intersection of these domains. Recognizing how AI can be used in nanotechnology is not just a theoretical pursuit but also a practical necessity. The speed in the discovery of new materials is enhanced; it has modified nanoscale manufacturing techniques, improved medical care, and helped create sustainable energy sources. A methodical investigation is necessary to completely understand these possibilities and negotiate the complexity of this progress. In order to increase awareness and encourage responsible innovation, it explores and outlines the moral, legal, and societal issues brought about by AI and nanotechnology collectively. In order to completely identify the idea of this convergence, the chapter highlights the significance of cooperation between domain-specific specialists, nanotechnologists, and AI researchers. The evaluation offers suggestions and information that can direct upcoming studies and educate decision-makers about the ethical advancement and application of AI in nanotechnology.

14.2 Robotics

The field of robotics incorporates a number of engineering and technology disciplines such as electronics, material science, mechanical engineering, and bioinformatics. All engineering branches and their technologies must coordinate for manufacture and operation. Applications of robots range from basic manufacturing to the military and defense industries. Diverse applications necessitate distinct operational methodologies and component kinds. Typically, all robots need to be strong, flexible, and long-lasting. They must also be lightweight for some applications, such as biological fields. These characteristics cause polymers, a large class of materials that compose of thermoplastics, nanomaterials, elastomers, etc., to function. The ability to modify polymers to achieve desired properties is one of its characteristics.The function of various polymers in the domain of robotics is briefly discussed in this chapter. Robotics is the branch of science that studies how to design, build, operate, and use robots. Robotics additionally encompasses the design and operation of the technological devices that control them. In his 1920 play, Rassum's Universal Robots, Czech novelist Karel Capek used the term "robot" for the first time. The Czech word for worker or servant is where the word "robot" first appeared. Robots are defined by the Oxford Dictionary as "machine capable of performing complex sequence of actions automatically, particularly those which can be programmed by a human."

14.3 Role of polymers in robotics

Robots are used in the polymers composition, and polymers and plastics are important components of robot manufacturing. Packaging, polymer manufacture, and other industries use commercial and industrial robots. Most of the time, polymers are inexpensive, lightweight, pliable, flexible, and simple to make. Because of these characteristics, they can be incorporated in numerous industries, for example, electronics, aerospace, and home products. Polymers can be molded into any shape, and their characteristics can be altered to suit the application. Compared to traditional robots, soft robotics are preferred these days due to their ease of contact with humans.

The most common polymer types used in robotic applications include thermoplastics, elastomers, polymeric gels, and materials based on epoxy and acrylic [18].

14.4 Advancements in nanotechnology

Molecular- and atomic-level engineering manipulation of valuable materials and devices is the focus of nanotechnology. Compared to existing developments in the chemical, biological, electronic, and related sectors, the development of nanotechnologies offers new and improved materials and devices that offer substantially higher overall performance, dependability, and efficiency.

Applications such as molecular electronics, tumors in which nanodevices can differentiate between cancerous and noncancerous ones and electro-mechanical computing devices are crucial for the advancement of whole new fields of research and technology. The gradual reduction of material size is a major benefit of materials science in nanotechnology. The primary benefit of precision nanodevices, which have been developed, is the unmatched precision they can attain.

However, other significant features of molecular interactions in related biological systems must be addressed in the development of nanodevices, as many of their operational characteristics are derived from molecular interactions between nanodevices, during use. The combination of computational and physical components is the other significant obstacle still facing nanodevices. Let's conclude by pointing out that a key component of contemporary advanced materials research is the process of assessing the quality of functional materials at microscopic to nanoscopic levels [19].

14.4.1 Nanomanufacturing and quality control

A paradigm change in nanomanufacturing and quality control is being driven by the combination of AI and nanotechnology [20, 21]. AI's incorporation into nanomanufacturing procedures has rendered it easier for scientists to precisely and efficiently assemble nanoscale structures, which is resulting in revolutionary advances across a range of industries: manufacturing processes are being revolutionized by AI-controlled robots with nanoscale precision [22, 23].

These robots' unmatched precision in manipulating nanomaterials makes it possible to assemble complex structures that were previously difficult to create. 3D nanoprinting methods are being optimized by AI algorithms. This approach opens up new possibilities in materials science, electronics, and medical devices by precisely creating intricate nanoscale architectures [24, 25]. Custom-designed nanomaterials and components are becoming possible thanks to AI-driven nanoscale additive manufacturing technologies [26]. This strategy has broad consequences for areas like aerospace, where there is a need for strong, lightweight nanomaterials (Figure 14.1).

AI plays an essential role in quality control, guaranteeing that devices and nanomaterials fulfill high requirements of dependability and efficiency [27, 28]. Machine vision systems driven by AI can examine materials and structures at the nanoscale with remarkable precision [29]. By identifying flaws and variations that the human eye might miss, these technologies improve the quality of the final product [30]. Real-time quality feedback is provided by AI-driven sensors and monitoring devices integrated into nanomanufacturing machinery [31]. This feature makes it possible to modify manufacturing procedures instantly, which reduces waste and mistakes. AI predicts maintenance requirements by analyzing data from nanomanufacturing equipment [32]. AI-driven preventative maintenance reduces downtime and guarantees constant production quality [33].

AI is used to optimize nanomanufacturing operations because of its ability to access huge data in order to determine the ideal factors for manufacturing of materials and guarantee high-grade nanomaterials, low consumption of resources, examine data from manufacturing processes [34, 35]. AI-enhanced materials characterization methods shed light on nanomaterial characteristics. This data guides modifications to production procedures, resulting in materials with the required qualities.

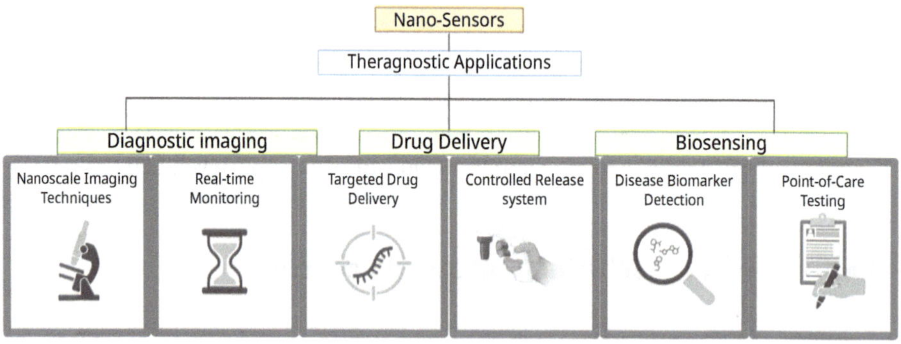

Figure 14.2: Additional developing techniques by novel methods.

14.4.2 Nanotechnology for biomedical soft robots

Since its introduction twenty years ago, soft robots have gained a lot of research interest; the biomedical field has been regarded as a crucial area of application [36]. Soft robots with promising features have been made possible by recent advancements in functional materials, particularly improved soft substances, which have significantly increased their prospective uses in the biomedical arena. Soft robots are divided into in vitro and in vivo categories on the basis of biomedical application characteristics. To help with difficult biomedical procedures and the replacement of human body parts with artificial organs, soft robots of in vivo category – that is, robots that are integrated with the human body [37] – are created [38]. For instance, research has been done on using soft robots to perform endoscopic procedures and minimally invasive surgery on humans, as well as to deliver medications under the skin or inside blood vessels. Conversely, soft robots of in vitro category – that is, robots that are cooperated outside the human body – evolved for a variety of biomedical procedures, ranging from precise surgery to physical rehabilitation. For instance, it has been stated that biomedical soft robots can help patients with rehabilitation by physically supporting them [39]. Biocompatibility is crucial in the design and construction of biomedical soft robotics because of the unique application conditions found in the human body. Specifically, to provide full functionality and good acceptability in application scenarios, the materials used in biomedical soft robots must entirely be compatible with organs and tissues [40].

14.4.3 Nanorobotics and medicine

Innovative developments in healthcare are being made possible by the intersection of nanorobotics and medicine. This convergence addresses medical issues in diagnosis, therapy, and medication distribution by utilizing the accuracy and versatility of nanoscale robots in the direction of AI algorithms [41]. The application of nanorobotics in medicine has enormous promise for diagnosis. These microrobots, which are frequently made to resemble biological objects, are extremely good at navigating the human body [42]. They can have sensors installed for cellular or molecular diagnostics and real-time monitoring. For instance, early detection and intervention may be made possible by the nanorobots' ability to identify infections, cancer cells, or disease-related biomarkers, strong classification performance, and reasoning. A neuroscience-inspired method called the complementary learning fuzzy neural network, suggested as computer-assisted intervention tool, is employed to analyze thermograms because of its superior classification performance, human-like reasoning, and high preliminary screening accuracy when compared to manual diagnostics [43]. Drug distribution is among the most exciting uses of nanorobotics in medicine. By avoiding healthy tissues and reducing adverse effects, these nanoscale robots can deliver medications straight to their intended locations [44]. Nanorobots that are guided by AI algorithms are able to traverse intricate biological settings and get past obstacles like the blood–brain barrier. By lowering the necessary drug dosage and improving therapeutic efficacy, this targeted drug delivery strategy minimizes toxicity and side effects. Additionally, minimally invasive surgery is being revolutionized by AI-driven nanorobotics. These robots have remarkable precision in executing complex tasks. Advanced sensors are responsible for the high precision of nanorobots, which influences their capacity to navigate the complex biological environment.

AI and nanorobotics together enhance autonomy and decision-making [45]. These nanorobots are able to make effective choices in response to shifting bodily situations by using real-time data from sensors. For instance, they can react rapidly to emergencies or modify medicine release rates in accordance with the patient's physiological parameters. By increasing the accuracy of targeted medicines, better diagnostics, and creative treatment approaches, the future of medicine is expected to see significant breakthroughs due to the combination of nanorobotics and AI. By utilizing modes of hybrid control such as light field trigger and magnetic field drive, bioactive microbots can execute targeted treatments like tumor removal with minimal or no adverse effects. AI integration has the potential to further optimize nanorobots, enabling a response to intricate biological circumstances that could basically revolutionize patient outcomes and clinical practice [46]. One of these is the advancement of personalized medicine techniques, in which nanorobots are customized to an individual's genes and medical requirements. Furthermore, AI-powered nanorobotics have the potential to transform regenerative medicine by repairing damaged organs or tissues at the cellular level [47].

14.4.4 Nanotechnology in biorobotics

The field of biorobotics is widely transdisciplinary in both science and technology. In order to improve our understanding of human physiology, the initial definitions from mid-1990s emphasized challenges of simulating and modeling biological systems [48]. Biological systems and robotics were conceptually related and viewed as "metaphor" in this viewpoint. It was also seen as an essential connection between AI-based technology and individuals.

The following are the main goals of biorobotics: (1) establishing skilled mechanical systems with improved functionality and performance; (2) using robotic materials to better understand specific animal (including human) mechanisms and behaviors; and (3) achieving safe and efficient interactions (like new low invasive therapies) with biological structures. Nowadays, nanotechnology is significant for achieving these objectives, providing previously unattainable possibilities [49].

Particularly for medical applications, nanorobots are regarded as one of the most promising systems. Nanorobots are intelligent systems composed of nanoscale component assemblies through overall dimensions below the range of micrometers [50]. Although it is generally acknowledged that medical nanorobots with true functionality will significantly improve healthcare, a number of problems must be resolved before the many functional components that constitute a nanorobot can be effectively developed and assembled. The physical principles that govern interactions at the nanoscale level are primarily responsible for these challenges. In addition to the significant scaling of effects of volume (related to weight, heat capacity, and inertia) that are superimposed by surface effects (related to forces of adhesion, heat transfer, and friction), chemical bonding forces play a larger character by decreasing the size, as do quantum effects. It highlights the necessity of producing precise assembly and manipulation techniques in order to accomplish the intended goals [51].

14.4.5 Nanotechnology for actuation

A vital part of each device is its actuator, which enables movement and, more generally, interaction with the environment. Artificial molecular machines represent an intriguing approach to nano-actuation. The manufacturing of such structures must be predicated on chemical synthesis and manipulation techniques [52]. Remote stimuli, which include light, can produce and govern single molecules as well as supramolecular systems (such as creating isomerization). Moreover, they exhibit a number of drawbacks [53].

These consist of the following: (i) these are only available in solution and are difficult to handle separately; (ii) For continuous rotations or alternatively, motions without a mechanical coupling with a load, they are not very helpful; and (iii) an operation's yield is often quite poor. A recent study emphasizes that structured molecular equipment assemblies, coupled with frameworks of metal and organic, are used as nanorobotic arms, in addition to outlining a roadmap (which excludes molecular switches due to their inability to generate useful work) [54].It also highlights the possibility of developing new kinds of enzyme-like mimics. These machines can exert forces well below 1 N, despite their promising size [55].

Other nanoscale actuators, often in tens of range of nanometers, are called biomotors. There has been some advancement in this sector since the initial attempts to use kinesin, a motor protein to create movement [56] and to bind to a surface F1-ATPase molecule in order to generate observable binding [57]. Nanoparticles have mostly been employed for nanoscale actuation [58]. Ten to fifteen years ago, CNTs were employed as main components that allowed actuation, demonstrating their potential [59, 60]. The performance of nanoparticles and nanocomposites is improving these days, particularly when combined with smart materials [61, 62].

Remote forces can also be used to guarantee actuators for nanoscale devices. When it pertains to magnetic fields, magnetized living things as well as nanoparticles with magnetic properties have been utilized, for example, magnetic resonance imaging [63]. Controlling the navigation of magnetic materials requires an understanding of their properties. Magnetic susceptibility (v), which is the ratio of the magnetic field applied (H) to the induced magnetization (M), is the crucial characteristic.

Micro-biorobots, also known as biohybrid microrobots, are devices that combine artificial and live components (such as cells or tissues), with the goal of small-scale exploitation of the living matter special properties within a combined system. According to this structure, the whole of living cells – for example, sperm or flagellated bacteria – are utilized to offer miniaturized robots their micro/nanoscale motility. They can be combined with magnetic nanoparticles, powders, or layers [64, 65], or used with bacterial cells that exhibit their own magnetic properties, which includes magnetotactic bacteria that are identified by organelles known as magnetosomes and constitute magnetic crystals [66, 67].

Although they have only been tested in vitro thus far, bio-hybrid inert microrobots derived from modified monocytes have been suggested [68]. To transport drugs to the right places, these microrobots must take advantage of monocytes' chemotactic activity and transmigrating motility. Moreover, as chemotactic gradients may only be recognized at extremely short distances, using chemotaxis, a single targeting technique, would indicate and restrict treatment efficacy. This method would greatly expand its potential if it were linked to a proper "pre-navigation" of the microrobot by utilizing additional technical techniques, thereby permitting a closer approach to the target. Nanoparticles can play a significant role in this paradigm, for example, by acting as intracellular transducers to increase muscle contractility [69, 70]. To progress this research, a number of nanotechnological techniques will be required, particularly if complex biorobots made up of heterotypic cell clusters – multiple cell types – are to be studied [71]. Resonant devices, in which nanomechanical element is activated in one of its resonant modes, make up the majority of actuation-related nanoelectromechanical systems (NEMS) [72]. Interesting examples of NEMS for optomechanics have recently been published; including NEMS controlled by optical gradient forces [73] and NEMS that can regulate photonic crystal cavities' optical biostability [74].

14.4.6 Nanotechnology for sensing

Technology for sensing is developing quickly. Since nanosensors would be tiny, ubiquitous, and undetectable, they are very intriguing. Above all, their sensitivity should surpass that of their macroscopic equivalents. At the moment, nanoparticles rule the nanosensor stage due to their many uses and operational mechanisms. By taking advantage of their altered conductivity upon exposure to particular compounds, nanotubes were first employed as sensors of the nanoscale [75, 76]. Presently, a number of nanoparticles are known to produce quantifiable results in response to specific response that is generated.

Gold nanoparticles are primarily used to detect chemical or biological targets by either modifying their surface to particular characteristic elements (such as antibodies, oligonucleotides, or reactant molecules) that release signals, which can be detected, like fluorescence, when they come in contact with the desired target or by taking advantage of changes in the nanoparticle's conductivity, redox characteristics, or surface plasmon resonance character [77]. By utilizing their potential to function as catalysts and improve the transfer of electron, additional nanoparticles of metal (Ag and Pt), oxide nanoparticles (TiO_2, SiO_2, ZrO_2, and MnO_2), and semiconductor nanoparticles (CdS and PbS) may additionally serve as the basis for electrochemical sensors and biosensors.

Furthermore, as was previously mentioned for gold nanoparticles [78, 79], they can be functionalized by particular ligands.

One of the most ensuring nanoparticles to accomplish multifunctional sensing at the nanoscale is CNTs [80]. They can be utilized to transduce various physical inputs by taking advantage of diverse detecting techniques such as their special mechanical and electrical characteristics. Single-walled CNTs' electromechanical characteristics are especially relevant for mechanical unit sensing. Covalent carbon–carbon bonds, for example, have a high degree of elasticity, making them suitable for use as nanoscale piezo resistors with gauge factors as high as 1,000 and the capacity to return to their starting state even after significant deformation.

Recently, polymeric nanoparticles have also been suggested. Polyacrylamide, poly (methyl methacrylate), poly(sodium styrene sulfonate), *N*-isopropylacrylamide, and other nondegradable and degradable materials are employed to create nanoparticles that are primarily used for chemical properties' dynamic monitoring, cell and tissue metabolism, and response to external stimuli. Because of their high biocompatibility, they are particularly helpful in bio-related applications. Although their conformational variations in response to certain action can also be utilized, they are typically functionalized with target-specific ligands for this purpose [81, 82]. These days, flexible polymeric matrices containing piezoelectric nanoparticles, for example, barium titanate, zinc oxide, and alkaline niobate are being used to create nanocomposites. These devices transform mechanical energy into electrical outputs that may be sensed and stored using piezoelectric materials. They have primarily been proposed as energy harvesting systems [83–85]. This is an intriguing biorobotics application that enables the creation of gadgets that, when in use, can partially recharge their batteries. However, these devices are also appealing as force and stretching/bending sensors through nanoscale characteristics due to their capacity to react to both pressure and deformation stimuli.

Biological molecules have been recommended into nanosensors; the most promising ones are deoxyribonucleic acid and proteins (enzymes, antigens, and antibodies) [86–88]. When a binding event occurs, these molecules' sensing mechanism is predicated on a conformational shift. For instance, the binding may cause an electrically active component and an electrode to move closer, enabling the detection of an electrical signal. Future integration of biological molecules with autonomous or semiautonomous miniature robots for cutting-edge medical treatments appears to have great potential. Nevertheless, there are currently no technologies for signal processing and collection that are as compressed as these sensor molecules [89].

14.5 Artificial intelligence in polymer science

In recent years, AI has grown as an essential platform to resolve scientific as well as technical issues in a range of fields. AI has enormous potential advantages, but it also poses significant risks and challenges. Scientific research will be more significantly and more rapidly impacted by this new trend. In order to comprehend the possible effects of AI, not only on their present activity but also on their fundamental mission as research and learning institutions, scientific institutions and universities everywhere urgently need to conduct strategic evaluations. This is crucial to maximize potential benefits and minimize potential drawbacks, as well as to make sure that new advancements are utilized to their fullest potential [90, 91].

AI is becoming a reality in daily life and is no longer merely a sci-fi "what might be" fantasy. This is a result of several recent discoveries and advancements that have been fueled in part by public and private funding as well as significant expenditures made by huge businesses. We are at the start of a new era that will profoundly alter many industries, particularly those where advancement has been impeded by the complexity of the data to be analyzed. Even though the scientific research industry is typically at the forefront of creating and utilizing new technology, it has demonstrated a startling lack of interest in the latest developments.

Our AI systems outperform the top human players at chess, go, and poker. These days, call center duties can be completed well by AIs without the callers realizing they are speaking to a machine. Currently, using a speech recognition program is less expensive than hiring a human assistant. The creation of systems that can enhance the capacity to solve complicated problems in a way similar to that of humans has demonstrated the validity of the Darwin challenge. The influence of these and similar discoveries on day-to-day operations has been minimal [92].

AI offers new opportunities for innovation by enabling data-driven forecasts, revealing hidden patterns, and automating time-consuming operations. However, several scientific disciplines, like polymer science, are not yet ready for its integration, and its rapid rise has proved disruptive. Adoption of AI has been particularly difficult due to the intricacies of polymer systems and the discipline's predominantly empirical character. Many polymer scientists continue to encounter major obstacles, such as a lack of transdisciplinary training and technical complexity. Since they were discovered more than a century ago, polymers have mostly relied on trial-and-error testing. This is a time-consuming procedure that entails generating monomers, refining macromolecular structures, and modifying formulation characteristics in order to determine the optimal combination for a particular application. Even though the strategy has definitely led to important advances, it is still essentially gradual. Numerous options remain undiscovered because of the intricate and nonlinear interactions between structure and characteristics, as well as the enormous combinatorial explosion of polymer design possibilities. What innovative materials could be concealed in this vast, unexplored design space?

In such a huge and complicated world, how can we speed up the discovery of latest materials? No matter how ambitious, human ability cannot accomplish such a huge undertaking. Nonhuman intelligence, also referred to as AI [93, 94], enters the picture here, not as a solution but rather as a transformational tool, and aids that changes the way we conduct scientific study [95–97]. Artificial intelligence has already opened doors in disciplines like biology [98, 99]. For instance, Deep Mind's Alpha Fold resolved a frequent protein folding problem [100]. This innovation demonstrates AI's remarkable capacity to identify unidentified sequences and generate assumptions that outperform human perception.

Is polymer science the next area where AI will have a revolutionary impact? Imagine an artificial assistant who works round the clock to find solutions to urgent industrial and societal demands, evaluate enormous volumes of information to find opportunities that have been missed, and reveal important gaps in the literature. Working on the discoveries, a similar assistant may suggest a novel polymer that is precisely suited for a certain use, suggest an economical and effective production route, and take the polymer's intended lifespan into consideration. Additionally, it might forecast how it will deteriorate and offer solutions to improve its sustainability and recyclability. However, this assistant wouldn't merely provide theoretical recommendations. To carry out the suggested synthesis, it could easily communicate with an automated system. It may repeatedly optimize the process by evaluating real-time experimental data, and dynamically modifying response parameters to accomplish predetermined goals, while maintaining total safety. Practical concerns like locating and acquiring required solvents and reagents would also be taken care of, guaranteeing a seamless transition between expected and implementation. Additionally, this technology could consolidate, preserve, and structure information at each stage, producing thorough data that monitors development and offers valuable information to direct researchers. Although this may seem like a lofty goal for "today," it is actually more realistic than we may realize. According to a recent study by Martin and colleagues describing "self-driving labs" of the future, the instruments required to enable seamless combination of AI, automation, and laboratory operations are either evolving or currently under development [101]. This vision is gradually becoming a reality due to rapid advances in robotics and AI. Perhaps, the scientists' creativity and capacity to fully utilize this technology are the only real constraints. Another intelligent assistant might soon be available to aid with this duty as well. The number of published papers has increased, reflecting the exponential expansion of AI applications in polymer science in the last ten years. Though the AI's ability in polymer science is clearly evident – many top-notch reviews have previously addressed this subject – the future is still unclear [102–107]. How does AI operate? Which resources should we acquire in order to start experimenting with and using AI in our studies?

14.6 AI as a novel paradigm in polymer science research

Fundamentally, AI is the machine's capacity to mimic intelligence of humans, including pattern recognition, prediction, and data-driven learning [108, 109]. The word "AI" in chemistry frequently arouses interest but also perplexity, particularly in contrast to molecular simulations or physical modeling techniques. Polymer research has traditionally relied on conventional approaches, which include molecular dynamics and density functional theory simulations that use explicit equations a well as physical principles to determine changes such as phase transformations, chain conformations, or mechanical events.

14.6.1 Key machine learning techniques

ML is a branch of AI that focuses on creating models that can learn from data and produce conclusions or predictions without explicit programming [110–112]. ML is being used more and more in polymer science to assist material discovery, optimize synthesis [113–125], and predict characteristics [126–133].

ML is divided into three ways, as shown in Figure 14.2: supervised learning (SL), unsupervised learning (UL), and RL. The kinds of challenges these approaches tackle and the way these models learn from data vary [134–139].

14.6.1.1 Supervised learning

Models in SL acquire knowledge using datasets with labels, in which every input has a corresponding output. This method is similar to conventional classroom instruction, in which a teacher guides students' learning by giving them both accurate and inaccurate examples. In contrast to human learning, though, the development of model is assessed throughout training in order to modify its characteristics as well as boost corrections. Until the model achieves a high-performance threshold, which guarantees accurate predictions, the learning process is repeated iteratively.

Polymer research has embraced SL in order to use extensive experimental datasets to answer difficult material issues. Predicting the compatibility of polymers with solvents is one such use [140–141].

Complex data analysis activities, especially those involving the processing of microscope images, have been automated with SL. The accurate localization and characterization of nanoparticles inside polymer matrices, which is typically accomplished by hand or with time-consuming image analysis techniques, is one of the primary challenges in polymer nanocomposite research. Qu et al. created a deep learning-based technique to identify and measure nanoparticles in transmission electron microscopy (TEM) pictures in order to solve this problem. They train a particular kind of neural network, CNNs, on labeled datasets of nanoparticle sizes and positions using an SL process [142]. An automated cropping and labeling technique (DOPAD) was used to extract 279,057 annotated subpictures from the 193 dataset, which comprises 72 TEM images.

When compared to manual approaches, the model greatly enhances the speed and accuracy of nanoparticle characterization by properly predicting the location and dimension of nanoparticles in newly formed TEM pictures. This method improves the study of polymer nanocomposite materials and makes it easier to conduct research on functional polymer-based nanotechnologies, advanced materials, and coatings.

14.6.1.2 Unsupervised learning

A powerful technique for identifying patterns in unlabeled data – that is, data without any predetermined outputs – is UL [143]. To uncover hidden correlations, clusters, or patterns, UL models autonomously analyze data, as opposed to SL, which depends upon certain input–output pairs. Stated differently, it is comparable to a student studying texts on their own to find recurring themes without the assistance of an instructor. As a result, UL is particularly helpful for interpreting complex polymer datasets where it may be challenging to identify or specify experimental labels. UL is specifically helpful in case of clustering, which groups polymers to similar structural or chemical characteristics together, and dimensionality reduction, which rectifies datasets of high-dimensional polymers through reserving important information [144, 145].

In polymer research, UL approaches have been effectively used to obtain insightful information from intricate datasets. Ziolek et al. [138] examined the micelles structure in nanoscale, made of linear block copolymers and four arms, using UL techniques. They found micelle structures groups with comparable corona arrangements by aggregating molecular conformations, and dimensionality reduction made the intricate structural changes easier to understand. Their approach allowed for a deeper comprehension of self-assembly mechanisms that are important for medications delivery and biomaterials creation.

14.6.1.3 Reinforcement learning

RL is an ML subset, which uses rewards for optimum behavior and interactions with the environment to teach models [146]. In contrast to SL, which trains models on datasets that are labeled, RL algorithms learn novel methods via trial and error. This is well-adopted for problems that call for sequential decision-making. Stated differently, it is like a toddler discovering that fire is dangerous only after coming in contact with it – the knowledge is acquired from actual experience rather than previous training.

Because RL entails long-term reward optimization, sequential decision-making, and exploration–exploitation trade-off, it is substantially more complex than supervised and UL. RL dynamically adapts methods based on ongoing input rather than learning from static datasets, necessitating sophisticated algorithms and substantial processing resources. Although these characteristics make RL an effective instrument for autonomous experimental control and improving polymerization processes, they also add to the complexity of its implementation and mathematics [147, 148]. Here are condensed examples to show how RL is used in polymer science.

With minimal real-world testing, Warren and associates [129] established an AI-driven closed-loop polymerization system that maximizes reversible-addition fragmentation chain transfer parameters to produce the required molecular weight and disparity. Their method incorporates concepts from RL, in which the system repeatedly tries reaction conditions, assesses the outcomes, and adjusts its strategy based on input from real-time analytical methods such as gel permeation chromatography and nuclear magnetic resonance. The system learns by engaging with the polymerization process rather than depending on preexisting datasets. It does this by methodically adjusting temperature and reaction time to maximize monomer conversion while minimizing disparity.

The main features of each ML strategy are compiled in Table 14.1 to offer a structured comparison, emphasizing the variations of information needs, analytical techniques, and applicability to research of polymer.

14.7 AI real-world tools

Platforms, cheminformatics tools, and programming libraries that can help researchers organize, analyze, and model polymer data are listed in Table 14.2. Open-source solutions that encourage openness, reproducibility, and accessibility in polymer research are especially beneficial since they enable a broader scientific community to participate in AI-driven materials discovery.

Python has become the preferred programming language for users who are just starting out with AI because of its ease of use, adaptability, and robust library ecosystem. User-friendly settings for writing and running Python code are offered by platforms like Jupyter Notebooks and Google Colab, which frequently do not require any

Table 14.1: Evaluation of approaches of key machine learning methods.

Characteristics	Supervised learning (SL)	Unsupervised learning (UL)	Reinforcement learning (RL)
Category	Labeled information (pairs of input–output)	Unlabeled data	No fixed classifications; develops by involvement
Aim	Determine results (classification/ regression)	Reduce dimensions or organize related data into clusters	Discover a series of steps to optimize rewards
Learning process	Memorize from explicit examples	Recognize hidden structures autonomously	Learns by experience and input from the surroundings
Optimization focus	Minimize loss (error)	Identifies clusters, patterns, and representations	Boost long-term benefits
Computational complexity	Average	Average–High	Extremely high (complex decision-forming)

installation or sophisticated computer power. These tools make it very easy for scientists for datasets loading, data cleaning, and preprocessing, deploying ML approaches. Open-source frameworks like Pandas and Numpy allow functional exploration along with large datasets alteration, including calculating averages, row filtering, and managing missing data.

Low-code or no-code platforms offer alternate points of entry for people who are afraid to start learning to code. Drag-and-drop interfaces for creating ML processes are provided by tools such as KNIME, which enable preparing data, training models, and assessing predictions, without requiring any code. Similar to this, Google's Teachable Machine streamlines categorization procedures, and through user-friendly online interfaces, systems such as Azure ML and Google AutoML allow learning about custom models on available datasets.

Engaging in presentation of molecules along with utilizing cheminformatics tools such as RDKit, BigSMILES is one of the most potent uses of AI in polymer research. Researchers can transform chemical structures into machine-readable representations, including molecular fingerprints or SMILES strings, using RDKit, which can then be used as inputs for AI models [149]. Researchers may now represent structural changes and unpredictability in polymer chains because of BigSMILES, which expands the ability through offering stochastic polymers using standardized nomenclature. Scientists may easily predict properties like dielectric constants or biodegradability when associated along with particular databases such as Polymer Genome that provides pre-trained models for prediction of properties. Rapid hypothesis testing is made possible by the tools that Polymer Genome offers for investigating connections between molecular descriptors and polymer characteristics.

Table 14.2: Science of polymers. FT stands for free-tier, OS for open source, and P for proprietary: Real-world AI tools.

Category	Tool (access type)	Functionality
General programming	Python (OS)	Data science open-source programming language
Coding and execution	Jupyter notebooks (OS)	A Python script running interactive coding environment
	Anaconda (FT)	Python distribution simplifies package management and deployment for scientific computing
Cloud-based execution	Google Colab (FT)	An installation-free cloud-based environment for executing Python code
Data processing	Numpy (OS)	Fast array manipulations are made possible by this numerical computing package
Data visualization	Matplotlib (OS)	Graphing library to create plots of publication quality
	Seaborn (OS)	Visualization of statistical data for scientific graphs with an integrated theme setting
No-code ML platforms	KNIME (OS)	Drag-and-drop ML workflow tool for non-coders
	Google AutoML (P)	Automated machine learning model builder
	Azure ML (P)	Large-scale AI projects using cloud-based machine learning services
	Teachable machine (FT)	User-friendly AI tool for random classification task without coding
Cheminformatic	RDKit (OS)	Transforms chemical structures into representations that AI can use
	BigSMILES (OS)	Standardized stochastic polymer representation
	Polymer Genome (FT)	Pretrained AI algorithms for predicting polymer properties

With AI technologies, navigating huge datasets – a frequent problem in data science – becomes much simpler. Consider, for example, a polymers dataset that includes their molecular weights, temperature stability, and mechanical properties. A researcher can use Python's Pandas module to observe connections between molecular weight and tensile strength, compute averages for specific categories, and filter polymers with Tg over a predefined threshold in a fraction of time that it would take conventional tools such as Excel. In addition to saving time, these approaches create new opportunities for data analysis at a level that was previously unthinkable.

It is now a matter of when rather than if AI will be included into polymer research. The hurdles to entry have never been lower, thanks to the availability of open-source libraries, user-friendly platforms, and pre-trained models. Researchers can take their first steps into this revolutionary subject by beginning with basic tools like KNIME for drag-and-drop workflows or Scikit-learn for developing prediction models. There are a lot of thorough and useful resources accessible for people who want to use these technologies right away. Frequently, customers only need to download and execute pre-written code with little preparation [150–156].

14.8 Challenges and future directions in the convergence of AI and nanotechnology

The growing integration of AI and nanotechnology expresses variety of difficulties as well as generate novel possibilities for research. The key to this convergence is interdisciplinary cooperation, which brings together specialists from several disciplines, including computer science, materials science, and medication. Although it could be highly challenging, closing the knowledge gap across various fields is necessary for the effective creation of novel solutions. Large, high-quality datasets are essential for AI systems to flourish. Obtaining enough data can be an issue in many uses of nanotechnology. For reliable AI-driven solutions, it is essential to guarantee data accessibility, diversity, and accuracy.

The ethical implementation of AI in nanotechnology creates a number of ethical concerns, consisting of intellectual property rights, data privacy, and equitable access to AI-driven innovations. Accounting all these ethical issues is necessary to guaranteeing the responsible and equitable use of emerging technology as well as in building public trust. Rapid advancements in AI and nanotechnology are often difficult for regulatory frameworks to keep up with. Governments and international institutions must modify as well as enforce guidelines to preserve safety, security, and moral standards in these developing businesses. Scaling up AI-driven nanomanufacturing processes and nanodevices for practical applications could be challenging and expensive. It's still difficult to achieve scalability without sacrificing accuracy and effectiveness. As AI and nanotechnology become more and more integrated into vital systems and applications, security and privacy issues become increasingly significant.

It is essential to mitigate security flaws and protect data privacy in order to stop abuse and online threats. The fusion of AI along with nanotechnology continues to influence science and technology in the future. Ethical norms, policies, and best practices for responsible usage of AI will be necessary for navigating the evolving world. Furthermore, new nanomaterials with specialized qualities will be made possible by AI-driven materials discovery, which will help a number of industries. AI and nanotechnology will make it possible to provide highly individualized healthcare on basis of patient's unique genetic characteristics and medical requirements. Additionally, researchers will concentrate on energy-efficient AI-driven manufacturing techniques and green nanomaterial synthesis methods as well as other sustainable nanomanufacturing procedures. Complex simulations, optimization methods, and cryptography solutions will all be advanced by quantum computing, combined with AI algorithms. In the post-quantum computing age, this will ensure data security by opening the door for quantum-safe AI. Additionally, quantum materials development by AI improves quantum technologies such as quantum sensors and computation.

14.9 Data resources: accessibility, availability, and difficulties

Collaborative coding platforms and structured, high-quality datasets are essential for AI convergence in polymer research. Several initiatives have been developed to support scholars through giving repositories for exchanging curated datasets, machine-readable polymer representations, and ML models. These tools help scientists to promote creativity as well as discovery by effectively training and refining AI models. Several specific databases have been designed to allow the use of AI in polymer research through offering pre-trained models as well as structured data. BigSMILES provides standardized notation system for presenting stochastic polymers, Materials Project contains computationally derived polymer-related data, Polymer Genome provides AI-driven polymer property predictions, and NIST Polymer Database collects experimentally verified polymer properties and serves as benchmark for applications of AI.

Beyond databases that are polymer-specific, the polymer science community can take advantage of a number of general platforms that facilitate data accessibility, AI model sharing, and collaborative coding. These platforms not only make it easier to collaborate across disciplines, but they also provide ideas for creating customized alternatives that are exclusive to polymer research, increasing accessibility along with awareness. Zenodo is an open-access repository for ML models and structured datasets that provide proper attribution with digital object identifiers, while Hugging Face is well-known for collection of pre-trained AI models, which includes tools exclusive to polymers. GitHub is still a necessary platform for version-controlled AI workflows, collaborative development, and dataset storage, enhancing transparency and reproducibility. Even while these materials are becoming more widely available, there are still major obstacles to data accessibility and standardization. Much research continue to have uneven, fragmented, or unavailable datasets, either with little metadata or with important information about experimental results, characterization methods, and synthesis conditions. Polymer science runs the risk of falling behind fields like materials science and biology, where open data standards have already facilitated the rapid implementation of AI, in the absence of established data-sharing norms. To improve reproducibility and speed up development, funding organizations and scientific journals must actively promote change. They can do this by requiring the publication of structured datasets alongside research publications. Creating interoperable datasets that form the basis of AI-driven research requires developing community standards for collection of data, annotation, and distribution.

14.10 Future outlook

More than simply a technical breakthrough, the incorporation of ML with polymer science heralds the start of a revolutionary age. ML has the capability to entirely transform the design, synthesis, and optimization of polymers, changing both research and industrial uses. ML-driven techniques, ranging from autonomous testing to predictive modeling, are expected to speed up discovery while rendering previously unobtainable materials accessible.

14.10.1 Artificial learning's growing application in polymer science

ML approaches will become more important in data-driven material discovery as polymer science experimental datasets continue to improve in size and complexity. Although SL has shown promise in forecasting important polymer characteristics, RL and autonomous learning will probably receive more attention in the future in order to support more independent and flexible research methodologies. It is anticipated that automated reaction optimization would benefit greatly from RL. RL allows an AI agent to interact with a polymerization process, continuously adjusting reaction parameters to maximize desired outcomes, in contrast to SL, which depends on historical data.

For instance, systems driven by RL could improve formulation procedures, speed up material discovery, and improve reaction conditions by continually gaining knowledge through real-time testing data. When combined with high-throughput experiments, the ML-based structures have the potential to investigate large chemical regions with previously unheard-of efficiency. The potential for UL for finding implicit structure-property correlations in polymer datasets has also been overlooked in polymer science. In polymer informatics, the absence of uniform labeling has been one of the main obstacles to its use, making SL techniques more instantly useful. But as structured polymer informatics frameworks develop further, autonomous learning will be an important tool for material design, latent pattern recognition, and molecular dataset clustering.

14.10.2 AI-powered experiments and self-selected labs

In the future of polymer science, which is anticipated to see the growth of AI-powered autonomous laboratories, standard synthesis, characterization, and optimization procedures will be managed by AI-guided robotic systems, greatly accelerating the research cycle. These self-operating labs will free up scientists to focus on more challenging scientific problems by transferring manual investigation to AI-assisted hypothesis-driven study. Through fusing AI to developed characterization methods and reaction monitoring systems, these labs will allow reactors that are self-optimizing to alter formulation parameters in real time and generate polymers through precise characteristics. Researchers will be able to effectively investigate novel polymer chemistries that are currently too difficult or resource-intensive to probe manually, thanks to the combination of AI, automation, and high-throughput testing. Real-time input from spectroscopy, chromatography, or other in situ monitoring methods could be used to continually improve reaction parameters in self-learning reactors. This strategy guarantees ideal material qualities while reducing trial-and-error testing. AI-powered formulation models have the potential to improve productivity and reproducibility in industrial settings by dynamically adjusting compositions of polymer for coatings, applications in biomedicine, and additive manufacturing.

14.10.3 Polymer science innovation and sustainability

ML is becoming more and more prominent as a crucial technique for creating ecologically friendly polymers due to the demand for sustainable materials. By forecasting integration in combinations of polymers and improving separation methods, ML models can revolutionize recycling operations, maximize the recyclability of polymer compositions, and assist in the design of biodegradable polymers with customized degradation profiles. ML-driven advancements in polymer science will be essential to achieving a circular economy by reducing energy requirements in reprocessing.

14.11 Conclusion

AI and nanotechnology are having a significant impact on the quantum computing field. AI helps to optimize quantum algorithms, but nanoscale components are crucial to the development of quantum processors. This convergence holds the potential to address challenging issues in domains like materials science and encryption and generate previously unparalleled computing resources. But with every technological development, there are issues that need to be resolved. To address the knowledge gaps between domain-specific scientists, nanotechnologists, and AI specialists, interdisciplinary cooperation is crucial. As this convergence develops, it is imperative to pay attention to legal usage of AI, data privacy, and scalability.

Future standards, suggested practices, and ethical principles ought to guide the proper use of AI in nanotechnology. Future developments in healthcare, environmental sustainability, and energy solutions will be greatly influenced by nanotechnology, while AI-driven materials research evolves to provide incredible discoveries. AI-powered quantum computing has the potential to completely transform a number of sectors. In summary, the combination of AI with nanotechnology seems to be evidence of human creativity and inventiveness. This convergence has the ability to change our world by addressing problems and moral dilemmas while pursuing state-of-the-art research as well as development. It can lead to new possibilities for knowledge and technological development and solve some of the most pressing issues of our day.

List of abbreviations

AI	Artificial intelligence
CNT	Carbon nanotubes
ML	Machine learning
SL	Supervised learning
UL	Unsupervised learning
RL	Reinforcement learning
TEM	Transmission electron microscopy

References

[1] Jaber W. Artificial intelligence and nanotechnology: Transforming the future. Artificial Intelligence in the Age of Nanotechnology, IGI Global, 2024;1–24. https://doi.org/10.4018/979-8-3693-0368-9.ch001.

[2] Natale S and Ballatore A. Imagining the thinking machine: Technological myths and the rise of artificial intelligence. Convergence, 2020;26(1):3–18. https://doi.org/10.1177/1354856517715164.

[3] Duan Y, Edwards JS and Dwivedi YK. Artificial intelligence for decision making in the era of big data–evolution, challenges and research agenda. International Journal of Information Management, 2019;48:63–71. https://doi.org/10.1016/j.ijinfomgt.2019.01.021.

[4] Dwivedi YK, Hughes L, Ismagilova E, Aarts G, Coombs C, Crick T, Duan Y, Dwivedi R, Edwards J, Eirug A and Galanos V. Artificial Intelligence (AI): Multidisciplinary perspectives on emerging challenges, opportunities, and agenda for research, practice and policy. International Journal of Information Management, 2021;57:101994. https://doi.org/10.1016/j.ijinfomgt.2019.08.002.

[5] Torfi A, Shirvani RA, Keneshloo Y, Tavaf N and Fox EA Natural language processing advancements by deep learning: A survey. arXiv preprint arXiv: 2003.01200. 2020 Mar 2. https://arxiv.org/abs/2003.01200.

[6] Sarker IH. Deep learning: A comprehensive overview on techniques, taxonomy, applications and research directions, SN Comput. Science, 2021;2(6):420. https://doi.org/10.1007/s42979-021-00815-1.

[7] Singh N and Sabrol H. Convolutional neural networks-an extensive arena of deep learning. A comprehensive study. Archives of Computational Methods in Engineering, 2021;28(7):4755–4780. https://doi.org/10.1007/s11831-021-09551-4.

[8] Patel JK, Patel A and Bhatia D. Introduction to Nanomaterials and nanotechnology. In: Emerging Technologies for Nanoparticle Manufacturing. 2021; (pp. 3–23). Cham: Springer International Publishing. https://doi.org/10.1007/978-3-030-50703-9_1.

[9] Roco MC, Hersam MC, Mirkin CA, Mirkin CA and Tuominen M. Synthesis, processing, and manufacturing of components, devices, and systems. Nanotechnology Research Directions for Societal Needs, 2011;2020:109–158. https://doi.org/10.1007/978-94-007-1168-6_4.

[10] Raj S, Khurana S, Choudhari R, Kesari KK, Kamal MA, Garg N, Ruokolainen J, Das BC and Kumar D. Specific Targeting Cancer Cells With Nanoparticles and Drug Delivery in Cancer therapy. In Seminars in Cancer Biology. 2021; (vol. 69, pp. 166–177). Academic Press. https://doi.org/10.1016/j semcancer.2019.11.002.

[11] Yetisgin AA, Cetinel S, Zuvin M, Kosar A and Kutlu O. Therapeutic nanoparticles and their targeted delivery applications. Molecules, 2020;25(9):2193. https://doi.org/10.3390/molecules25092193.

[12] Shivalkar S, Gautam PK, Chaudhary S, Samanta SK and Sahoo AK. Recent development of autonomously driven micro/nanobots for efficient treatment of polluted water. Journal of Environmental Management, 2021;281:111750. https://doi.org/10.1016/j.jenvman.2020.111750.

[13] Singh AV, Ansari MH, Laux P and Luch A. Micro-nanorobots: Important considerations when developing novel drug delivery platforms. Expert Opinion on Drug Delivery, 2019;16(11):1259–1275. https://doi.org/10.1080/17425247.2019.1676228.

[14] Mitchell S, Qin R, Zheng N and P´erez-ramírez J. Nanoscale engineering of catalytic materials for sustainable technologies. Nature Nanotechnology, 2021;16(2):129–139. https://doi.org/10.1038/s41565-020-00799-8.

[15] Manickam P, Mariappan SA, Murugesan SM, Hansda S, Kaushik A, Shinde R and Thipperudraswamy SP. Artificial intelligence (AI) and internet of medical things (IoMT) assisted biomedical systems for intelligent healthcare. Biosensors, 2022;12(8):562. https://doi.org/10.3390/bios12080562.

[16] Rane N. Integrating leading-edge artificial intelligence (AI), internet of things (IOT), and big data technologies for smart and sustainable architecture, engineering and construction (AEC) industry: Challenges and future directions. Engineering and construction (AEC) industry: Challenges and future directions. Srrn, 2023; https://doi.org/10.2139/ssrn.4616049.

[17] ˇLuby S, Lubyova M, Siffalovi ˇ ˇc P, Jergel M and Majkov´ A E. A brief history of nanoscience and foresight in nanotechnology. In Nanomaterials and Nanoarchitectures: A Complex Review of Current Hot Topics and Their Applications. 2015; (pp. 63–86). Netherlands: Springer. https://doi.org/10.1007/978-94-017-9921-8_4.

[18] Reghunadhan A, Krishna A, Jose AJ. Polymers in robotics. InPolymer science and innovative applications 2020 Jan 1 (pp. 393–421). Elsevier.

[19] Chantipmanee N and Xu Y. Nanofluidic manipulation of single nanometric objects: Current progress, challenges, and future opportunities. Engineering, 2024; https://doi.org/10.1016/j.eng.2024.01.001.

[20] Konstantopoulos G, Koumoulos EP and Charitidis CA. Digital innovation enabled nanomaterial manufacturing; machine learning strategies and green perspectives. Nanomaterials, 2022;12(15):2646. https://doi.org/10.3390/nano12152646.

[21] Rahman MA, Saleh T, Jahan MP, McGarry C, Chaudhari A, Huang R, Tauhiduzzaman M, Ahmed A, Mahmud AA, Bhuiyan MS and Khan MF. Review of intelligence for additive and subtractive manufacturing: Current status and future prospects. Micromachines, 2023;14(3):508. https://doi.org/10.3390/mi14030508.

[22] Huang Z, Shen Y, Li J, Fey M and Brecher C. A survey on AI-driven digital twins in industry 4.0: Smart manufacturing and advanced robotics. Sensors, 2021;21(19):6340. https://doi.org/10.3390/s21196340.

[23] Zhou F, Ni H, Zhu G, Bershadsky L, Sha R, Seeman NC and Chaikin PM. Toward three-dimensional DNA industrial nanorobots. Science Robotics, 2023;8(85):eadf1274. https://doi.org/10.1126/scirobotics.adf1274.

[24] Mahmood MA, Visan AI, Ristoscu C and Mihailescu IN. Artificial neural network algorithms for 3D printing. Materials, 2020;14(1):163. https://doi.org/10.3390/ma14010163.

[25] Ruberu K, Senadeera M, Rana S, Gupta S, Chung J, Yue Z, Venkatesh S and Wallace G. Coupling machine learning with 3D bioprinting to fast track optimisation of extrusion printing. Applied Materials Today, 2021;22:100914. https://doi.org/10.1016/j.apmt.2020.100914.

[26] Sui X, Downing JR, Hersam MC and Chen J. Additive manufacturing and applications of nanomaterial-based sensors. Mat Today: Proceedings, 2021;48:135–154. https://doi.org/10.1016/J.MATTOD.2021.02.001.

[27] Hoffmann R and Reich C. A systematic literature review on artificial intelligence and explainable artificial intelligence for visual quality assurance in manufacturing. Electronics, 2023;12(22):4572. https://doi.org/10.3390/electronics12224572.

[28] Magarelli G, Freire AM and Silva LP. Electrochemical sensors coupled with machine learning for food safety and quality inspection. In: Food Quality Analysis. 2023; (pp. 171–200). Academic Press. https://doi.org/10.1016/b978-0-323-95988-9.00001-1.

[29] Kim H, Han J and Han TY. Machine vision-driven automatic recognition of particle size and morphology in SEM images. Nanoscale, 2020;12(37):19461–19469. https://doi.org/10.1039/d0nr04140h.

[30] Liu YN, Wu FY, Tian RY, Shi YX, Xu ZQ, Liu JY, Huang J, Xue FF, Liu BY and Liu GQ. The bHLH-zip transcription factor SREBP regulates triterpenoid and lipid metabolisms in the medicinal fungus Ganoderma lingzhi. Communications Biology, 2023;6(1):1. https://doi.org/10.1038/s42003-022-04154-6.

[31] Kwon YT, Kim H, Mahmood M, Kim YS, Demolder C and Yeo WH. Printed, wireless, soft bioelectronics and deep learning algorithm for smart human–machine interfaces. ACS Applied Materials and Interfaces, 2020;12(44):49398–49406. https://doi.org/10.1021/acsami.0c14193.

[32] Arinez JF, Chang Q, Gao RX, Xu C and Zhang J. Artificial intelligence in advanced manufacturing: Current status and future outlook. Journal of Manufacturing Science and Engineering, 2020;142(11):110804. https://doi.org/10.1115/1.4047855.

[33] Hoffmann MA and Lasch R. Tackling industrial downtimes with artificial intelligence in data-driven maintenance. ACM Computing Surveys, 2023;56(4):1–33. https://doi.org/10.1145/3623378.

[34] Wang S, Shen Z, Shen Z, Dong Y, Li Y, Cao Y, Zhang Y, Guo S, Shuai J, Yang Y and Lin C. Machine-learning micropattern manufacturing. Nano Today, 2021;38:101152. https://doi.org/10.1016/J.NANTOD.2021.101152.

[35] Mekki-Berrada F, Ren Z, Huang T, Wong WK, Zheng F, Xie J, Tian IP, Jayavelu S, Mahfoud Z, Bash D and Hippalgaonkar K. Two-step machine learning enables optimized nanoparticle synthesis. Npj Computational Materials, 2021;7(1):55. https://doi.org/10.1038/s41524-021-00520-w.

[36] Hsiao J, Chang J and Cheng C. Soft medical robotics: Clinical and biomedical applications, challenges, and future directions. Adv Robot, 2019;33(21):1099e1111. https://doi.org/10.1080/01691864.2019.1679251.

[37] Wang H, Zhang R, Chen W, Wang X and Pfeifer R. A cable-driven soft robot surgical system for cardiothoracic endoscopic surgery: Preclinical tests in animals. Surgical Endoscopy, 2017;31:3152e3158. https://doi.org/10.1007/s00464-016-5340-9.

[38] McCandless M, Perry A, DiFilippo N, Carrol A, Billatos E and Russo S. A soft robot for peripheral lung cancer diagnosis and therapy. Soft Robot, 2022;9(4):754e766. https://doi.org/10.1089/soro.2020.0127.

[39] Paez-Granados D, Yamamoto T, Kadone H and Suzuki K. Passive flow control for series inflatable actuators: Application on a wearable soft-robot for posture assistance. IEEE Robotics and Automation Letters, 2021;6(3):4891e4898. https://doi.org/10.1109/LRA.2021.3070297.

[40] Yang Y and Jiao P. Nanomaterials and nanotechnology for biomedical soft robots. Materials Today Advances, 2023 Mar 1;17:100338.

[41] Dong H, Lin J, Tao Y, Jia Y, Sun L, Li WJ and Sun H. AI-enhanced biomedical micro/nanorobots in microfluidics. Lab Chip, 2024;24(5):1419–1440. https://doi.org/10.1039/d3lc00909b.

[42] Kong X, Gao P, Wang J, Fang Y and Hwang KC. Advances of medical nanorobots for future cancer treatments. Journal of Hematologic Oncology, 2023;16(1):74. https://doi.org/10.1186/s13045-023-01463-z.

[43] Tan TZ, Quek C, Ng GS and Ng EY. A novel cognitive interpretation of breast cancer thermography with complementary learning fuzzy neural memory structure. Expert Systems with Applications, 2007;33(3):652–666. https://doi.org/10.1016/j.eswa.2006.06.012.

[44] Agrahari V, Agrahari V, Chou ML, Chew CH, Noll J and Burnouf T. Intelligent micro-/nanorobots as drug and cell carrier devices for biomedical therapeutic advancement: Promising development opportunities and translational challenges. Biomaterials, 2020;260:120163. https://doi.org/10.1016/j.biomaterials.2020.120163.

[45] Nistor MT and Rusu AG. Nanorobots With Applications in medicine. In Polymeric Nanomaterials in Nanotherapeutics. 2019; (pp. 123–149). Elsevier. https://doi.org/10.1016/b978-0-12-813932-5.00003-0.

[46] Chen Y, Chen D, Liang S, Dai Y, Bai X, Song B, Zhang D, Chen H and Feng L. Recent advances in field-controlled micro–nano manipulations and micro–nano robots. Advanced Intelligent Systems, 2022;4(3):2100116. https://doi.org/10.1002/aisy.202100116.

[47] Nosrati H and Nosrati M. Artificial intelligence in regenerative medicine: Applications and implications. Biomimetics, 2023;8(5):442. https://doi.org/10.3390/biomimetics8050442.

[48] Dario P, Guglielmelli E, Allotta B and Carrozza MC. Robotics for medical applications. IEEE Robotics and Automation Magazine, 1996 Sep;3(3):44–56.

[49] Dario P, Hannaford B and Takanishi A. Guest editorial special issue on biorobotics. IEEE Transactions on Robotics, 2008 Feb 25;24(1):3–4.

[50] Ferreira A and Martel S. Guest editorial: Special issue on nanorobotics. IEEE Transactions on Robotics, 2014 Feb 3;30(1):1–2.

[51] Dong L and Nelson BJ. Tutorial-Robotics in the small part II: Nanorobotics. IEEE Robotics and Automation Magazine, 2007 Sep;14(3):111–121.

[52] Balzani V, Credi A, Raymo FM and Stoddart JF. Artificial molecular machines. Angewandte Chemie International Edition, 2000 Oct 2;39(19):3348–3391.

[53] Requicha AA. Nanorobots NEMS, and nanoassembly. Proceedings of the IEEE, 2004 Nov 8;91 (11):1922–1933.

[54] Coskun A, Banaszak M, Astumian RD, Stoddart JF and Grzybowski BA. Great expectations: Can artificial molecular machines deliver on their promise?. Chemical Society Reviews, 2012;41(1):19–30.

[55] Badjic JD, Balzani V, Credi A, Silvi S and Stoddart JF. A molecular elevator. Science, 2004 Mar 19;303 (5665):1845–1849.

[56] Dennis JR, Howard J and Vogel V. Molecular shuttles: Directed motion of microtubules along nanoscale kinesin tracks. Nanotechnology, 1999 Sep 1;10(3):232.

[57] Montemagno C and Bachand G. Constructing nanomechanical devices powered by biomolecular motors. Nanotechnology, 1999 Sep 1;10(3):225.

[58] Raguse B, Müller KH and Wieczorek L. Nanoparticle actuators. Advances in Materials, 2003 Jun 5;15 (11):922–926.

[59] Baughman RH, Cui C, Zakhidov AA, Iqbal Z, Barisci JN, Spinks GM, Wallace GG, Mazzoldi A, De Rossi D, Rinzler AG, Jaschinski O. Carbon nanotube actuators. Science. 1999 May 21;284(5418):1340–4.

[60] Fennimore AM, Yuzvinsky TD, Han WQ, Fuhrer MS, Cumings J and Zettl A. Rotational actuators based on carbon nanotubes. nature, 2003 Jul 24;424(6947):408–410.
[61] Terasawa N, Hayashi Y, Koga T, Higashi N and Asaka K. High-performance polymer actuators based on poly (ethylene oxide) and single-walled carbon nanotube–ionic liquid-based gels. Sensors and Actuators B: Chemical, 2014 Oct 31;202:382–387.
[62] Ye Z and Sitti M. Dynamic trapping and two-dimensional transport of swimming microorganisms using a rotating magnetic microrobot. LOC, 2014;14(13):2177–2182.
[63] Martel S, Felfoul O, Mathieu JB, Chanu A, Tamaz S, Mohammadi M, Mankiewicz M and Tabatabaei N. MRI-based medical nanorobotic platform for the control of magnetic nanoparticles and flagellated bacteria for target interventions in human capillaries. The International Journal of Robotics Research, 2009 Sep;28(9):1169–1182.
[64] Sitti M. Voyage of the microrobots. Nature, 2009 Apr 30;458(7242):1121–1122.
[65] Magdanz V, Sanchez S and Schmidt OG. Development of a sperm-flagella driven micro-bio-robot. Advances in Materials, 2013 Dec;25(45):6581–6588.
[66] Martel S, Tremblay CC, Ngakeng S, Langlois G. Controlled manipulation and actuation of micro-objects with magnetotactic bacteria. Applied Physics Letters. 2006 Dec 4;89(23):233904.
[67] Martel S, Mohammadi M, Felfoul O, Lu Z and Pouponneau P. Flagellated magnetotactic bacteria as controlled MRI-trackable propulsion and steering systems for medical nanorobots operating in the human microvasculature. The International Journal of Robotics Research, 2009 Apr;28(4):571–582.
[68] Park SJ, Lee Y, Choi YJ, Cho S, Jung HE, Zheng S, Park BJ, Ko SY, JO P and Park S. Monocyte-based microrobot with chemotactic motility for tumor theragnosis. Biotechnology & Bioengineering, 2014 Oct;111(10):2132–2138.
[69] Ricotti L, Fujie T, Vazão H, Ciofani G, Marotta R, Brescia R, Filippeschi C, Corradini I, Matteoli M, Mattoli V, Ferreira L. Boron nitride nanotube-mediated stimulation of cell co-culture on micro-engineered hydrogels. PloS one. 2013 Aug 14;8(8):e71707.
[70] Ricotti L, Das Neves RP, Ciofani G, Canale C, Nitti S, Mattoli V, Mazzolai B, Ferreira L and Menciassi A. Boron nitride nanotube-mediated stimulation modulates F/G-actin ratio and mechanical properties of human dermal fibroblasts. Journal of Nanoparticle Research, 2014 Feb;16:1–4.
[71] Kamm RD and Bashir R. Creating living cellular machines. Annals of Biomedical Engineering, 2014 Feb;42:445–459.
[72] Ekinci KL. Electromechanical transducers at the nanoscale: actuation and sensing of motion in nanoelectromechanical systems (NEMS). small. 2005 Aug;1(8–9):786–797.
[73] Cai H, Xu KJ, Liu AQ, Fang Q, Yu MB, Lo GQ, Kwong DL. Nano-opto-mechanical actuator driven by gradient optical force. Applied Physics Letters. 2012 Jan 2;100(1): 013108.
[74] Tian F, Zhou G, Du Y, Chau FS. Applications of nanoelectromechanical actuators in nano optomechanics. In2013 International Conference on Optical MEMS and Nanophotonics (OMN) 2013 Aug 18 (pp. 173–174). IEEE.
[75] Kong J, Franklin NR, Zhou C, Chapline MG, Peng S, Cho K and Dai H. Nanotube molecular wires as chemical sensors. science, 2000 Jan 28;287(5453):622–625.
[76] Cui Y, Wei Q, Park H and Lieber CM. Nanowire nanosensors for highly sensitive and selective detection of biological and chemical species. science, 2001 Aug 17;293(5533):1289–1292.
[77] Saha K, Agasti SS, Kim C, Li X and Rotello VM. Gold nanoparticles in chemical and biological sensing. Chemical Reviews, 2012 May 9;112(5):2739–2779.
[78] Luo X, Morrin A, Killard AJ and Smyth MR. Application of nanoparticles in electrochemical sensors and biosensors. Electroanalysis: An International Journal Devoted to Fundamental and Practical Aspects of Electroanalysis, 2006 Feb;18(4):319–326.
[79] Sakar MS, Neal D, Boudou T, Borochin MA, Li Y, Weiss R, Kamm RD, Chen CS and Asada HH. Formation and optogenetic control of engineered 3D skeletal muscle bioactuators. LOC, 2012;12 (23):4976–4985.

[80] Hierold C, Jungen A, Stampfer C and Helbling T. Nano electromechanical sensors based on carbon nanotubes. Sensors and Actuators A: Physical, 2007 May 1;136(1):51–61.

[81] Motornov M, Roiter Y, Tokarev I and Minko S. Stimuli-responsive nanoparticles, nanogels and capsules for integrated multifunctional intelligent systems. Progress in Polymer science, 2010 Jan 1;35(1–2):174–211.

[82] Ray A and Kopelman R. Hydrogel nanosensors for biophotonic imaging of chemical analytes. Nanomedicine, 2013 Nov 1;8(11):1829–1838.

[83] Park KI, Lee M, Liu Y, Moon S, Hwang GT, Zhu G, Kim JE, Kim SO, Kim DK, Wang ZL and Lee KJ. Flexible nanocomposite generator made of BaTiO3 nanoparticles and graphitic carbons. Advances in Materials, 2012 Jun 12;24(22):2999–3004.

[84] Xu S, Yeh YW, Poirier G, McAlpine MC, Register RA and Yao N. Flexible piezoelectric PMN–PT nanowire-based nanocomposite and device. Nano Letters, 2013 Jun 12;13(6):2393–2398.

[85] Jeong CK, Park KI, Ryu J, Hwang GT and Lee KJ. Large-area and flexible lead-free nanocomposite generator using alkaline niobate particles and metal nanorod filler. Advanced Functional Materials, 2014 May;24(18):2620–2629.

[86] Zhao HQ, Lin L, Li JR, Tang JA, Duan MX and Jiang L. DNA biosensor with high sensitivity amplified by gold nanoparticles. Journal of Nanoparticle Research, 2001 Aug;3:321–323.

[87] Hauser CA, Maurer-Stroh S and Martins IC. Amyloid-based nanosensors and nanodevices. Chemical Society Reviews, 2014;43(15):5326–5345.

[88] Dolatabadi JE and De la Guardia M. Nanomaterial-based electrochemical immunosensors as advanced diagnostic tools. Analytical Methods, 2014;6(12):3891–3900.

[89] Ricotti L and Menciassi A. Nanotechnology in biorobotics: Opportunities and challenges. Journal of Nanoparticle Research, 2015 Feb;17:1–0.

[90] Iu Zaman BU. Transforming education through AI benefits risks and ethical considerations. 2024. https://doi.org/10.20944/preprints202401.0123.v1

[91] Arif H, Kumar A, Fahad M and Hussain HK. Future Horizons: AI-Enhanced Threat Detection in Cloud Environments: Unveiling Opportunities for Research. International Journal of Multidisciplinary Sciences and Arts, 2024;3(1):242–251. https://doi.org/10.47709/ijmdsa.v3i4.

[92] Thakur A and Kumar A. Recent advances on rapid detection and remediation of environmental pollutants utilizing nanomaterials-based (bio) sensors. Science of The Total Environment, 2022; https://doi.org/10.1016/j.scitotenv.2022.152345.

[93] Russell SJ, Norvig P. Artificial intelligence: A modern approach;[the intelligent agent book]. Prentice hall; 1995.

[94] LeCun Y, Bengio Y, Hinton G. Deep learning. nature. 2015 May 28;521(7553):436–44.

[95] Ramprasad R, Batra R, Pilania G, Mannodi-Kanakkithodi A, Kim C. Machine learning in materials informatics: recent applications and prospects. npj Computational Materials. 2017 Dec 13;3(1):54.

[96] Anosova O, Kurlin V, Senechal M. The importance of definitions in crystallography. IUCrJ. 2024 Jul 1;11(4):453–63.

[97] Rajkomar A, Dean J, Kohane I. Machine learning in medicine. New England Journal of Medicine. 2019 Apr 4;380(14):1347–58.

[98] Sanchez-Lengeling B, Aspuru-Guzik A. Inverse molecular design using machine learning: Generative models for matter engineering. Science. 2018 Jul 27;361(6400):360–5.

[99] Vamathevan J, Clark D, Czodrowski P, Dunham I, Ferran E, Lee G, Li B, Madabhushi A, Shah P, Spitzer M, Zhao S. Applications of machine learning in drug discovery and development. Nature reviews Drug discovery. 2019 Jun;18(6):463–77.

[100] Jumper J, Evans R, Pritzel A, Green T, Figurnov M, Ronneberger O, Tunyasuvunakool K, Bates R, Žídek A, Potapenko A, Bridgland A. Highly accurate protein structure prediction with AlphaFold. nature. 2021 Aug 26;596(7873):583–9.

[101] Park H, Li Z, Walsh A. Has generative artificial intelligence solved inverse materials design?. Matter. 2024 Jul 3;7(7):2355–67.

[102] Chen L, Pilania G, Batra R, Huan TD, Kim C, Kuenneth C, Ramprasad R. Polymer informatics: Current status and critical next steps. Materials Science and Engineering: R: Reports. 2021 Apr 1;144:100595.

[103] Ignacz G, Bader L, Beke AK, Ghunaim Y, Shastry T, Vovusha H, Carbone MR, Ghanem B, Szekely G. Machine learning for the advancement of membrane science and technology: A critical review. Journal of Membrane Science. 2025 Jan 1;713:123256.

[104] Gormley AJ, Webb MA. Machine learning in combinatorial polymer chemistry. Nature Reviews Materials. 2021 Aug;6(8):642–4.

[105] Li Y, Bai N, Chang Y, Liu Z, Liu J, Li X, Yang W, Niu H, Wang W, Wang L, Zhu W. Flexible iontronic sensing. Chemical Society Reviews. 2025;54(10):4651–700.

[106] Martin TB, Audus DJ. Emerging trends in machine learning: a polymer perspective. ACS Polymers Au. 2023 Jan 18;3(3):239–58.

[107] Wang J, Tian K, Li D, Chen M, Feng X, Zhang Y, Wang Y, Van der Bruggen B. Machine learning in gas separation membrane developing: Ready for prime time. Separation and Purification Technology. 2023 May 15;313:123493.

[108] Lichtenthaler U. Five maturity levels of managing AI: From isolated ignorance to integrated intelligence. Journal of Innovation Management. 2020 Mar 30;8(1):39–50.

[109] Shanker S. Turing and the Origins of AI. Philosophia Mathematica. 1995 Jan 1;3(1):52–85.

[110] Hollingsworth SA, Dror RO. Molecular dynamics simulation for all. Neuron. 2018 Sep 19;99 (6):1129–43.

[111] Garrigue L. Hohenberg–Kohn theorems for interactions, spin and temperature. Journal of Statistical Physics. 2019 Nov;177(3):415–37.

[112] Jordan MI, Mitchell TM. Machine learning: Trends, perspectives, and prospects. Science. 2015 Jul 17;349(6245):255–60.

[113] Ravaioli G, Domingos T, Teixeira RF. A framework for data-driven agent-based modelling of agricultural land use. Land. 2023 Mar 27;12(4):756.

[114] Zdeborová L. New tool in the box. Nature Physics. 2017 May;13(5):420–1.

[115] Wu S, Kondo Y, Kakimoto MA, Yang B, Yamada H, Kuwajima I, Lambard G, Hongo K, Xu Y, Shiomi J, Schick C. Machine-learning-assisted discovery of polymers with high thermal conductivity using a molecular design algorithm. Npj Computational Materials. 2019 Jun 21;5(1):66.

[116] Lu Y, Yalcin D, Pigram PJ, Blackman LD, Boley M. Interpretable machine learning models for phase prediction in polymerization-induced self-assembly. Journal of Chemical Information and Modeling. 2023 May 19;63(11):3288–306.

[117] Kuenneth C, Schertzer W, Ramprasad R. Copolymer informatics with multitask deep neural networks. Macromolecules. 2021 Jun 29;54(13):5957–61.

[118] Monteith S, Glenn T. Automated decision-making and big data: concerns for people with mental illness. Current psychiatry reports. 2016 Dec;18(12):112.

[119] Aoki Y, Wu S, Tsurimoto T, Hayashi Y, Minami S, Tadamichi O, Shiratori K, Yoshida R. Multitask machine learning to predict polymer–solvent miscibility using Flory–Huggins interaction parameters. Macromolecules. 2023 Jul 10;56(14):5446–56.

[120] Sutliff BP, Goyal S, Martin TB, Beaucage PA, Audus DJ, Orski SV. Correlating near-infrared spectra to bulk properties in polyolefins. Macromolecules. 2024 Feb 28;57(5):2329–38.

[121] Al-Garadi MA, Varathan KD, Ravana SD, Ahmed E, Mujtaba G, Khan MU, Khan SU. Analysis of online social network connections for identification of influential users: Survey and open research issues. ACM Computing Surveys (CSUR). 2018 Jan 31;51(1):1–37.

[122] Zhou Y, Wen J, Nie Y. Microscopic mechanisms of self-healing in polymers revealed by molecular simulations and machine learning. Macromolecules. 2024 Mar 22;57(7):3258–70.

[123] Bera A, Akash TS, Ishraaq R, Pial TH, Das S. Hydrogen bonding inside anionic polymeric brush layer: Machine Learning-driven exploration of the relative roles of the polymer steric effect, charging, and type of screening counterions. Macromolecules. 2024 Feb 10;57(4):1581–92.
[124] Chandrasekaran A, Kim C, Venkatram S, Ramprasad R. A deep learning solvent-selection paradigm powered by a massive solvent/nonsolvent database for polymers. Macromolecules. 2020 Jun 11;53 (12):4764–9.
[125] Miccio LA, Schwartz GA. Mapping chemical structure–glass transition temperature relationship through artificial intelligence. Macromolecules. 2021 Feb 5;54(4):1811–7.
[126] Schneider L, de Pablo JJ. Combining particle-based simulations and machine learning to understand defect kinetics in thin films of symmetric diblock copolymers. Macromolecules. 2021 Oct 28;54 (21):10074–85.
[127] Park NH, Zubarev DY, Hedrick JL, Kiyek V, Corbet C, Lottier S. A recommender system for inverse design of polycarbonates and polyesters. Macromolecules. 2020 Nov 27;53(24):10847–54.
[128] Wheatle BK, Fuentes EF, Lynd NA, Ganesan V. Design of polymer blend electrolytes through a machine learning approach. Macromolecules. 2020 Oct 21;53(21):9449–59.
[129] Knox ST, Parkinson SJ, Wilding CY, Bourne RA, Warren NJ. Autonomous polymer synthesis delivered by multi-objective closed-loop optimisation. Polymer Chemistry. 2022;13(11):1576–85.
[130] Bone JM, Childs CM, Menon A, Poczos B, Feinberg AW, LeDuc PR, Washburn NR. Hierarchical machine learning for high-fidelity 3D printed biopolymers. ACS Biomaterials Science & Engineering. 2020 Nov 20;6(12):7021–31.
[131] Zhou T, Qiu D, Wu Z, Alberti SA, Bag S, Schneider J, Meyer J, Gámez JA, Gieler M, Reithmeier M, Seidel A. Compatibilization efficiency of graft copolymers in incompatible polymer blends: Dissipative particle dynamics simulations combined with machine learning. Macromolecules. 2022 Sep 1;55(17):7893–907.
[132] Meleties M, Britton D, Katyal P, Lin B, Martineau RL, Gupta MK, Montclare JK. High-throughput microrheology for the assessment of protein gelation kinetics. Macromolecules. 2022 Feb 9;55 (4):1239–47.
[133] Richbourg NR, Peppas NA. High-throughput FRAP analysis of solute diffusion in hydrogels. Macromolecules. 2021 Nov 9;54(22):10477–86.
[134] Mendels D, Byléhn F, Sirk TW, de Pablo JJ. Systematic modification of functionality in disordered elastic networks through free energy surface tailoring. Science Advances. 2023 Jun 7;9(23):eadf7541.
[135] Khajeh A, Schweigert D, Torrisi SB, Hung L, Storey BD, Kwon HK. Early prediction of ion transport properties in solid polymer electrolytes using machine learning and system behavior-based descriptors of molecular dynamics simulations. Macromolecules. 2023 Jun 23;56(13):4787–99.
[136] Wilson AN, St John PC, Marin DH, Hoyt CB, Rognerud EG, Nimlos MR, Cywar RM, Rorrer NA, Shebek KM, Broadbelt LJ, Beckham GT. PolyID: Artificial intelligence for discovering performance-advantaged and sustainable polymers. Macromolecules. 2023 Oct 19;56(21):8547–57.
[137] Shi J, Rebello NJ, Walsh D, Zou W, Deagen ME, Leao BS, Audus DJ, Olsen BD. Quantifying Pairwise Similarity for Complex Polymers. Macromolecules. 2023 Sep 6;56(18):7344–57.
[138] Ziolek RM, Smith P, Pink DL, Dreiss CA, Lorenz CD. Unsupervised learning unravels the structure of four-arm and linear block copolymer micelles. Macromolecules. 2021 Apr 14;54(8):3755–68.
[139] FERJI K. Basic Concepts and Tools of Artificial. science.;25:26.
[140] Chollet F. Deep learning with Python. simon and schuster; 2021 Dec 21.
[141] FERJI K. Basic Concepts and Tools of Artificial. science.;25:26.
[142] Qu EZ, Jimenez AM, Kumar SK, Zhang K. Quantifying nanoparticle assembly states in a polymer matrix through deep learning. Macromolecules. 2021 Apr 2;54(7):3034–40.
[143] Shoukat S, Nawaz S, Rasheed MM, Javaid A, Sami HA. Efficiency of OLS and Huber M estimator in case of Outliers. Journal of Excellence in Social Sciences. 2024 Jul 5;3(3):55–60.

[144] Solorio-Fernández S, Carrasco-Ochoa JA, Martínez-Trinidad JF. A review of unsupervised feature selection methods. Artificial Intelligence Review. 2020 Feb;53(2):907–48.
[145] Saxena A, Prasad M, Gupta A, Bharill N, Patel OP, Tiwari A, Er MJ, Ding W, Lin CT. A review of clustering techniques and developments. Neurocomputing. 2017 Dec 6;267:664–81.
[146] Botvinick M, Ritter S, Wang JX, Kurth-Nelson Z, Blundell C, Hassabis D. Reinforcement learning, fast and slow. Trends in cognitive sciences. 2019 May 1;23(5):408–22.
[147] Ma R, Zhang H, Luo T. Exploring high thermal conductivity amorphous polymers using reinforcement learning. ACS Applied Materials & Interfaces. 2022 Mar 28;14(13):15587–98.
[148] Li H, Collins CR, Ribelli TG, Matyjaszewski K, Gordon GJ, Kowalewski T, Yaron DJ. Tuning the molecular weight distribution from atom transfer radical polymerization using deep reinforcement learning. Molecular Systems Design & Engineering. 2018;3(3):496–508.
[149] Yan C, Feng X, Wick C, Peters A, Li G. Machine learning assisted discovery of new thermoset shape memory polymers based on a small training dataset. Polymer. 2021 Feb 1;214:123351.
[150] Cahill ST, Young JE, Howe M, Clark R, Worrall AF, Stewart MI. Assignment of Regioisomers Using Infrared Spectroscopy: A Python Coding Exercise in Data Processing and Machine Learning. Journal of Chemical Education. 2024 Jun 20;101(7):2925–32.
[151] Lafuente D, Cohen B, Fiorini G, García AA, Bringas M, Morzan E, Onna D. A Gentle introduction to machine learning for chemists: an undergraduate workshop using python notebooks for visualization, data processing, analysis, and modeling. Journal of Chemical Education. 2021 Aug 13;98(9):2892–8.
[152] Qiu R, Lin Z, Yang Z, Gao L. Machine Learning for Infrared Spectral Classification of Polyvinyl butyrals with Identical Chemical Groups: An Example for Undergraduate Chemistry Classes. Journal of Chemical Education. 2024 Jan 3;101(2):328–36.
[153] FERJI K. Basic Concepts and Tools of Artificial. science.;25:26.
[154] Berhe TA, Su WN, Hwang BJ. Halide perovskites' multifunctional properties: coordination engineering, coordination chemistry, electronic interactions and energy applications beyond photovoltaics. Inorganics. 2024 Jun 28;12(7):182.
[155] Thrall ES, Martinez Lopez F, Egg TJ, Lee SE, Schrier J, Zhao Y. Rediscovering the Particle-in-a-Box: Machine Learning Regression Analysis for Hypothesis Generation in Physical Chemistry Lab. Journal of Chemical Education. 2023 Nov 29;100(12):4933–40.
[156] Peña-Martínez J, Beltrán-Martínez J, Cano-Ortiz A, Rosales-Conrado N. Teaching the Nature of Science Through Biodiesel Synthesis from Waste Cooking Oil: A Literature Review with Experimental Insights. Sustainable Chemistry. 2025 May 19;6(2):15.

Editor's biography

Dr. Rishabha Malviya completed his B.Pharm. from Uttar Pradesh Technical University and M.Pharm. (pharmaceutics) from Gautam Buddha Technical University, Lucknow, Uttar Pradesh. His Ph.D. (pharmacy) work was in the area of novel formulation development techniques. He has 15 years of research experience and has been working as a professor in the Department of Pharmacy, School of Medical and Allied Sciences, Galgotias University, for past 12 years. His areas of interest include formulation optimization, nanoformulation, targeted drug delivery, localized drug delivery, and characterization of natural polymers as pharmaceutical excipients. He has authored more than 200 research/review papers for national/international reputed journals. He has 58 patents (19 grants, 38 published, and 1 filed) and publications in reputed national and international journals with a total of 350 cumulative impact factors. He has also received an Outstanding Reviewer Award from Elsevier. He has authored/edited/editing 58 books (Wiley, CRC Press/Taylor and Francis, Springer, River Publisher, IOP Publishing, and OMICS publication) and authored 150 book chapters. His name has been included in the World's Top 2% Scientists list for the years 2020, 2021, 2022, 2023, and 2024 by Elsevier BV and Stanford University. He is reviewer/editor/editorial board member of more than 50 national and international reputed journals.

Sonali Sundram completed her B.Pharm. and M.Pharm. (pharmacology) from AKTU, Lucknow. She has worked as a research scientist in the project of ICMR in King George's Medical University, Lucknow. Then, she joined BBDNIIT and currently she is working in Galgotias University, Greater Noida. Her Ph.D. (pharmacy) work was in the area of neurodegeneration and nanoformulation. Her areas of interest include neurodegeneration, clinical research, and artificial intelligence. She has edited/authored/editing 32 books (Wiley-Scrivener Publishing, CRC Press/Taylor and Francis, IOP Publishing, Springer Nature, Apple Academic Press/Taylor and Francis Group, and River Publishers) and authored 45 book chapters. She has attended as well as organized more than 15 national and international seminars/conferences/workshops. She has more than eight patents, national and international, to her credit.

Dr. Sathvik Belagodu Sridhar is currently working as an associate dean and professor at RAK College of Pharmacy (RAKCOP), RAK Medical and Health Sciences University, Ras Al Khaimah, United Arab Emirates (UAE). He completed his Ph.D. in clinical pharmacy from Rajiv Gandhi University of Health Sciences, Bengaluru, India. He received an FIP fellowship for higher training in renal clinical pharmacy at Royal Adelaide Hospital and Queen Elizabeth Hospital in Adelaide, South Australia. He worked as an assistant professor and clinical pharmacist at JSS College of Pharmacy and JSS Medical College Hospital, Mysore, India, for 8 years and later joined International Medical University, Malaysia, as a senior lecturer in the Department of Pharmacy Practice from 2008 to 2010. He joined as an assistant professor in the Department of Clinical Pharmacy at RAKCOP, RAK Medical and Health Sciences University, in 2010 and worked as a professor and chairperson of the Department of Clinical Pharmacy and Pharmacology from 2018 to 2023. He has been honored with a prestigious teacher award by the Society of Pharmaceutical Education and Research (SPER). He has also received training in pharmacovigilance at the Uppsala Monitoring Centre, Uppsala, Sweden. He has 25 years of experience in teaching and practicing clinical pharmacy in India, Malaysia, and the UAE. His research interests include medication safety,

https://doi.org/10.1515/9783111681870-015

pharmacovigilance, pharmacoepidemiology, and renal and psychiatry clinical pharmacy. He has more than 75 research papers published in reputed national and international journals, along with book chapters. He has presented more than 75 research papers at numerous national and international conferences and has been honored with many awards. He also serves as a reviewer for various national and international pharmacy practice-related journals.

Shivam Rajput has completed his B.Pharm. from Pt. B.D. Sharma University of Health Sciences, Rohtak, and M.Pharm. from Galgotias University, Greater Noida. He has authored/edited 10 books with Scrivener Publishing/Wiley/Apple Academic Press/Taylor and Francis Group, and Springer Nature. He has published more than 15 Scopus-indexed papers with reputed publishers such as Elsevier and Bentham Science. He has attended more than 10 international conferences.

Index

https://doi.org/10.1515/9783111681870-016